D1222192

CHEMICAL ZOOLOGY

Volume X

AVES

Contributors to This Volume

R. G. BOARD
ALAN H. BRUSH
DONALD S. FARNER
BLAINE R. FERRELL
B. M. FREEMAN
F. REED HAINSWORTH
D. J. HORNSEY
SHMUEL HURWITZ
JÜRGEN JACOB
ALBERT H. MEIER
C. PAUL
A. G. SCHNEK
C. VANDECASSERIE
LARRY L. WOLF

CHEMICAL ZOOLOGY

Edited by MARCEL FLORKIN
LABORATOIRES DE BIOCHIMIE
INSTITUT LÉON FREDERICQ
LIÈGE, BELGIUM

BRADLEY T. SCHEER
PROFESSOR EMERITUS
DEPARTMENT OF BIOLOGY
COLLEGE OF LIBERAL ARTS
EUGENE, OREGON

Volume X

AVES

Edited by ALAN H. BRUSH
THE BIOLOGICAL SCIENCES GROUP
THE COLLEGE OF LIBERAL ARTS AND SCIENCES
THE UNIVERSITY OF CONNECTICUT
STORRS, CONNECTICUT

ACADEMIC PRESS New York San Francisco London 1978
A Subsidiary of Harcourt Brace Jovanovich, Publishers

ACADEMIC PRESS, INC.
111 Fifth Avenue, New York, New York 10003

United Kingdom Edition published by
ACADEMIC PRESS, INC. (LONDON) LTD.
24/28 Oval Road, London NW1 7DX

Library of Congress Cataloging in Publication Data

Florkin, Marcel.
Chemical zoology.

Includes bibliographies.
CONTENTS:—v. 1. Protozoa, edited by G. W. Kidder.—v. 2. Porifera, Coelenterata, and Platyhelminthes.—v. 3. Echinodermata, Nematoda, and Acanthocephala. [etc.]
1. Biological chemistry. I. Scheer, Bradley Titus, Date joint author. II. Kiddler, George Wallace, Date ed. III. Title. [DNLM: 1. Amphibia. 2. Reptiles. W1CH276 v. 9 1974 / QL641 A526 1974]
QP514.F528 591.1'9'2 67-23158
ISBN 0-12-261040-7

PRINTED IN THE UNITED STATES OF AMERICA

Contents

Chapter 1. **Introduction**

Donald S. Farner

Chapter 2. **Plasma and Egg White Proteins**

R. G. Board and D. J. Hornsey

Chapter 3. **Chemical Embryology**

B. M. FREEMAN

Chapter 4. **Feather Keratins**

ALAN H. BRUSH

Chapter 5. **Avian Pigmentation**

ALAN H. BRUSH

Chapter 6. **Uropygial Gland Secretions and Feather Waxes**

JÜRGEN JACOB

List of Contributors

Numbers in parentheses indicate the pages on which the authors' contributions begin.

R. G. BOARD (37), School of Biological Sciences, Bath University, Claverton Down, Avon, England

ALAN H. BRUSH (117, 141), The Biological Sciences Group, The College of Liberal Arts and Sciences, University of Connecticut, Storrs, Connecticut

DONALD S. FARNER (1), Department of Zoology, University of Washington, Seattle, Washington

BLAINE R. FERRELL (213), Department of Zoology and Physiology, Louisiana State University, Baton Rouge, Louisiana

B. M. FREEMAN (75), Houghton Poultry Research Station, Houghton, Huntingdon, Cambs, England

F. REED HAINSWORTH (307), Department of Biology, Syracuse University, Syracuse, New York

D. J. HORNSEY (37), School of Biological Sciences, Bath University, Claverton Down, Avon, England

SHMUEL HURWITZ (273), Agricultural Research Organization, The Volcani Center, Bet Dagan, Israel

JÜRGEN JACOB (165), Biochemisches Institut für Umweltcarcinogene, Ahrensburg/Holst, West Germany

ALBERT H. MEIER (213), Department of Zoology and Physiology, Louisiana State University, Baton Rouge, Louisiana

C. PAUL (359), Faculté des Sciences, Chimie Generale I, Université Libre de Bruxelles, Brussels, Belgium

A. G. SCHNEK (359), Faculté des Sciences, Chimie Generale I, Université Libre de Bruxelles, Brussels, Belgium

C. VANDECASSERIE (359), Faculté des Sciences, Chimie Generale I, Université Libre de Bruxelles, Brussels, Belgium

LARRY L. WOLF (307), Department of Biology, Syracuse University, Syracuse, New York

Preface

There has been a recent spate of books on birds that are not traditional ornithological texts. This activity is certainly justifiable as avian models provide incisive thinking and experimental substrates for a wide variety of investigations at almost all levels of biological organization. In a sense, this book is also different. It is not a comprehensive ornithology textbook. Both physiological and biochemical topics are considered with a balance between systems orientation and development of specific problems or functions. It covers aspects of avian function and molecular organizations not treated elsewhere or considered only as an afterthought. There are approximately 8600 species of birds, classified into about 160 families. Their variety is limited only by the most inhospitable climatic conditions and the physiological and biochemical restraints imposed by flight, oviparity, and size. While they share many of the biochemical and physiological mechanisms of mammals due to a common ancestry, they are unique in many ways. This has led to a burgeoning of adaptive responses, and many of these demands, controls, and mechanisms are touched on in the chapters of this volume.

A major function of this book is to introduce ornithologists, field biologists, and investigators from other experimental areas to the chemical aspects of avian biology. In this role the volume complements other available sources. This approach also serves to single out birds as a group worthy of study both in their own right and in relation to other vertebrate groups. Birds are well known taxonomically and have served many studies on ecology, population dynamics, reproductive biology, and speciation. Their unique physiological and structural requirements and environmental interactions are often ultimately based in adaptations of their biochemical machinery. These are the aspects of the biology of birds that each author was asked to explore.

Each contributor was asked to concentrate on recent results and new interpretation of experimental data and theoretical advances. Much of the earlier work has been covered by reference to reviews and monographs. An attempt has been made to emphasize the broader aspects of the topics, but some are so "typically" avian as to exclude other groups. In fact, several problems are essentially limited to birds and therefore represent a very special interest.

ALAN H. BRUSH

Contents of Other Volumes

Volume I: PROTOZOA

Volume II

Section I: PORIFERA

Section II: COELENTERATA, CTENOPHORA

Section III: PLATYHELMINTHES, MESOZOA

Volume III

Section I: ECHINODERMATA

Volume IV: ANNELIDA, ECHIURA, AND SIPUNCULA

Volume V: ARTHROPODA Part A

Volume VI: ARTHROPODA Part B

Volume VII: MOLLUSCA

Volume VIII: DEUTEROSTOMIANS, CYCLOSTOMES, AND FISHES

Section I: DEUTEROSTOMIANS

Section II: VERTEBRATES (CRANIATA)

Volume IX

Section I: AMPHIBIA

Section II: REPTILIA

CHAPTER 1

Introduction

Donald S. Farner

I. Origin and General Characteristics

The class Aves includes approximately 8600 living species. This number is now affected only slightly by the discovery of species truly new to science, recently at a rate of no more than a few per decade. It is affected more by progress in the field of avian systematics, which often combines into single species forms that previously were assigned individual specific status. Although much remains to be learned about birds, the knowledge available on distribution, natural history, and general biology is far more extensive than for any other class of animals. It is because of this that it has been possible to use birds so effectively in the study of the general aspects of evolutionary biology, comparative physiology, ethology, ecology, environmental physiology, and other contemporary fields of biology.

The class Aves can be characterized with greater ease and less controversy than most classes of animals. Indeed, the common noun, bird, has the same precision as the scientific name. The identifying characteristics of birds are largely associated, in one way or another, with flight. All modern birds are capable of flight or were derived from flying ancestors. Briefly, birds may be characterized as bipedal vertebrates with forelimbs modified for flight and with unique diagnostic integumentary structures, the feathers. They are exclusively oviparous, and, with the exceptions of temporary heterothermy in some species, are homeothermal. The complex lung–air–sac system is a unique adaptation to flight.

Birds share with at least some reptiles such phylogenetically significant features as a monocondylous skull, quadrate articulation of the mandible, a single sound-transmitting bone (columella) in the middle ear, epidermal scales, uncinate costal processes, intertarsal ankle joint, nucleated erythrocytes, uricotelic nitrogen metabolism, renal portal system, and numerous features of embryonic development. Birds differ from reptiles mainly in features that accompany the development of flight. These include homeothermy, complete double circulatory system, complex lung–air–sac system, feathers, and wings that are different in structure from those of the pterosaurs and other flying reptiles. Avian and mammalian homeothermy, although similar in many respects, are clearly convergent phenomena. The reptilian affinities of birds are especially conspicuous in earlier fossil forms, including *Archeopteryx* which, however, seems too specialized and to have evolved too late to have been ancestral to contemporary groups. Although it long seemed clear that the ancestor of birds was to be sought among pseudosuchian reptiles of the Lower Triassic, the avian fossil record is so sparse that a satisfactory lineage cannot be constructed. A careful and thoughtful recent review of the available evidence lead Ostrum (1976) to the conclusion that a more immediate ancestry occurred among the coelosauerian dinosaurs. It should be noted that birds were contemporary with an extensively developed and diversified post-Triassic group of flying reptiles, the pterosaurs, from their origin until the decline of the latter in the late Cretaceous. Although there were interesting convergences between birds and pterosaurs, it is obvious that the former were not derived from the latter. Brodkorb (1971) has reminded us that "The Age of Reptiles (Mesozoic Era) saw not only the origin of birds but also more than half of their history." He argues, doubtless controversially, that the avifaunae of the Pleistocene, Pliocene and Miocene epochs were richer in species than that of the Recent. Brodkorb estimated that the total number of avian species ". . . past and present, is approximately 154,000 . . ." and suggests, therefore, that the ". . . less than 9600 living and fossil birds presently known still represent a tiny fraction of those that existed. . . ."

Birds range in size over five orders of magnitudes of weight. Among the smallest species, the Bee Hummingbird, *Mellisuga helenae,* has a length of about 5 cm, about half of which is tail. In contrast, the largest living bird the Ostrich, *Struthio camelus*, may weigh as much as 135 kg and stand as high as 2.4 m. The largest known species is the flightless fossil Elephantbird, *Aepyornis titan*, which may have weighed as much as 450 kg and stood as much as 3 m tall. The Giant Moa, *Diornis maxima*, also flightless, was probably of similar size. Among living flying birds the greatest weights, up to 17 kg, are attained in the Trumpeter Swan, *Cygnus buccinator*, and in the Great Bustard, *Otis tarda*, of western and central Asia. *Cygnus buccinator* and the White Pelican, *Pelecanus erythrorhynchos* (weight, ca. 5–7 kg), have wing spans of up to about 3 m; but the greatest record wing span among living birds (3.5 m) is that of the Wandering Albatross, *Diomeda exulans* (weight, ca. 8–10 kg). Among the living flightless, swimming species the Emperor Penguin, *Aptenodytes forsteri* (length, 1.2 m; body weight, 42.5 kg), is the largest.

The class Aves is doubtless monophyletic. As in other classes of animals, phylogenetic relationships are assessed differently by different specialists. The expression of relationships in terms of orders and families, and their serial arrangement, are subjective. A recent classification by Storer (1971) recognizes 28 orders with contemporary species. (The number of contemporary species is indicated in parentheses.)

Tinamiformes—Tinamous (42 species)
Rheiformes—Rheas, nandus (2)
Struthioniformes—Ostrich (1)
Casuariiformes—Emus, cassowaries (5)
Dinornithiformes—Kiwis (3)
Podicipediformes—Grebes (18)
Sphenisciformes—Penguins (17)
Procellariiformes—Albatrosses, shearwaters, fulmars, petrels (ca. 95)
Pelecaniformes—Tropic birds, boobies, cormorants, pelicans, frigate birds (57)
Anseriformes—Screamers, ducks, geese, swans (153)
Phoenicopteriformes—Flamingos (6)
Ciconiiformes—Herons, bitterns, boatbills, whale-headed storks, hammerheads, ibises, spoonbills, storks (116)
Falconiformes—New World vultures, ospreys, Old World vultures, kites, hawks, eagles, secretarybirds, falcons (226)
Galliformes—Curassows, chachalacas, hoatzin, megapodes, guineafowl, pheasants, turkeys (275)
Gruiformes—Rails, coots, sungrebes, Kagu, sunbittern, roatelos, buttonquails, plainswanderer, cranes, Limpkin, trumpeter, seriemas, bustards (204)
Charadriiformes—Jacanas, painted-snipes, oystercatchers, ibisbill, stilts, avocets, crabplover, thick-knees, pratincoles, coursers, plovers, lapwings, woodcocks, snipes, sandpipers, turnstones, phalaropes, seedsnipes, sheathbills, sandgrouse, skuas, jaegers, gulls, skimmers (313)
Gaviiformes—Loons (4)
Columbiformes—Pigeons, doves (306)

Psittaciformes—Parrots, lories (339)
Cuculiformes—Plantaineaters, turacos, cuckoos (150)
Strigiformes—Barn owls, owls (145)
Caprimulgiformes—Oilbird, frogmouths, owlet-frogmouths, potoos, nightjars (95)
Apodiformes—Crested swifts, swifts, hummingbirds (410)
Coliiformes—Colies, mousebirds (6)
Trogoniformes—Trogons (34)
Coraciiformes—Kingfishers, todies, motmots, bee-eaters, rollers, groundrollers, cuckoo-rollers, hoopoe, wood-hoopoes, hornbills (192)
Piciformes—Jacamars, puffbirds, barbets, honeyguides, toucans, woodpeckers, piculets, wrynecks (391)
Passeriformes—Songbirds and related species. The largest and most diverse of the orders. Storer recognizes 58 families (5044)

Alternative classifications are available, and much work is still needed in systematic theory and the methodology of taxomonic studies. Our understanding of classification is necessarily tied to an understanding of the genetic similarity and phylogenetic histories of the organisms involved.

Birds have evolved an almost bewildering array of trophic adaptations. The trophic dimension is often the most important in the definition of the niche of a species. Adaptations have evolved that allow the exploitation of most potential food items on the surface of the earth, in the air over its surface, and in the upper layers of its waters. Only the fossorial mode of feeding appears not to have been evolved although some species obtain food items from the upper few centimeters of the surface of the earth. Birds occur in the second (herbivore), third (carnivore including insectivore), and fourth (secondary carnivore) trophic levels. Some species (omnivores) have roles in more than one level; these roles may change seasonally. Some species are involved in symbiotic and commensal relationships.

II. The Integument

The integument of birds is characterized by the production of feathers of a substantial variety, scaly coverings of legs and feet, claws, horny coverings of legs and feet, claws, uropygial glands, horny coverings of the beak (ramphothecae) and, in various systematic groups, such structures as spurs, webs, wattles, combs, anal glands, and outer ear glands. The integument and its derivatives have been reviewed recently by Stettenheim (1972) and analyzed in much greater detail by Lucas and Stettenheim (1972).

As in other vertebrates, the skin consists of two principal layers. These are the epidermis which is derived from ectoderm, and the dermis, which is derived from mesoderm through mesenchyme in contact with skin ectoderm. Proliferation of cells, differentiation, and surface sloughing continue in the epidermis throughout the life of the bird.

Feathers distinguish birds from all other vertebrates and are the most complex derivatives of the integument in the entire phylum. On structural bases several kinds of feathers can be recognized (Chandler, 1916; Stettenheim, 1972). These include the contour feathers, semiplumes, down feathers, bristles, and filoplumes. Although the mature feather is almost exclusively an epidermal derivative, the follicle in which it is formed contains both dermal and ectodermal elements. The pulp of the developing feather is of mesodermal origin. Follicles develop in a regular pattern that is responsible for the definitive feather tracts of the adult birds. Follicles, feather muscles, and surrounding dermis have a rich, complex innervation and vascularization. The development and structure of follicles are described concisely, with extensive references to the literature, by Stettenheim (1972).

Feathers are replaced in temporal patterns that differ extensively among the major systematic groups (Stresemann and Stresemann, 1966; Voitkevich, 1962; Palmer, 1972; references to much of the very extensive literature can be found in these three treatises). Despite much research, the mechanisms that control molt are poorly understood (Payne, 1972; see also Chapter 4, this volume). The widely accepted assertion that the thyroid gland plays a major role (reviews by Assenmacher, 1958; Voitkevich, 1962) has not been sustained by careful experimental investigation (e.g., Tanabe and Katsuragi, 1962). Indeed it seems possible that the increase in thyroid activity often observed at the time of molt may be a thermoregulatory adjustment to decreased insulation.

Feathers appear to have evolved as insulative structures in the development of homeothermy and thermoregulation, and the specialization of flight feathers came later (Ostrom, 1974). In addition, feathers, by virtue of color and pattern, have important roles in protective coloration. They are also often involved in behavioral displays, frequently by alteration of feather position by means of subcutaneous striated muscles, nonstriated pennamotor muscles, and at least in some cases by nonstriated apterial muscles and tensor muscles.

The contour feathers are arranged in tracts called pterylae. The intervening areas, apterae, may have down feathers in varying density. Characteristic patterns of pterylae and apterae can be recognized at the level of families or higher groups. As early as 1840 Nitzsch and Burmeister proposed a highly useful and logical nomenclature for the pterylae. This has been adopted, with some changes, in the treatise of Lucas and Stettenheim (1972).

Unhydrolyzed feathers contain about 90% protein, mostly keratin, the nature of which is presented in Chapter 4. A variety of nitrogenous nonkeratin compounds occurs in feathers. These include water-soluble proteins, water-soluble polypeptides with amino acid compositions similar to protein, free amino acids, ammonia, uric acid, and purines. The feather keratin is structurally different from that in the beak or scale.

In several species of flying birds, feathers contain about 2% lipids which appear to be characteristically different from lipids in other vertebrate tissues. The differences reside in the high content of cholestanol and free fatty acids (Bolliger and Varga, 1961). Cholestanol has been suggested as having a role in waterproofing. However, there is the possibility that the cholestanol is derived from the uropygial gland (oil gland) and is not intrinsic to the feathers (Spearman, 1966; see also, Chapter 6, this volume).

The plumages of birds have a great variety of colors and color patterns. The literature devoted to the myriad aspects of this field is extensive. The colors of feathers may be produced by pigmentary colors, structural colors, or combinations thereof. They are discussed at length in Chapter 5. The control mechanisms responsible for pattern production are complex and virtually unknown.

The skin of birds is basically of the vertebrate type with an epidermis of ectodermal origin and dermis derived from the mesoderm. Bird skin is relatively thin, elastic, and often translucent. It has relatively few attachments to body muscles, but, unlike other vertebrates, has numerous attachments to the skeleton. The avian skin, except for the uropygial gland (Chapter 6), was long considered essentially glandless. However the investigations of Varićak (1938), Lucas (1968), and Lucas and Stettenheim (1972) suggest that it is extensively a sebaceous secretory organ. Sweat glands, however, are lacking.

III. The Skeletomotor System

The primary characteristics of this system, in comparison with other classes of vertebrates, are reflections of the essential requirements for flight. Secondary characteristics are associated with the types of flight and other modes of locomotion, the nature of food taken, and the mode of food procurement. The literature on the anatomy of the avian skeletomuscular system and its adaptations is vast. Most of the major contributions were reviewed by Bock (1974). Flying birds alternate the support of body weight between the wings and legs or in many species, between the wings and whole-body buoyancy in water. The degree to which any system serves this function varies greatly. Terrestrial flightless birds show adaptations to weight-bearing by legs alone. Penguins alternate between support by buoyancy and by legs. Swimming is effected by use of legs as in ducks and grebes or by use of wings as in the penguins. The vertebral column of the trunk is relatively short and rigid. The ribs bear distinctive uncinate processes. Apart from birds such processes are known only in a few reptiles including the tuatara *(Sphenodon)*. A prominent ventral keel of the sternum serves for insertion of flight muscles. The skeleton is relatively light and certain bones contain air sacs which are components of a complex system associated with the respiratory tract although they have no role in external respiration. Air sacs are often reduced in diving birds in which increased density reduces bouyancy.

Adaptation of the forelimb for flight has resulted in an extensive reduction of the number and the mobility of carpal and metacarpal bones and phalanges. The pelvic bones are fused and, in turn, fuse with the synsacrum.

The femur is relatively short and sometimes contains an air sac. The fibula is reduced drastically. Some of the tarsal bones are fused with the distal end of the tibia forming a tibiotarsus. The remaining three tarsal bones are joined with three fused metatarsal bones to form a single tarsometatarsus. Thus the "ankle joint" between tibiotarsus and tarsometatarsus is relatively further from the end of the limb than is the true ankle joint in most other terrestrial vertebrates. The feet display an extensive array of modifications as adaptations for mode of locomotion (aquatic, terrestrial, cursorial, etc.) and, as in raptors, also for procurement of food.

The skull is relatively light, a further adaptation to flight. In most species the bones that make up the brain case are at least partially "pneumatized," that is, they contain extensive air spaces between the inner and outer surface which serve to optimize strength with minimum weight. As in the reptiles, the mandible consists of several bones and hinges on movable quadrate bones, thus providing a double-jointed, wide-gaping articulation. In many species there is also a movable articulation of the maxilla with the quadrate bones. The relatively large eyes displace the brain dorsally and posteriorly. As in the reptiles the skull articulates with the first vertebra through a single condyle.

The literature on the skeletal muscles of birds is extraordinarily extensive. The most useful review of general and comparative morphology is that of George and Berger (1966) which contains an extensive list of references. General functional aspects are treated extensively by Bock (1974). Histologically and ultrastructurally avian skeletal muscle conforms basically to the general plan of skeletal muscle of terrestrial vertebrates. There are conspicuous differences in the amount of mass invested in the various groups of skeletal muscles. In active, powerful fliers, such as hummingbirds and swallows, 25–35% of the body mass consists of flight muscle (Hartman, 1961). Conversely, only 2% or less occurs in the leg musculature. In ducks and geese equal fractions of body mass, about 30% each, occur in the flight and leg musculature. In addition to the muscles with origins and/or insertions on the skeleton, there are sheets or bands of subcutaneous striated muscles beneath the loose areas of skin. A complex system of smooth muscles is involved in the movement of the feathers (Stettenheim, 1972).

The occurrence of red and white muscle fibers in the pectoralis muscle, the principal flight muscle of birds, has long been known. The extensive studies by George and his associates (see George and Berger, 1966, for summary) have contributed significant information concerning their relative distribution and metabolic features. They may be regarded as the two extremes of metabolic and structural specialization, but intermediate types occur. Some pertinent structural and biochemical features are outlined in Table I. In the domestic fowl, and possibly in other galliform species, the pectoralis contains primarily white and intermediate fibers. In the domestic pigeon, and perhaps other columbiform species, the Japanese quail *(Coturnix coturnix)*, and species in several other orders only white and red fibers occur. A variety of good flying species have only red and intermediate fibers. Many other good fliers have only red fibers. There appears to be a correlation between sustained flight and red fibers, and between brief rapid flight and white fibers, but this is certainly an oversimplification. In migratory species, which use lipids as fuel for flight, the pectoralis appears to be composed largely, or exclusively, of red fibers. George and Berger (1966) describe the activity of red fibers as "sustained" in contrast to white fibers as "rapid." Although it may be semantic, it seems better to designate the activity of the white fibers as limited to a brief period of rapid contractions supported by anaerobic glycolysis. In my view, Bock (1974) correctly asserted that the dichotomies of red and white fibers and that of twitch and tonus fibers differentiated by their physiological properties, innervation, and sarcotubular systems, i.e., "Fibrillenstruktur und Felderstruktur" (Krüger, 1952), be maintained independently, at least until further information becomes available.

TABLE I
SELECTED DIFFERENCES BETWEEN THE RED AND WHITE TYPES OF AVIAN MUSCLE[a]

	Red	White
Color	Red	White or pinkish
Diameter	Small	Large
Blood supply	Rich	Scant
Position of nuclei	Within fiber	Perimeter of fiber
Myoglobin content	High	Low
Sarcoplasm	Granular	Clear
Mitochondria	Abundant	Sparse
Fat content	High	Low
Glycogen content	Low	High
Lipase activity	High	Low
Activities of glycogenic and glycolytic enzymes	Low	High
Lactic dehydrogenase activities		
Cytoplasmic	Low	High
Mitochondrial	High	Low
Other oxidative enzymes	High	Low
Alkaline phosphatase	Fairly high	Moderate
Acid phosphatase	Moderate	Fairly high
ATPase (acid)	Low	High
ATPase (neutral)	High	Low
Acetylcholinesterase	High	Low
Butyrylcholinesterase	Low	High

[a] Adapted from George and Berger (1966)

IV. Digestion and Nutrition

The digestive apparatus of birds adheres to the basic vertebrate pattern but is overlaid with conspicuous adaptations for flight and dietary regime. The primary adaptation to flight has been a posterior shift of much of the cephalic phase of the physical processes of digestion. This has involved the development of two stomachs, an anterior secretory proventriculus, and posterior grinding stomach, gizzard, or ventriculus. In some groups there is an additional chamber between the ventriculus and the small intestine. The thickness of the muscle of the gizzard and its koilin lining varies in correlation with the nutritional regime. In many species an evagination of the esophagus, the crop, has a storage function. In some species storage is effected in the ventriculus, proventriculus, or in an unmodified esophagus. The central position of the relatively large ventriculus, as well as the more central temporary storage of food, and the relatively lighter head, contribute to a more central center of gravity that is favorable to flight.

The koilin of the lining of the ventriculus is a secreted polysaccharide–protein (Luppa, 1959; Webb and Colvin, 1964) of apparently unknown composition.

The morphology of the remainder of the digestive system does not differ conspicuously from the general vertebrate plan (for details, see Ziswiler and Farner, 1972). Paired caeca occur at the junction of the small and large intestine in most, but not all, groups. These are especially well developed in some galliform species for which there is evidence that the caeca may be the site of microbial digestion of cellulose and the absorption of the resulting volatile fatty acids (e.g., Gasaway, 1976a,b,c). The terminal chamber, the cloaca, is shared with the reproductive and excretory systems.

The enzymatic hydrolysis of fats, carbohydrates, and proteins in the avian digestive system differs in only minor ways from other vertebrate classes. Gastric juice, including pepsin, is produced in the proventriculus, whereas pepsin-catalyzed hydrolysis of peptide bonds occurs in the ventriculus. There is evidence of a gastric chitinolytic system in some species (Jeauniaux, 1962). It involves two enzymes, a chitinase and a chitobiase (*N*-acetylglucosaminidase), which together hydrolyze chitin to acetylglucosamine. Whether acetylglucosamine can be used metabolically is unknown. Even in insectivorous birds, it would be only a trivial source of energy. There is the possibility, however, that the system frees chitin-bound protein that would otherwise be unavailable.

The crops of doves and pigeons (Columbiformes) are distinctive in the production by both sexes of a milk that is fed to the young (see Fisher, 1972, for a brief review and references). Crop milk contains about 12% protein and 9% lipid. In the Greater Flamingo, *Phoenicopterus ruber,* and probably in other species of this genus, the esophagus produces a fluid that is initially the sole source of nutrition for the young. It contains about 8% protein and 18% fat (Lang *et al.*, 1962, Lang, 1963; Wackernägel, 1964). The male Emperor Penguin, *Aptenodytes forsteri,* which incubates the single egg and performs the initial care of the chick, produces a nutrient fluid in the crop during the brooding period (Prévost, 1961; Prévost and Vilter, 1963). On this sole source of nutrition the chick grows from a hatching weight of 300 gm to about 600 gm before it takes other food. This is a remarkable mobilization and transfer of energy in light of the fact that the male has no food intake throughout incubation and this period of the care of the young during which his weight decreases 12–15 kg from an initial 35 kg. The dry content of this fluid consists of 59% protein, 29% lipid, and 5.5% carbohydrate. Some procellariiform species apparently secrete an oil in the proventriculus that is used nutritionally by the young. But there is still some uncertainty about its origin and functions (for review, see Fisher, 1972).

The relatively sparse information on the processes of transfer of chemical species from the lumen of the gut into epithelium and thence into the circulatory system suggests that transport mechanisms are generally similar to those of mammals which have been investigated more extensively (for brief summaries, see Hudson *et al.*, 1971; Ziswiler and Farner, 1972). The absorption of calcium, which involves both passive and active transport, is discussed in Chapter 8 of this volume.

An interesting and important derivative of the digestive system is the bursa of Fabricius. First described in 1621 (Fabricius ab Aguapendente, 1621), the bursa arises embryologically as a dorsal, cranially directed outgrowth of the coprodeum (Romanoff, 1960). Its development can be arrested or prevented by treatment of the early embryo with sex steroids. Normally regression occurs during the first year of life. The bursa has a critical role in the production of immunoglobulin and immunologic competence during the early life of the bird. It is thought that it serves as the source of immunologically competent cells for other lymphatic tissues and that the cells then proliferate in the recipient tissues and become independent of the bursa (for review, see Payne, 1971). But the alternate hypothesis of a bursal hormone that induces immunocompetence in other lymphoid tissues has not been precluded.

V. Excretion, Osmoregulation, and Ionoregulation

These interrelated homeostatic processes must be viewed in birds first from the general aspect of the uricotelic nitrogen metabolism which is characteristic of the entire class. Uric acid constitutes about 80% of the urinary nitrogen; ammonia is the next most abundant nitrogenous constituent. It seems most likely that uricotelism evolved primarily as an embryonic adaptation that converts nitrogenous wastes into a relatively nontoxic form. Uric acid also has relatively low osmotic activity. Although the conversion of α-amino nitrogen into uric acid has a significant energy cost, it provides a means by which the developing embryo can exist with its accumulating waste nitrogen. In the adult bird a continued uricotelic nitrogen metabolism, although with a smaller but significant excretion of ammonia nitrogen, apparently provides some basis for water conservation. A urine with a low water : nitrogen ratio may also be of some advantage in flight. For a useful review of these functions consult Shoemaker (1972).

The kidney is the major organ of nitrogen excretion and it also plays significant roles in osmo- and ionoregulation. The avian kidney possesses two basic types of nephrons. The "reptilian" type, which occurs near the surface of the cortex, lacks the loop of Henle. The second or "mammalian" type, which occurs in the medulla, has a loop of Henle between the proximal and distal tubules. Because the avian kidney has a renal portal vein, the peritubular capillaries receive blood from three sources. These are branches of the renal portal vein, the efferent glomerular arterioles, and occasional arterioles without glomerular involvement (Siller and Hindle, 1969). It is clear that the flow rates from these sources are adjustable. This suggests that the ratio of uric acid excreted via filtration to that removed by tubular secretion can be varied and that a very low ratio would conserve water. This could provide the basis for physiological control or for evolutionary adaptation to the environmental availability of water. Similarly, an increase in the ratio of functioning mammalian-type tubules to reptilian-type could provide the basis for increased water conservation, either as a mechanism for physiologic control or a basis for evolutionary adaptation (Dantzler, 1966). A further adaptive development for conservation of water may be that of increased length of the loop of Henle as indicated by the relative thickness of the medulla (Johnson, 1974). If the arrangement of the loops of Henle and the collecting ducts constitutes a countercurrent multiplier, then longer loops of Henle could provide higher osmolality in the medulla and greater water recovery (Poulson, 1965; Skadhauge, 1973, 1975). There is as yet insufficient physiologic information to assess these suggestions derived from anatomical studies. It should be emphasized that the concentrating capacities of avian kidneys are generally not as great as those of mammals (Johnson, 1974; Skadhauge, 1973, 1975).

A major control of kidney function is exerted via the hypothalamic (posterior pituitary) hormone, arginine vasotocin, which increases the permeability to water of the distal tubules and collecting ducts (Shoemaker, 1973; Skadhauge, 1973). Urine is moved through the ureter by peristalsis (Gibbs, 1929). Uric acid occurs as a precipitate in the collecting ducts and ureters (McNabb and Poulson, 1970). At low flow rates urine is viscous because of the presence of mucus (Shoemaker, 1972). Ureteral urine is further modified by reabsorption of salt and water in the cloaca and large intestine (Skadhauge, 1973, 1975). Clearly the cloaca and large intestine are important organs in both iono- and osmoregulation (Shoemaker, 1973).

The occurrence of paired nasal glands within or around the orbits of birds has been known for three centuries. Although Schildmacher (1932) demonstrated a correlation between their size and the intake of sea water, it remained for Schmidt-Nielsen *et al.* (1958) to demonstrate their function as osmo- and ionoregulatory organs. These "salt glands" secrete hypertonic salt solutions as adaptations to the intake of sea water or to desert conditions in which body fluids tend to become hypertonic. Ensuing investigations by Schmidt-Nielsen and his colleagues, and others, indicate that functional salt glands occur in species of at least nine orders of birds but are conspicuously absent among passerine birds. The function of the salt gland is the subject of reviews by Peaker (1975) and Peaker and Linzell (1975).

VI. The Respiratory System

The avian respiratory system has three major functions: gas exchange in external respiration, thermoregulation, and vocalization. Morphologically the system is unique among all air-breathing vertebrates. At the bifurcation of the trachea to form the two primary bronchii, there is a small chamber, the syrinx, which is the primary organ of vocalization. The paired lungs are relatively small and inexpansible. A very complex system of air sacs whose detailed morphology varies extensively among the species of birds is connected with the lungs and bronchi by multiple pathways. Some of the air sacs occur within bones. They are membranous, elastic, and lightly vascularized. Although the air sac system has no significant role in respiratory gas exchange, the posterior air sacs (abdominal and postthoracic) have an important role as bellows that drive air through the lung. The air sac system is described in detail in the excellent monograph of Duncker (1971).

The lungs are rigidly extended systems of inelastic tubes of constant volume. The finest (tertiary) bronchi, the parabronchi, constitute an anastamosing system connected to the secondary and primary bronchi in such a manner as to make the avian lung a "through-flow" system in contrast with the "dead end" lungs of other terrestrial vertebrates. The parabronchi give rise to numerous fine (diameter 3–10 μm) anastamosing air capillaries which are interlaced with blood capillaries that provide the surfaces for gas exchange (e.g., West *et al.*, 1977). The exchange surface per unit volume is about 10 times that of the mammalian lung. The careful studies of Duncker (1971) indicate that the lung actually consists of two parts, an extensive "paleopulmo" in which the flow of air is doubtless anteriorly unidirectional and relatively constant, as first proposed by Hazelhoff (1943) and later by Bretz and Schmidt-Nielsen (1971), and a smaller laterocaudal part, the "neopulmo," whose parabronchial network is interposed between the primary bronchus and the posterior air sacs. In the neopulmo the air flow reverses with inspiration and expiration. Duncker (1971) suggests that in the higher orders of birds resting respiration can be sustained by the neopulmo alone and that the paleopulmo becomes functional only with greater activity, such as flight.

Contrary to the situation in mammals, gas exchange in the avian lung, because of the constant unidirectional flow of air, at least in the paleopulmo, never approaches diffusion equilibrium. This doubtless accounts for the fact that the avian lung extracts a higher fraction of oxygen from air than does the mammalian lung and is important for adequate respiration at high altitudes and in intensive activity. Experiments by Scheid and Piiper (1972) and Piiper and Scheid (1973) argue most forcefully that the physical arrangement of the blood capillaries to the air capillaries is that of crosscurrent relationship. This conclusion is supported by recent anatomical studies (Duncker, 1974; Abdulla and King, 1975; West *et al.*, 1977). West *et al.* (1977) suggest that an auxiliary countercurrent arrangement may exist. Although crosscurrent diffusion is not as effective as countercurrent diffusion it nevertheless provides a system in which an equilibrium such as occurrs in the mammalian alveolus between inspiratory phases is avoided. If crosscurrent diffusion is generally applicable in the avian lung it would mean that the reversal of the direction of air flow proposed by Duncker (1971) for the neopulmo can occur without affecting gas exchange appreciably.

The oxygen-carrying capacity of avian blood is similar to that of mammals. Older reports of a lower capacity are the result of the failure to recognize that the nucleated erythrocytes of birds have a high rate of oxygen consumption (Lutz *et al.*, 1974) so that undetected losses of oxygen occurred in the samples in the process of collection and analysis.

VII. Circulation

In principle, the circulatory systems of birds and mammals are similar. But to a substantial extent this represents evolutionary convergence incident to the development of homeothermy and high metabolic rates. The important differences between birds and mammals in both the morphology and physiology of the circulatory system are discussed in a review by Jones and Johansen (1972).

As in all nonmammalian vertebrates the avian erythrocyte is nucleated. Blood also contains three types of granulocytes, lymphocytes, monocytes, and thrombocytes. The thrombocytes are fragile mononucleated cells that perform the same role in coagulation of blood as the platelets of mammals.

The adult avian heart is completely four-chambered. The sinus venosus is greatly reduced and subdivided and does not function as a chamber. The major arteries arise directly from the ventricles. The left atrioventricular valve consists of two cusps as in mammals, but in contrast the right is monocuspid. Avian hearts are proportionately larger than those of mammals. The possible advantages of this are discussed by Jones and Johansen (1972). The contractile cardiac muscle fibers lack T-tubules and have much smaller diameters than those of mammals. The sarcoplasmic tubule system is well developed. Not unexpectedly, avian cardiac muscle is very rich in mitochondria.

Generally, the morphology of the arterial system varies extensively among the species, but conforms in principle with the basic plan of higher vertebrates. The aorta is derived from the right embryonic aortic arch. The carotid arteries vary in their origin and adult morphology (Glenny, 1955; Baumel and Gerchman, 1968).

The legs of many species of birds contain relatively compact networks of veins and arteries, the "rete mirabile," which function as heat exchangers that reduce heat loss in the unfeathered parts of the leg (e.g., Irving and Krog, 1954; Steen and Steen, 1965). Another rete, the "rete mirabile ophthalmicum," occurs in the temporal region of the head of many, if not all species. It appears also to function as a heat transfer organ that allows the temperature of the brain to remain lower than body temperature, especially during hyperthermia (Kilgore *et al.*, 1976).

The venous system includes three portal systems. The hypophyseal portal system carries blood from the hypothalamus to the pars distalis. Most commonly it consists of two separate groups of portal veins that connect the distinct anterior and posterior parts of the median eminence of the hypothalamus with the cephalic and caudal lobes of the hypothalamus, respectively (Vitums *et al.*, 1964; Dominic and Singh, 1969). The renal portal system permits a variable flow of blood from the posterior part of the body through the peritubular capillaries of the kidney. Cephalic and caudal renal portal veins arise from the external iliac veins and supply blood to the peritubular capillaries. But the caudal renal portal veins also anastomose with the coccygeomesenteric vein. Other veins from the body and legs join the system. At is junction with the renal vein the external iliac vein has a valve, the renal portal valve, which contains doubly innervated smooth muscle which can control the rate of flow of blood into the renal portal veins. However, blood in the posterior renal portal veins may flow to the renal capillaries or directly to the liver via the coccygeomesenteric vein. Blood from the anterior renal portal veins may flow into the renal capillaries or into the vertebral venous sinus and thence to the jugular vein (Akester, 1967). The third portal system, the hepatic, constitutes a major blood supply to the liver. In the domestic fowl, at least, there are two hepatic portal veins. The larger right hepatic vein receives blood from most of the digestive tract and the spleen, whereas the smaller left portal vein receives blood from the proventriculus and gizzard. The coccygeomesenteric vein from the large intestine and, as noted above, from the renal portal system flows into the right heaptic portal vein (Akester, 1971).

VIII. Flight

Birds as flying machines have been the subject of numerous investigations. A good nontechnical general review is that of Dorst (1974). The mechanics and aerodynamics of flapping flight are discussed in detail by Oehme (1963), Pennycuick (1975), and Greenewalt (1975). Considerable information is also available on other aspects such as steering (Oehme, 1976a,b,c), hovering flight (Greenewalt, 1975; Hainsworth, Chapter 9, this volume), gliding (Nachtigall, 1975), and soaring (Pennycuick, 1975). Pennycuick suggests that in large birds the cost of operating the tonic muscles in soaring is of the order of magnitude of standard metabolic power and therefore represents a substantial saving of energy in comparison with the cost of flapping flight. It is of interest to note that several species, such as the White Stork *(Ciconia ciconia)*, Black Stork *(Ciconia nigra)*, and some species of eagles perform long-distance migration almost exclusively by soaring. Albatrosses and shearwaters execute long-distance foraging flights primarily by soaring. The energy cost of flight is discussed briefly in Section IX and more extensively by Berger and Hart (1974) who also discuss other physiologic adjustments during flight.

IX. Metabolism and Thermoregulation

Birds as homeotherms uniformally have high rates of energy intake and metabolism. Metabolic rate (metabolic power) whether standard (or basal), existence level, or at any level of activity, is a function of about the 0.7th power of body mass. The following approximate relationship holds:

$$\dot{H}_{sm} = aM^b,$$

where $\dot{H}_{sm}$ is standard metabolic power in watts, M is weight in kilograms. For passerine birds a is about 6.6 and b is about 0.72; for nonpasserine birds a = ca. 4.3, b = ca. 0.73 (Dawson and Hudson, 1970; Aschoff and Pohl, 1970; Calder and King, 1974; Calder, 1974; see also Chapter 9). These values do not differ greatly from those of mammals. Under natural conditions $\dot{H}$ is affected by thermoregulatory requirements, motor activity, and other factors. In summit metabolism, $\dot{H}_{max}$, i.e., the highest metabolic power that can be generated under standard conditions to maintain body temperature in low environmental temperature, a = ca. 20 and b = ca. 0.65. In flight a is probably in excess of 50 with b = ca. 0.73. This indicates a tenfold difference between standard metabolic power and metabolic power in flight but this value certainly varies extensively according to species and with flight conditions (e.g., Bernstein *et al.*, 1973; Farner, 1970; Greenewalt, 1975; Kespaik, 1968; Lyuleeva, 1962, 1970; Tucker, 1968, 1972, 1973).

The methods by which the energy costs of flight have been estimated are briefly outlined by Farner (1970). Among these are the ingenious wind tunnel experiments of Tucker (summaries and reviews, 1974, 1975) and Bernstein *et al.* (1973) in which the most precise measurements can be made. Tucker himself (e.g., 1975) recognized that the carrying of a face mask and tube and flight with no forward progress may impose an increment of energy cost over the natural level of utilization of energy in flight. The analyses of Greenewalt (1975) suggests that calculations based on wind tunnel experiments may give results that are as much as 2.5 times higher than the cost of natural flight. When the energy costs of flying for birds and running in mammals are compared as functions of body weight, the cost of flight is lower, however swimming costs even less (Schmidt-Nielsen, 1972).

Some strides have been made in the development of an understanding of the uses of energy under natural conditions (King, 1974). These activities include incubation (Drent, 1973, 1975; Kendeigh, 1963, 1973), reproduction (e.g., Kendeigh, 1952, 1973; King, 1973; Ricklefs, 1974), migration (e.g., Hussel, 1969; Nisbett *et al.*, 1963; Raveling and LeFebvre, 1967), molt (e.g., Dolnik, 1965; Kendeigh, 1973), growth of young (e.g., Westerterp, 1973; Drent, 1977), and cost of procurement of food (see Chapter 9, this volume). Nevertheless, the difficult field of the partition of energy under natural conditions through the course of the annual cycle with variable climate conditions is still in an early stage of development, although such problems have been investigated more thoroughly in birds than in any other group of animals.

However energy is partitioned among basal metabolism, growth or replacement of tissue components, transport functions, thermogenesis in thermoregulation, internal communication and regulation, or manifold modes of external work through striated muscle, it must come ultimately from the food intake. Although an overwhelming fraction of the investigations on the intermediary metabolism of birds has been effected on the domestic fowl there appear, at least at this time, to be no significant qualitative differences among the species of birds in nutritional requirements (Fisher, 1972) or in general pattern of intermediary metabolism, which are in most aspects similar to those of mammals (Hazelwood, 1972). The characteristically avian features of intermediary metabolism have been summarized for carbohydrates by Hazelwood (1972, 1976) and Pearce and Brown (1971). Similar summaries are available for amino acids and proteins, including the formation of uric acid from α-amino nitrogen (Boorman and Lewis 1971; Griminger, 1976a), and for lipids (Griminger, 1976b; Annison, 1971; Hazelwood, 1972). Again this work is based primarily on domestic species.

Energy released in metabolism is used both for internal work (synthesis, transport, communication, etc.) and external work (locomotion, food intake, nest construction, etc.). In such functions, 70% or more of the energy is thermodynamically unavailable and "lost" as heat. Birds as homeotherms maintain a heat content sufficient to hold body temperature within relatively restricted limits which appear to be almost independent of body size (King and Farner, 1961; Calder and King, 1974). Environmental temperatures below the thermoneutral zone require additional chemical thermogenesis. Above the thermoneutral zone additional heat is produced in the work of hyperventilation, by increased rate of circulation of blood and by the general elevation in metabolic rate of tissues. The extensive physiological and behavioral mechanisms of thermoregulation are reviewed by King and Farner (1961, 1964), Calder and King (1974), Richards (1975), and Shilov (1968). Temporary hypothermia or torpor occurs in species of the orders Apodiformes and Caprimulgiformes. Heat loss is reduced through reductions in the temperature gradient between body and environment. Energy is also spared by the concomitant reduction in metabolic rate incident to the lower body temperature. It is probable that nocturnal hypothermia, regulated at a lower limit, is a constant feature of the metabolism of most species of hummingbirds for much of the year. Continuous homeothermy is mitigated against by the high surface–volume ratio of small body size. The Poorwill, *Phalaenoptilus nuttalii*, performs true hibernation when the availability of food becomes too low to support homeothermy (Jaeger, 1948, 1949). The avian hypothermic state is described in detail by Calder and King (1974).

X. Integration and Coordination

As in all vertebrates integration occurs at intracellular and organismal levels, the latter being effected by nervous and endocrine systems. The separation of these systems is largely a matter of convenience which has arisen historically because of methodologies and the aspects which investigators have assumed in their research. There are numerous functional relationships between the systems and, indeed, they share a common organ in the hypothalamus. In this volume this traditional separation is maintained with the separate treatment of the endocrine system (Chapter 7).

Despite innumerable hiatuses in our knowledge, the literature on the avian nervous system is vast and suffers extensively from the lack of a standardized nomenclature. A useful general review is that of Cohen and Karten (1974). In general, however, the organization and basic physiology of the avian nervous system conform to the general plan of the nervous system of higher vertebrates. In external morphology it is similar to that of modern reptiles. There are 12 cranial nerves, the XIIth being derived ontogenically from the first somites of the spinal cord. The number of pairs of spinal nerves is variable. For example, there are 51 in the Ostrich, 38 in the domestic pigeon, 36 in the domestic fowl. The number of pairs of cervical nerves is most variable and corresponds with the number of cervical vertebrae. Somatic afferent innervation and somatic and visceral efferent innervation were reviewed briefly by Bennett (1974).

As in other vertebrate animals, the *autonomic* nervous system consists of those efferent nerve fibers that have synapses in ganglia outside of the central nervous system. In birds the postganglionic fibers of the autonomic system innervate smooth muscle, many secretory cells, cardiac muscle and the intrinsic striated muscle of the eye. The traditionally categorical separation of these fibers into sympathetic and parasympathetic components, and especially the designation of them, respectively, as adrenergic and cholinergic, although convenient, now often proves to be misleading (Campbell, 1970; Bennett, 1974).

The preganglionic fibers of the *cephalic autonomic system* leave the brain via cranial nerves III through XII. Additional fibers come from the cranial autonomic ganglionic. Generally the preganglionic fibers that leave the brain in cranial nerves III–IX synapse with postganglionic neurons whose axons innervate appropriate cephalic effectors. Because of numerous anastomoses, both the fine anatomy and physiology of this part of the autonomic nervous system are incompletely understood. A large component of preganglionic fibers pass peripherally in the vagus which also contains afferent visceral fibers. The ganglia supplied by these fibers generally lie near or within the cervical, thoracic, and abdominal organs that are innervated by the postganglionic fibers. From the cervical through the lumbar segments relatively short preganglionic autonomic fibers pass from the spinal cord via the ventral roots. In the thoracolumbar segments these fibers synapse in the paravertebral chain or in the prevertebral ganglia. In the cervical region the preganglionic fibers pass rostrad to synapses in the superior cervical ganglionic which also receives fibers that leave the brain in cranial nerves. Postganglionic fibers are relatively long and pass either to visceral plexus or directly to effectors. Possibly uniquely avian is the ganglionated nerve of Remak which extends from the caudal end of the rectum to the cephalic end of the duodenum. According to Bennett (1974) it is continuous at its cephalic end with the celiac and mesenteric plexus and also receives vagal autonomic fibers. Caudally, it is continuous with the lower extensions of the paravertebral trunks and with the ganglionic pelvic plexuses. Remak's nerve receives numerous fibers from the paravertebral chains of ganglia and has numerous connections with the gut and its intrinsic ganglia.

As a gross generalization the afferent vagal fibers tend to provide direct communication from the brain to specific groups of effectors. Autonomic fibers from the spinal cord, because of their numerous ganglionic connections lack this specificity, and as a system have a more general and integrative function.

Cholinergic, adrenergic, noradrenergic, and noncholinergic–nonadrenergic fibers have been demonstrated in the avian autonomic nervous system. Morphologic and functional relationships of this system, including the intrinsic components of visceral organs, have been ably and concisely reviewed by Bennett (1974) on which this summary is largely based.

As in reptiles, the spinal cord is approximately of the same length as the neural canal with the consequence that the spinal nerves extend laterally from the cord. The internal organization of the cord conforms with general plan for higher vertebrates although much remains to be learned about its functional organization. A brief and useful description has been assembled by Pearson (1972).

A characteristically avian structure is the glycogen body of the spinal cord (Imhof, 1905). It is a pear-shaped formation of glycogen-filled modified glial cells located within the vertebral column dorsal to the spinal cord at the level of the emergence of the roots of the spinal cord. Gage (1917), Terni (1924), and Romanoff (1960) have summarized the early investigations on this unique organ. Its embryology and vascularization have been presented in some detail by Doyle and Watterson (1949) and Watterson (1949). Its biochemical characteristics have been reviewed briefly by De Gennaro (1959), Hazelwood (1972), and Benzo *et al.* (1975). The last demonstrated that utilization of glycogen via the pentose phosphate cycle is functionally significant. The rates of glycogenesis and glycolysis are relatively low. Despite its position in relation to the spinal cord, its functional significance remains enigmatic.

Generally the avian brain conforms in basic principle with the general vertebrate organization, but has become significantly specialized with the extensive development of the eyes and vestibular apparatus in conjunction with the evolution of flight. Although literature is extensive it is likewise fragmentary and plagued by differences in nomenclature. Important among the general treatises are the monumental anatomical contributions of Kappers (1921, 1947) and Kappers *et al.* (1936). The more functionally oriented treatises of Ten Cate (1936, 1965) and the more recent integrated accounts of Pearson (1972) and especially of Cohen and Karten (1974) are also useful. Some more specific aspects are considered by Portmann and Stingelin (1961), and Stingelin (1965); van Tienhoven and Juhász (1962) and Karten and Hodos (1967) have produced useful stereotaxic atlases. Generally the striking morphological characteristics include much reduced olfactory lobes, relatively large cerebral hemispheres, large optic lobes, and a relatively very large cerebellum. Unlike the case in mammals the increased size of the cerebrum is due to an extensive development of the corpus striatum which performs many of the functions associated with the mammalian cerebral cortex. The cortical components, including the hippocampus, perimygdalar area, and prepyriform areas constitute a relatively small fraction of the cerebral mass.

The development, morphology and function of the avian forebrain, including the olfactory bulbs and lobes, the cerebral hemispheres, the thalamus, and hypothalamus have been subjected to detailed reviews by Pearson (1972) and Cohen and Karten (1974) which provide citations of much of the important literature. Additional information concerning the morphological and functional aspects of the hypothalamus can be found in Wingstrand (1951) and Oksche and Farner (1974). Its general functions, and those of the thalamus, have been reviewed by Pearson (1972). The structure and function of the pineal body, which varies remarkably in morphology among the families and orders, have been reviewed recently by Menaker and Oksche (1974). A notable recent development is the demonstration of an apparently essential role in overt circadian periodicities in the House Sparrow, *Passer domesticus* (Gaston, 1971; Menaker and Zimmerman, 1976). Its functions as an endocrine gland are discussed in Chapter 7 of this volume.

The avian mesencephalon is characterized by large optic lobes whose supraventricular walls constitute the highly striated optic tectum. Although much of the integrative activity of the mesencephalon has, in the evolution of higher vertebrates, shifted anteriorly into the forebrain, the optic tectum remains the site of termination of most of the retinal fibers and it is well developed and intricately organized in a manner similar to that of the reptiles. The lateral mesencephalic nucleus receives fibers from vestibulocochlear system. The tectum has extensive connections both caudally and to the forebrain.

The avian cerebellum is a relatively large organ with distinct transverse folia. The degree of development varies extensively among families and orders. Its size is doubtless to be correlated with its importance in motor coordination and equilibrium necessary for flight. The cerebellum receives sensory information, directly and indirectly, from the vestibulocochlear system, retina, and tactile receptors. It has extensive connections with more anterior parts of the brain and with the spinal cord. The anatomy and functional anatomy of the cerebellum have been reviewed by Pearson (1972).

The medulla oblongata is the most posterior part of the brain. Posteriorly it extends with no clear morphologically identifiable boundary into the spinal cord, which, as in all vertebrates may be considered as its posterior extension. However, the medulla oblongata, as in other vertebrates, has a distinctive morphology and characteristic functions. For birds these have been most recently reviewed by Pearson (1972). The older treatise of Ten Cate (1936) remains useful.

During the past two decades experimental investigations have resulted in concepts of the functional organization of the avian brain. Despite its different origin, the external striatum is clearly functionally and organizationally comparable to the mammalian neocortex. The necessity of a thorough reconsideration of ideas drawn from the classical anatomical investigation is presented eloquently and compellingly by Cohen and Karten (1974).

XI. The Senses and Sense Organs

Birds have the same complement of senses and sense organs as the other higher vertebrates. The evolution of flight was certainly only possible with the simultaneous evolution of highly developed visual and vestibular sensory systems, and a central neural system capable of processing and integrating the information received from them.

The importance of the eyes is emphasized by their relatively large size. The eyes of some hawks and owls are at least as large as human eyes. The eye of the Ostrich is the largest among the terrestrial vertebrates. In some species the combined weight of the eyes exceeds that of the brain. Although avian eyes have the same basic structure and functional organization as those of other vertebrates, there are features that are highly specialized or unique to birds. These have been discussed in considerable detail in the very useful treatises of Walls (1942), Rochon-Duvigneaud (1943), Duke-Elder (1958), Pumphrey (1961), Pearson (1972), and Sillman (1973).

Notable as a unique feature is the pecten, a structure of highly variable morphology that extends into the vitreous body from a base directly over the optic disc. Known to science for three centuries, its functional significance is still controversial. The more than thirty hypotheses have been reviewed and evaluated by Wingstrand and Munk (1965). Their own experiments, and those of others, suggest that by diffusion or transport into the vitreous body, the functions of the pecten may be, possibly among others, nutritive and respiratory. Its great morphological variation leaves open the question of qualitative differences in functions among the families and orders of birds.

Encephalic photoreception must be added to the visual use of light as a source of external information. First demonstrated by Benoit (1935, 1938) in the domestic Mallard, *Anas platyrhynchos*, it was more recently found in *Passer domesticus* by Menaker and his colleagues (e.g., Menaker and Keatts, 1968; Menaker *et al.*, 1970; McMillan *et al.*, 1975a,b); in the White-crowned Sparrow, *Zonotrichia leucophrys gambelii* by Gwinner *et al.* (1971) and Yokoyama and Farner (1976); in the Golden-crowned Sparrow, *Z. atricapilla* by Gwinner *et al.* (1971); and in the domesticated Japanese quail, *Coturnix coturnix* by Kato *et al.* (1967) and Homma and Sakakibara (1971). The photoreceptive elements appear to be diffusely distributed in the region of the infundibular nucleus of the hypothalamus (K. Yokoyama, D. S. Farner, A. Oksche, and T. R. Darden, unpublished). Using this system circadian rhythms in motor activity can be entrained by 24-hour in light–dark cycles (Menaker, 1971; Menaker and Zimmerman, 1976). In species in which it constitutes information for the growth and development of the gonads, day length is "measured" by a system that includes a circadian oscillation in photosensitivity (for review, see Farner, 1975).

As in other terrestrial vetebrates the avian ear provides information on sound and change in position in three coordinates. The avian vestibular apparatus adheres anatomically to the general vertebrate system of three semicircular canals with the associated saculus, utriculus, and lagena and provides information on change in position. The sound-receiving apparatus involves a tympanic membrane; a single middle ear bone, the columella, as in amphibians and reptiles; and, in the inner ear, a noncoiled cochlea with a basal membrane that oscillates in accordance with the frequency and amplitude of sound that impinges on the tympanum. The transduction of mechanical energy into information in the form of nerve impulses, and the transmission and processing thereof, together with proprioreception by the skin, are reviewed by Schwartzkopff (1973) and Ilyichev (1972).

In addition to those of the inner ear, mechanoreceptors are widely distributed in the avian body. These include free nerve fiber endings which are especially abundant in the feather papillae. Two tactile corpuscles, those of Merkel and Grandry, occur in subepithelial connective tissue. Herbst corpuscles, which resemble the mammalian corpuscles of Pacini, occur in large numbers in the bills of all birds. Although not extensively studied, proprioreceptors occur within skeletal muscles and visceral organs. The somatoreceptors have been reviewed recently by Schwartzkopff (1973).

Although chemoreception has generally been regarded as poorly developed in birds, it is by no means insignificant (Wenzel, 1973). With the exception of differences in central processing of information which have not yet been assessed fully, this sensory modality differs from those of mammals only in the relatively reduced overall importance as a source of external information. However Papi and his associates (Papi, 1976) have obtained extensive experimental evidence that olfaction is involved in orientation of homing pigeons.

To the sensory modalities listed, the use of geomagnetism as information in orientation and homing must be added. Although suggested for at least a half century, until recently most experimental approaches have given negative or ambivalent results, probably because of the use of overly strong fields. Many of the early experiments have been reviewed by Emlen (1975). The major recent impetus in the recognition of this sensory modality has come from experiments by Merkel and Wiltschko (1965), Merkel and Fromme (1958), Fromme (1961), Merkel *et al.* (1964), Wiltschko (1972), Merkel (1971), Wiltschko and Merkel

(1966, 1971), and Wiltschko and Wiltschko (1972, 1975a,b, 1976). In general, these experiments have involved the effects of modification of the direction and intensity of the magnetic field on the orientation of migratory behavior *(Zugunruhe)* in captivity. Emlen (1975) reviewed critically the results of these investigations, of those on Indigo Buntings, *Passerina cyanea,* in collaboration with the Wiltschkos, and of those on homing pigeons carrying small magnets or Helmholtz coils that disturb the magnetic environment of the bird (e.g., Keeton, 1971, 1974; Walcott and Green, 1974). Emlen regards the results as "tantalizing" evidence that geomagnetism is used as information in establishing direction of flight. This "tantalizing" evidence and the results of subsequent investigations (Wiltschko and Wiltschko, 1976) indicate that perception of magnetism is indeed a sensory modality, at least in some species of birds, even though nothing is known about the sensory apparatus.

XII. Reproduction

All species of birds are oviparous. The structures and function of the reproductive systems of both sexes are subjects of a vast literature reviewed recently by Lofts and Murton (1973). Control of reproductive functions are reviewed and discussed by Farner (1975), Farner and Lewis (1971), and Chapter 7 of this volume. With the exception of the megapodes, which use appropriately controlled environmental heat (e.g., Frith, 1962), the developing embryo is maintained at a relatively constant temperature by "incubation" by adult birds. Incubation usually involves transfer of body heat from the incubating bird to the egg; but under elevated environmental temperatures heat may pass from the egg to the incubating birds for dissipation (Drent, 1973). The complex physiological and behavioral processes of incubation have been reviewed by Drent (1973, 1975) and by White and Kinney (1974). The development of the avian embryo has been described in great detail by Romanoff (1960). Aspects of functional development have been reviewed most usefully by Freeman and Vince (1974), and biochemical aspects are discussed in Chapter 3 of this volume. Incubating adults of most species develop a highly vascularized, edematous incubation patch that furthers heat exchange. Incubation is normally performed by one or both parents. But a considerable number of species have evolved brood parasitism in which the eggs are incubated and the young reared by the host species (Chance, 1940; Friedmann, 1929, 1948, 1955, 1960; Nicolai, 1964).

The avian egg is telolecithal and contains all the nutrient materials required for the entire embryonic life. The yolk is surrounded by layers of albumen, two shell membranes, and a hard shell composed of minerals, mostly calcium as calcite, laid down in a protein matrix, all produced by the oviduct. Egg size varies among species from 2–27% of body weight and is a power function of body size. Rahn *et al.* (1975) have shown for a substantial number of families and orders that $W = a B^b$, where W and B are egg and body weights, respectively, in grams. For all groups examined b was about 0.67 whereas a varied and was characteristic of each group. In the Anatidae and Phasianidae, at least, the weight of the clutch is proportional to the square root of body weight. This may hold also for the Fringillidae.

The eggshell provides mechanical protection and prevents invasion by microorganisms. In conjunction with the shell membranes, it retards the loss of water (Rahn and Ar, 1974; Ar *et al.*, 1974). The rates of diffusion of gases, oxygen, carbon dioxide, and water through the shell are proportional to their diffusion constants times the ratio of pore area to shell thickness (Ar *et al.*, 1974). The conductance of the shell for these gases is also proportional to $W^{0.78}$ where W is egg weight. Since 0.78 is very close to the power function that relates body size and standard metabolic rate, this suggests that shell thickness and geometry of the pores are adapted closely to the metabolic rate of the embryo (Wangensteen and Rahn, 1971; Ar *et al.*, 1974). As eggs become larger, shell thickness increases as a linear function of $W^{0.46}$. This is presumably an adaptation to provide adequate mechanical strength of the shell. But this increased thickness requires an increase in pore area so that pore area is proportional to $W^{1.24}$ (Ar *et al.*, 1974). Thus, a 1000 gm egg has about 5100 times as much pore area as a 1 gm egg. The older literature on the composition and structure of the eggshell have been reviewed by Romanoff and Romanoff (1949). More recent investigations include those of Tyler (1956, 1964, 1966, 1969) and of Schmidt (1963, 1964, 1968, 1970). A brief, useful review was presented by Gilbert (1971).

Many eggshells are white. By deposition of porphyrins and cyanins in varying amounts and combinations a huge array of species- or subspecies-specific colors and color patterns have evolved. The size of the clutch varies from one to more than twenty and is, within limits, fixed within the species or population. Selection of clutch size in evolution is complex but fundamentally clutch size, or the total number of eggs laid in a season by double or multiple-brooded species, must be related to the minimum number young required to maintain the population in steady state, to the survival rate of eggs and young, to the cost of the reproductive effort in time and energy, and to the availability of energy and other resources (King, 1973; Ricklefs, 1974). In nidifugous species it may be limited by the capacity of the female to produce eggs. Among the many extensive treatises that bear on the evolution of clutch size are those of Lack (1968), von Haartman (1951, 1971), and Cody (1971). The incubation period is only roughly a function of egg size (Heinroth, 1922; Drent, 1973; Rahn and Ar, 1974). It ranges from as little as 10 days in the Great Spotted Woodpecker, *Dendrocopus major*, to 80 days in the Royal Albatross, *Diomeda epomophora*. The internal temperature of the incubated egg appears to be essentially independent of both egg size and body size and body temperature of the incubating adult (Drent, 1975).

Among the orders of birds, young are hatched at different stages of development. The extremes range from the naked, totally dependent, typically altricial young of passerine species to the essentially independent precocial young of the megapodes. Correspondingly, there is a similar spectrum in the extent of posthatching care provided by adults. A useful survey of parental care and its evolution is provided by Kendeigh (1952).

XIII. Migration

Although some species of insects, fish, and mammals undertake migratory movements, this phenomenon has evolved most conspicuously in birds. Migration is the annual movement of the members of a species or a population between discrete breeding and wintering (or nonbreeding) areas. It is an adaptation that permits breeding in areas with seasonally abundant trophic resources, usually at mid or high latitudes with long summer days. The nonbreeding area is usually one with milder climate than that of the winter season in the breeding area. Trophic resources on the wintering grounds are often as abundant or more abundant than in summer in the breeding area but are often divided among more species and more individuals. Because many species have migratory and nonmigratory populations, because many genera have migratory and nonmigratory species, and because none of the families consist entirely of migratory species, it follows that migration probably has evolved independently many times. The failure to recognize this multiple origin of migration has led to numerous pointless controversies. The most useful of the general treatises of migration are those of Dorst (1956), Schüz (1971), and Dolnik (1975).

The high energy requirement for migration is met through the accumulation of fat in specific fat organs (Farner, 1955; King, 1972; Berthold, 1975; Dolnik, 1975, 1976). These fat reserves are the result of programmed hyperphagia that causes an increased intake of food up to a level of fat deposition that is consistent, with a safety margin, with the duration of the migratory flight. Fat is an ideal storage medium because of its high caloric yield and the water produced in hydrolysis. In species that make long uninterrupted flights over water or desert areas, the initial fat reserve may be as much as 50% of body weight. In species that fly over land, migration is often interrupted for brief periods during which the fat reserves are restored. After migration the appestat is reset to a lower level and the fat reserves return to minimal levels. The still poorly understood endocrine aspects of fat deposition, and migration, in general, are reviewed by Meier and Ferrell in Chapter 7 of this volume.

The mechanisms by which birds follow a precise route in migration are the subject of a huge and frequently confusing and controversial literature which has been ably reviewed by Gwinner (1971), Emlen (1975), and Dolnik (1975). Much confusion has been injected into the literature by the assumptions, implicit at least, that the migrating individual uses a single source of information or that there is common route-following mechanism for all migratory species. The results of experimental investigations of the past two decades suggest that navigation during migration involves the use of a variety of sources of external information that must be processed in reference to endogenous entrained chronometers and to a "map" which may be learned or inherited, or both (see Emlen, 1975; Wallrath, 1974; Wiltschko and Wiltschko, 1976, for extensive discussions). There is a growing body of evidence that supports a concept of *Ortstreue* both in the breeding area and in the wintering. The basis for this must involve the navigation system as well as a detailed "map." The fact that *Ortstreue* seems invariably better developed in adults than in first-year birds argues for components of learning and memory in the development of the "map." The external sources of information used in migratory navigation and in homing have been reviewed recently by Emlen (1975), Wallrath (1974), and Wiltschko and Wiltschko (1976). The most prominent sources are topographic features, sun position, star positions, and the intensity, direction, and declination of the earth's magnetic field.

Migration usually follows a precise time schedule. This suggests that the system uses some external periodically recurring information. There is an abundance of evidence that the lengthening days of spring induce vernal hyperphagia, migratory fattening, and migratory behavior (for reviews, see Farner and Lewis, 1971; Berthold, 1975; Dolnik, 1975). The postnuptial molt and the ensuing hyperphagia, fattening, and late-summer migration appear to be physiologic sequelae of the photoperiodically induced vernal events (Farner, 1964; Dolnik, 1976) but a completely different explanation has been developed by Meier and his colleagues (Chapter 7, this volume).

In recent years Gwinner, Berthold, and their colleagues have accumulated compelling evidence for the existence of endogenous circennial periodicities in gonadal function, molt, fattening, and migratory behavior in several species (Berthold, 1975; Dolnik, 1975; Gwinner, 1975). These periodicities are imprecise and obviously must be entrained by an external "Zeitgeber" which could be some phase of the annual photocycle. The question remains open as to whether, for these species, the vernal increase in day length is primary information for the control system or has a role as a *Zeitgeber*. Our experience with White-crowned Sparrows, *Zonotrichia leucophrys gambelii*, points to the latter. But it must be kept in mind that migration doubtless had multiple evolutionary origins. It is possible also that the apparent difference between these two types of control is superficial (King and Farner, 1974).

REFERENCES

Abdulla, M. A., and King, A. S. (1975). *Respir. Physiol.* **23**, 267–290.

Akester, A. R. (1967). *J. Anat.* **101**, 569–594.

Akester, A. R. (1971). *In* "Physiology and Biochemistry of the Domestic Fowl" (D. J. Bell and B. M. Freeman, eds.), Vol. 2, pp. 783–839. Academic Press, New York.

Annison, E. F. (1971), *In* "Physiology and Biochemistry of the Domestic Fowl" (D. J. Bell and B. M. Freeman, eds.), Vol. 1, pp. 321–327. Academic Press, New York.

Ar, A., Paganelli, C. V., Reeves, R. B., Greene, D. G., and Rahn, H. (1974). *Condor* **76**, 153–158.

Aschoff, J., and Pohl, H. (1970). *J. Ornithol.* **111**, 38–47.

Assenmacher, I. (1958). *Alauda* **26**, 241–289.

Baumel, J. J., and Gerchman, L. (1968). *Am. J. Anat.* **122**, 1–18.

Bennett, T. (1974). *In* "Avian Biology" (D. S. Farner and J. R. King, eds.), Vol. IV, **4**, pp. 1–79. Academic Press, New York.

Benoit, J. (1935). *C. R. Seances Soc. Biol. Ses Fil.* **120**, 133–136.

Benoit, J. (1938). *C. R. Seances Soc. Biol. Ses Fil.* **127**, 909–914.

Benzo, C. A., De Gennaro, L. D., and Stearns, S. B. (1975). *J. Exp. Zool.* **193**, 161–166.

Berger, M., and Hart, J. S. (1974) *In* "Avian Biology" (D. S. Farner and J. R. King, eds.), Vol. IV, pp. 416–478. Academic Press, New York.

Bernstein, M. H., Thomas, S. P., and Schmidt-Nielsen, K. (1973). *J. Exp. Biol.* **58**, 401–410.

Berthold, P. (1975). *In* "Avian Biology" (D. S. Farner and J. R. King, eds.), Vol. V, pp. 77–128. Academic Press, New York.

Bock, W. J. (1974). *In* "Avian Biology" (D. S. Farner and J. R. King, eds.), Vol. IV, pp. 120–259. Academic Press, New York.

Bolliger, A., and Varga, D. (1961). *Nature (London)* **190**, 1125.

Boorman, K. N., and Lewis, D. (1971). *In* "Physiology and Biochemistry of the Domestic Fowl" (D. J. Bell and B. M. Freeman, eds.) Vol. **1**, pp. 339–372. Academic Press, New York.

Bretz, W. L., and Schmidt-Nielsen, K. (1971). *J. Exp. Biol.* **54**, 103–118.

Brodkorb, P. (1971). *In* "Avian Biology" (D. S. Farner and J. R. King, eds.), Vol. I, pp. 20–55. Academic Press, New York.

Calder, W. A., and King, J. R. (1974). *In* "Avian Biology" (D. S. Farner and J. R. King eds.), Vol. IV, pp. 260–415. Academic Press, New York.
Calder, W. A. (1974). *In* "Avain Energetics" (R. A. Paynter, ed.), pp. 86–151. Nuttall Ornithol. Club, Cambridge, Massachusetts.
Campbell, G. (1970). *In* "Smooth Muscle" (E. Bülbring *et al.*, eds.), pp. 418–450. Arnold, London.
Chance, E. P. (1940). "The Truth about the Cuckoo." Country Life, London.
Chandler, A. C. (1916). *Univ. Calif., Berkeley, Publ. Zool.* **13**, 243–446.
Cody, M. L. (1971). *In* "Avian Biology" (D. S. Farner and J. R. King, eds.), Vol. I, pp. 462–513. Academic Press, New York.
Cohen, D. H., and Karten, H. J. (1974). *In* "Birds: Brain and Behavior" (I. J. Goodman and M. W. Schein, eds.), pp. 29–73. Academic Press, New York.
Dantzler, W. H. (1966). *Am. J. Physiol.* **210**, 640–646.
Dawson, W. R., and Hudson, J. W. (1970). *In* "Comparative Physiology of Thermoregulation" (G. C. Whittow, ed.), Vol. **1**, pp. 224–310. Academic Press, New York.
De Gennaro, L. D. (1959). *Growth* **23**, 235–249.
Dolnik, V. R. (1965). *In* "Novosti Ornitologii," (E. I. Gavrilov, N. A. Gladnov, G. P. Dementiev, I. A. Dolgushinn, M. N. Korelov, M. A. Kuzmina, and A. K. Rustamov, eds.), pp. 124–126. "Nauk," Alma Alta.
Dolnik, V. R. (1975). "Migratsionnoe Sostoyanie Ptits." "Nauka," Moscow.
Dolnik, V. R. (1976). *In* "Fotoperiodizm Zhivotnykh i Rastenii (O. A. Skarlato and V. A. Zaslavskii, eds.), pp. 47–81. Akad. Nauk SSSR, Leningrad.
Dominic, C. J., and Singh, R. M. (1969). *Gen. Comp. Endocrinol.* **13**, 22–26.
Dorst, J. (1956). "Les Migrations des Oiseaux." Payot, Paris.
Dorst, J. (1974). "The Life of Birds." Columbia Univ. Press, New York.
Doyle, W. L., and Watterson, R. L. (1949). *J. Morphol.* **85**, 391–404.
Drent, R. (1973). *In* "Breeding Biology of Birds" (D. S. Farner, ed.), pp. 262–322. Natl. Acad. Sci., Washington, D.C.
Drent, R. (1975). *In* "Avian Biology" (D. S. Farner and J. R. King, eds.), Vol. V, 333–420. Academic Press, New York.
Drent, R. (1977). *Vogelwarte* **30** (in press).
Duke-Elder, S. (1958). *In* "System of Ophthalmology" (S. Duke-Elder, ed.), Vol. 1, pp. 397–427. Kimpton, London.
Duncker, H. R. (1971). *Ergeb. Anat.* **45**, No. 6.
Duncker, H. R. (1974). *Respir. Physiol.* **14**, 44–63.
Emlen, S. T. (1975). *In* "Avian Biology" (D. S. Farner and J. R. King, eds.), Vol. V, pp. 129–220. Academic Press, New York.
Fabricius ab Aquapendente, H. (1621). "De Formatio Ovi et Pulli." A. Bencii, Patavii.
Farner, D. S. (1955). *In* "Recent Studies in Avian Biology" (A. Wolfson, ed.), pp. 198–237. Univ. of Illinois Press, Urbana.
Farner, D. S. (1964). *Am. Sci.* **52**, 137–156.
Farner, D. S. (1970). *Fed. Proc., Fed. Am. Soc. Exp. Biol.* **29**, 1649–1663.
Farner, D. S. (1975). *Am. Zool.* **15**, Suppl., 117–135.
Farner, D. S., and Lewis, R. A. (1971). *Photophysiology* **6**, 325–370.
Fisher, H. (1973). *In* "Avian Biology" (D. S. Farner and J. R. King, eds.), Vol. II, pp. 431–471. Academic Press, New York.
Freeman, B. M., and Vince, M. A. (1974). "Development of the Avian Embryo." Chapman & Hall, London.
Friedmann, H. (1929). "The Cowbirds." Thomas, Springfield, Illinois.
Friedmann, H. (1948). "The Parasitic Cuckoos of Africa," Monogr. No. 1. Washington Acad. Sci., Washington, D.C.
Friedmann, H. (1955). *U.S., Natl. Mus., Bull.* **208**, 1–292.

Friedmann, H. (1960). *U.S., Natl. Mus., Bull.* **223**, 1–196.
Frith, H. J. (1962). "The Mallee-fowl." Angus & Robertson, Sydney, Australia.
Fromme, H. G. (1961). *J. Tierpsychol.* **18**, 205–220.
Gage, S. H. (1917). *J. Comp. Neurol.* **27**, 451–466.
Gasaway, W. C. (1976a). *Comp. Biochem. Physiol. A* **53**, 109–114.
Gasaway, W. C. (1976b). *Comp. Biochem. Physiol. A* **53**, 115–121.
Gasaway, W. C. (1976c). *Comp. Biochem. Physiol. A* **54**, 179–182.
Gaston, S. (1971). *In* "Biochronometry" (M. Menaker, ed.), pp. 541–548. Natl. Acad. Sci., Washington, D.C.
George, J. C., and Berger, A. J. (1966). "Avian Myology." Academic Press, New York.
Gibbs, O. S. (1929). *Am. J. Physiol.* **87**, 594–601.
Gilbert, A. B. (1971). *In* "Physiology and Biochemistry of the Domestic Fowl" (D. J. Bell and B. M. Freeman, eds.), Vol. 3, pp. 1379–1399. Academic Press, New York.
Glenny, F. H. (1955). *Proc. U.S. Natl. Mus.* **104**, 525–621.
Greenewalt, C. H. (1975). *Trans. Am. Philos. Soc.* **65**, 1–67.
Grimm, P. (1976a). *In* "Avian Physiology" (P. D. Sturkie, ed.), 3rd ed., pp. 233–251 Springer-Verlag, New York, Heidelberg, Berlin.
Grimm, P. (1976b). *In* "Avian Physiology" (P. D. Sturkie, ed.), 3rd ed., pp. 252–262. Springer-Verlag, New York, Heidelberg, Berlin.
Gwinner, E. (1971). *In* "Grundriss der Vogelzugskunde" (E. Schüz, ed.), pp. 299–348. Parey, Berlin.
Gwinner, E. (1975). *In* "Avian Biology" (D. S. Farner and J. R. King, eds.), Vol. V, pp. 221–285. Academic Press, New York.
Gwinner, E., Turek, F. W., and Smith, S. D. (1971). *Z. Vergl. Physiol.* **75**, 323–331.
Hartman, F. A. (1961). *Smithson. Misc. Coll.* **143(1)**.
Hazelhoff, E. H. (1943). *Versl. Gewone Vergad. Afd. Natuurkd., K. Ned. Adad. Wet.* **52**, 391–400.
Hazelwood, R. L. (1972). *In* "Avian Biology" (D. S. Farner and J. R. King, eds.), Vol. II, pp. 472–526. Academic Press, New York.
Hazelwood, R. L. (1976). *In* "Avian Physiology" (P. D. Sturkie, ed.), 3rd. ed., pp. 210–232. Springer–Verlag, New York–Heidelberg–Berlin.
Heinroth, O. (1922). *J. Ornithol.* **70**, 172–285.
Homma, K., and Sakakibara, Y. (1971). *In* "Biochronometry" (M. Menaker, ed.), pp. 333–341. Natl. Acad. Sci., Washington, D.C.
Hudson, D. A., Levin, R. J., and Smyth, D. H. (1971). *In* "Physiology and Biochemistry of the Domestic Fowl" (D. J. Bell and B. M. Freeman, eds.), Vol. 1, pp. 52–72. Academic Press, New York.
Hussell, D. J. T. (1969). *Auk* **86**, 75–83.
Ilyichev, V. (1972). "Bioakustika Ptits." University of Moscow.
Imhof, G. (1905). *Arch. Mikrosk. Anat.* **65**, 498–610.
Irving, L., and Krog, J. (1954). *J. Appl. Physiol.* **6**, 667–680.
Jaeger, E. C. (1948). *Condor* **50**, 45–46.
Jaeger, E. C. (1949). *Condor* **51**, 105–109.
Jeuniaux, C. (1962). *Ann. Soc. R. Zool. Belg.* **92**, 27–45.
Johnson, O. W. (1974). *J. Morphol.* **142**, 277–284.
Jones, D. R., and Johansen, K. (1972). *In* "Avian Biology" (D. S. Farner and J. R. King, eds.), Vol. II, 158–287. Academic Press, New York.
Kappers, C. U. A. (1921). "Vergleichende Anatomie des Nervensystems." Bohn, Haarlem.
Kappers, C. U. A. (1947). "Anatomie Comparée du Système Nerveux." Masson, Paris.
Kappers, C. U. A. Huber, G. C., and Crosby, E. C. (1936). "The Comparative Anatomy of the Nervous System of Vertebrates, Including Man." Macmillan, New York.

Karten, H. J., and Hodos, W. (1967). "A Stereotaxic Atlas of the Brain of the Pigeon, *Columba livia*." Johns Hopkins Press, Baltimore, Maryland.

Kato, M., Kato, Y., and Oishi, T. (1967). *Proc. Jpn. Acad.* **43**, 220–223.

Keeton, W. T. (1971). *Proc. Natl. Acad. Sci. U.S.A.* **68**, 102–106.

Keeton, W. T. (1974). *Adv. Study Behav.* **5**, 47–132.

Kendeigh, S. C. (1952). *Ill. Biol. Monogr.* **22**, 1–356.

Kendeigh, S. C. (1963). *Proc. Int. Ornithol. Congr., 13th, 1962* pp. 884–904.

Kendeigh, S. C. (1973). *In* "Breeding Biology of Birds" (D. S. Farner, ed.), pp. 111–117. Natl. Acad. Sci., Washington, D.C.

Kespaik, Y. (1968). *Izv. Akad. Nauk Est. SSR, Biol.* **17**, 179–190.

Kilgore, D. L., Bernstein, M. H., and Hudson, D. M. (1976). *J. Comp. Physiol. B* **110**, 209–216.

King, J. R. (1972). *Proc. Int. Ornithol. Congr., 15th, 1970* pp. 200–236.

King, J. R. (1973). *In* "Breeding Biology of Birds" (D. S. Farner, ed.), pp. 78–107. Natl. Acad. Sci., Washington, D.C.

King, J. R. (1974). *In* "Avian Energetics" (R. A. Paynter, ed.), pp. 4–85. Nuttall Ornithol. Club, Cambridge, Massachusetts.

King, J. R., and Farner, D. S. (1961). *In* "Biology and Comparative Physiology of Birds" (A. J. Marshall, ed.), pp. 215–288. Academic Press, New York.

King, J. R., and Farner, D. S. (1964). *In* "Handbook of Physiology," (D. B. Dill, E. F. Adolph, and C. G. Wilber, eds.), Sect. 4 "Adaptation to the Environment," pp. 603–624. American Physiological Society, Washington.

King, J. R., and Farner, D. S. (1974). *In* "Chronobiology" (L. E. Scheving, F. Halberg, and J. E. Pauly, eds.), pp. 625–629. Igaku Shoin Ltd., Tokyo.

Krüger, P. (1952). "Tetanus und Tonus der quergestreiften Skelettmuskeln der Wierbeltiere und des Menschen." Akad. Verlagsges., Leipzig.

Lack, D. (1968). "Ecological Adaptations for Breeding in Birds." Methuen London.

Lang, E. M. (1963). *Experientia* **19**, 532.

Lang, E. M., Thiersch, A., Thommen, H., and Wackernagel, H. (1962). *Ornithol. Beob.* **59**, 173–176.

Lofts, B., and Murton, R. K. (1973). *In* "Avian Biology" (D. S. Farner and J. R. King, eds.), Vol. III, pp. 1–107. Academic Press, New York.

Lucas, A. M. (1968). *Anat. Rec.* **160**, 386–387.

Lucas, A. M., and Stettenheim, P. (1972). *U.S., Dep. Agric., Agric. Handb.* **362.**

Luppa, H. (1959). *Acta Anat.* **39**, 51–81.

Lutz, P., Longmuir, I. S. and Schmidt-Nielsen, K. (1974). *Respir. Physiol.* **20**, 325–330.

Lyuleeva, D. S. (1962). *Mater. Vses. Ornitol. Konf., 3rd, 1962* **2**, pp. 79–81.

Lyuleeva, D. S. (1970). *Dokl. Akad. Nauk SSSR*, **190**, 1467–1470.

McMillan, J. P., Keatts, H. C., and Menaker, M. (1975a). *J. Comp. Physiol.* **102**, 251–256.

McMillan, J. P., Elliott, J. and Menaker, M. (1975b). *J. Comp. Physiol.* **102**, 257–262.

McNabb, F. M. A., and Poulson, T. L. (1970). *Comp. Biochem. Physiol.* **33**, 933–939.

Menaker, M. (1971). *In* "Biochronometry" (M. Menaker, ed.), pp. 315–332. Natl. Acad. Sci., Washington, D.C.

Menaker, M., and Keatts, H. (1968). *Proc. Natl. Acad. Sci. U.S.A.* **60**, 146–151.

Menaker, M., and Oksche, A. (1974). *In* "Avian Biology" (D. S. Farner and J. R. King, eds.), Vol. IV, pp. 80–119. Academic Press, New York.

Menaker, M., and Zimmerman, N. (1976). *Am. Zool.* **16**, 45–55.

Menaker, M., Roberts, R., Elliott, J., and Underwood, H. (1970). *Proc. Natl. Acad. Sci. U.S.A.* **67**, 320–325.

Merkel, F. W. (1971). *Ann. N.Y. Acad. Sci.* **188**, 283–294.

Merkel, F. W., and Fromme, H. G. (1958). *Naturwissenscha flen* **45**, 499–500.

Merkel, F. W., and Wiltschko, W. (1965). *Vogelwarte* **23**, 71–77.
Merkel, F. W., Fromme, H. G., and Wiltschko, W. (1964). *Vogelwarte* **22**, 168–173.
Nachtigall, W. (1975). *J. Ornithol.* **116**, 1–38.
Nicolai, J. (1964). *Z. Tierpsychol.* **21**, 129–204.
Nisbett, I. C. T., Drury, W. H., and Baird, J. (1963). *Bird-Banding* **34**, 107–138.
Nitzsch, C. L., and Burmeister, C. C. H. (1840). "System der Pterylographie." Halle.
Oehme, H. (1963). *Biol. Zentralbl.* **82**, 413–454.
Oehme, H. (1976a). *Beitr. Vogelk.* **22**, 58–66.
Oehme, H. (1976b). *Beitr. Vogelk.* **22**, 67–72.
Oehme, H. (1976c). *Beitr. Vogelk.* **22**, 73–82.
Oksche, A., and Farner, D. S. (1974). *Ergetb. Anat. Entwicklungsgesch.* **48**, 1–136.
Ostrom, J. H. (1974). *Q. Rev. Biol.* **49**, 27–47.
Ostrom, J. H. (1976). *Biol. J. Linn. Soc.* **8**, 91–182.
Palmer, R. S. (1972). *In* "Avian Biology" (D. S. Farner and J. R. King, eds.), Vol. II, pp. 65–102. Academic Press, New York.
Papi, F. (1976). *Verh. Dtsch. Zool. Ges.* **1976**, 184–205.
Payne, L. N. (1971). *In* "Physiology and Biochemistry of the Domestic Fowl" (D. J. Bell and B. M. Freeman, eds.), Vol. 2, 985–1038. Academic Press, New York.
Payne, R. B. (1972). *In* "Avian Biology" (D. S. Farner and J. R. King, eds.), Vol. II, 103–155. Academic Press, New York.
Peaker, M. (1975). *Symp. Zool. Soc. London* **35**, 107–128.
Peaker, M., and Linzell, J. L. (1975). "Salt Glands in Birds and Reptiles." Cambridge Univ. Press, London and New York.
Pearce, J., and Brown, W. O. (1971). *In* "Physiology and Biochemistry of the Domestic Fowl" (D. J. Bell and B. M. Freeman, eds.), Vol. 1, pp. 295–320. Academic Press, New York.
Pearson, R. (1972). "The Avian Brain." Academic Press, New York.
Pennycuick, C. J. (1975). *In* "Avian Biology" (D. S. Farner and J. R. King, eds.), Vol. V, 1–76. Academic Press, New York.
Piiper, J., and Scheid, P. (1973). *In* "Comparative Physiology, Locomotion, Respiration, Transport and Blood." (Bolis, S. H. P. Maddrell and K. Schmidt-Nielsen, eds.), pp. 161–185. North-Holland, Amsterdam.
Portmann, A., and Stingelin, W. (1961). *In* "Biology and Comparative Physiology of Birds" (A. J. Marshall, ed.), Vol. 2, pp. 1–36. Academic Press, New York.
Poulson, T. L. (1965). *Science* **148**, 389–391.
Prévost, J. (1961). *Actual. Sci. Ind.* **1291**, 2.
Prévost, J., and Vilter, V. (1963). *Proc. Int. Ornithol. Congr., 13th, 1962* Vol. 2, pp. 1085–1094.
Pumphrey, R. J. (1961). *In* "Biology and Comparative Physiology of Birds" (A. J. Marshall, ed.), Vol. 2, pp. 55–68. Academic Press, New York.
Rahn, H., and Ar, A. (1974). *Condor* **76**, 147–152.
Rahn, H., Paganelli, C. V., and Ar, A. (1975). *Auk* **92**, 750–765.
Raveling, D. G., and LeFebvre, E. A. (1967). *Bird-Banding* **38**, 97–113.
Richards, S. A. (1975). *Symp. Zool. Soc. London* **35**, 65–96.
Ricklefs, R. E. (1974). *In* "Avian Energetics" (R. A. Paynter, ed.), pp. 152–297. Nuttall Ornithol. Club, Cambridge, Massachusetts.
Rochon-Duvigneaud, A. (1943). "Les yeux et la vision des vertébrés." Masson, Paris.
Romanoff, A. L. (1960). "The Avian Embryo." Macmillan, New York.
Romanoff, A. L., and Romanoff, A. J. (1949). "The Avian Egg." Wiley, New York.
Scheid, P., and Piiper, J. (1972). *Respir. Physiol.* **16**, 304–312.
Schildmacher, H. (1932). *J. Ornithol.* **80**, 293–299.

Schmidt, W. J. (1963). *Z. Zellforsch. Mikrosk. Anat.* **54,** 848–880.
Schmidt, W. J. (1964). *Z. Zellforsch. Mikrosk. Anat.* **62,** 53–60.
Schmidt, W. J. (1968). *Z. Zellforsch. Mikrosk. Anat.* **86,** 444–452.
Schmidt, W. J. (1970). *Z. Zellforsch. Mikrosk. Anat.* **107,** 119–122.
Schmidt-Nielsen, K. (1972). *Science* **177,** 222–228.
Schmidt-Nielsen, K., Barker-Jørgensen, C., and Osaki, H. (1958). *Am. J. Physiol.* **193,** 101–107.
Schüz, E. (1971). "Grundriss der Vogelzugkunde." Parey, Berlin.
Schwartzkopff, J. (1973). *In* "Avian Biology" (D. S. Farner and J. R. King, eds.), Vol. III, pp. 417–477. Academic Press, New York.
Shilov, I. A. (1968). "Regulyatsiya teploobema u ptits." Moscow University Press.
Shoemaker, V. H. (1972). *In* "Avian Biology" (D. S. Farner and J. R. King, eds.), Vol. II, pp. 527–574. Academic Press, N.Y.
Siller, W. G., and Hindle, R. M. (1969). *J. Anat.* **104,** 117–135.
Sillman, A. J. (1973). *In* "Avian Biology" (D. S. Farner and J. R. King, eds.), Vol. III, 349–387. Academic Press, N.Y.
Skadhauge, E. (1973). *Dan. Med. Bull.* **20,** 1–82.
Skadhauge, E. (1975). *Symp. Zool. Soc. London* **35,** 97–106.
Spearman, R. I. C. (1966). *Biol. Rev. Cambridge Philos. Soc.* **41,** 59–96.
Steen, I., and Steen, J. B. (1965). *Acta Physiol. Scand.* **63,** 285–291.
Stettenheim, P. (1972). *In* "Avian Biology" (D. S. Farner and J. R. King, eds.), Vol. II, pp. 1–63. Academic Press, New York.
Stingelin, W. (1965). *Bibl. Anat.* **66,** 1–116.
Storer, R. W. (1971). *In* "Avian Biology" (D. S. Farner and J. R. King, eds.), Vol. I, pp. 1–18. Academic Press, New York.
Stresemann, E., and Stresemann, V. (1966). *J. Ornithol.* **107,** Suppl., 1–445.
Tanabe, Y., and Katsuragi, T. (1962). *Bull. Natl. Inst. Agric. Sci., Ser. G* **21,** 49–59.
Ten Cate, J. (1936). *Ergeb. Biol.* **13,** 93–173.
Ten Cate, J. (1965). *In* "Avian Physiology" (P. D. Sturkie, ed.), 2nd ed., pp. 697–751. Cornell Univ. Press (Comstock), Ithaca, New York.
Terni, T. (1924). *Arch. Ital. Anat. Embriol.* **21,** 55–86.
Tucker, V. A. (1968). *J. Exp. Biol.* **48,** 67–87.
Tucker, V. A. (1972). *Respir. Physiol.* **14,** 75–82.
Tucker, V. A. (1973). *J. Exp. Biol.* **58,** 689–709.
Tucker, V. A. (1974). *In* "Avian Energetics" (R. A. Paynter, ed.), pp. 298–333. Nuttall Ornithol. Club, Cambridge, Massachusetts.
Tucker, V. A. (1975). *Symp. Zool. Soc. London* **35,** 49–64.
Tyler, C. (1956). *J. Sci. Food Agric.* **7,** 483–493.
Tyler, C. (1964). *Proc. Zool. Soc. London* **142,** 547–583.
Tyler, C. (1966). *J. Zool.* **150,** 413–425.
Tyler, C. (1969). *J. Zool.* **158,** 395–412.
van Tienhoven, A., and Juhász, L. P. (1962). *J. Comp. Neurol.* **118,** 185–197.
Varićak, T. D. (1938). *Z. Mikrosk.-Anat. Forsch.* **44,** 119–130.
Vitums, A., Mikami, S.-I., Oksche, A., and Farner, D. S. (1964). *Z. Zellforsch. Mikrosk. Anat.* **64,** 541–569.
Voitkevich, A. A. (1962). "Pero Ptitsy." Akad. Nauk USSR, Moscow.
von Haartman, L. (1951). *Acta Zool. Fenn.* **67,** 1–60.
von Haartman, L. (1971). *In* "Avian Biology" (D. S. Farner and J. R. King, eds.), Vol. I, pp. 392–461. Academic Press, New York.
Wackernagel, H. (1964). *Int. Z. Vitaminforsch.* **34,** 141–143.
Walcott, C., and Green R. P. (1974). *Science* **184,** 180–182.

Wallrath, H. G. (1974). "Das Navigationssystem der Vögel." R. Oldenbourg Verlag, Munich.
Walls, G. L. (1942). *Bull. Cranbrook Inst. Sci.* **19**, 1–785.
Wangensteen, O. D., and Rahn, H. (1971). *Respir. Physiol.* **11**, 31–45.
Watterson, R. L. (1949). *J. Morphol.* **85**, 337–390.
Webb, T. E., and Colvin, J. R. (1964). *Can. J. Biochem.* **42**, 59–70.
Wenzel, B. M. (1973). *In* "Avian Biology" (D. S. Farner and J. R. King, eds.), Vol. III, pp. 389–416. Academic Press, New York.
West, N. H., Bamford, O. S., and Jones, D. R. (1977). *Z. Zellforsch. Mikrosk. Anat.* **176**, 553–564.
Westerterp, K. (1973). *Ardea* **62**, 137–158.
White, F. N., and Kinney, J. L. (1974). *Science* **186**, 107–115.
Wiltschko, W. (1972). *In* "Animal Orientation and Navigation," (S. R. Baller, K. Schmidt-Koenig, G. C. Jacobs, and R. E. Belleville, eds.), Sym. NASA SP-262, pp. 569–578. U.S. Govt. Printing Office, Washington, D.C.
Wiltschko, W., and Merkel, F. W. (1966). *Verh. Dsch. Zool. Ges.* **1965**, 362–367.
Wiltschko, W., and Merkel, F. W. (1971). *Vogelwarte* **26**, 245–249.
Wiltschko, W., and Wiltschko, R. (1972). *Science* **176**, 62–64.
Wiltschko, W., and Wiltschko, R. (1975a). *Z. Tierpsychol.* **37**, 337–355.
Wiltschko, W., and Wiltschko, R. (1975b). *Z. Tierpsychol.* **39**, 265–282.
Wiltschko, W., and Wiltschko, R. (1976). *J. Ornithol.* **117**, 362–387.
Wingstrand, K. G. (1951). "The Structure and Development of the Avian Pituitary." G. W. K. Gleerup, Lund.
Wingstrand, K. G., and Munk, O. (1965). *K. Dan. Vidensk. Selsk., Biol. Skr.* **14**, 1–64.
Yokoyama, K., and Farner, D. S. (1976). *Gen. Comp. Endocrinol.* **30**, 528–533.
Ziswiler, V., and Farner, D. S. (1972). *In* "Avian Biology" (D. S. Farner and J. R. King, eds.), Vol. II, 343–430. Academic Press, New York.

CHAPTER 2

Plasma and Egg White Proteins

R. G. Board and D. J. Hornsey

I. General Introduction

If a class of vertebrates is not included regularly in the studies leading to the elucidation of a fundamental aspect of physiology, then there is a tendency for the occasional investigation to attempt to establish links between the neglected class and the main body of information. This tendency is obvious with the plasma proteins of birds and is reflected in this chapter where organization and interpretation have been influenced by analogy to the plasma proteins of mammals.

An entirely different situation occurs with egg white proteins mainly because an elementary skill—the ability to break the shell and harvest the albumen without contamination with yolk material—is the only one to be mastered before a solution of proteins with no cellular organization, negligible enzyme activity, and little contamination with nonprotein material is obtained. Thus it is little wonder that eggs were a popular source of proteins at every phase in the evolution of methods for the isolation and characterization of proteins. An additional bonus, of course, is to have a solution of proteins, many of which have some easily demonstrated property or function. The egg white proteins exhibit these features (Table I). There are no grounds for complaint

TABLE I

SOME BIOLOGICAL AND PHYSICAL PROPERTIES OF THE MAIN PROTEINS OF THE ALBUMEN OF THE HEN'S EGG

Protein	Amount (%) in albumen	MW	p*I*	Biological property	Methods of assay
1. Ovalbumin	54	46,000	4.5	—	—
2. Ovotransferrin	12	76,600	6.05	Binding [2 atoms(mole^{-1})] of Fe^{3+}, Cu^{2+}, Mn^{2+} Co^{2+}, Cd^{2+}, Zn^{2+}, Ni^{2+}	Spectral studies (Tan and Woodworth, 1968; Phelps and Antonini, 1975); growth studies with microbes (Theodore and Schade, 1965a,b)
				Binding of bicarbonate	
3. Ovomucoid	12	28,000	4.1	Inhibition of proteases	Standard biochemical techniques
				Anaphylactoid response in rats	Injection of albino rats
4. Lysozyme	3.4	14,300	10.7	Hydrolysis of $\beta(1-4)$ glycosidic bond in peptidoglycans	Spectral studies with particulate substrate (cell walls of *Micrococcus lysodeikticus*)
				Scaffolding of yolk/embryo through electrostatic interaction with ovomucin	Rheological methods
5. Ovomucin	3.5	*See 4* (Above)	4.5–5.0	Scaffolding of yolk/embryo through interaction with lysozyme	Rheological methods
				Virus antihemagglutination	Serological methods
6. Ovoinhibitor	1.5	44/49,000	5.1	Inhibition of proteases	Standard biochemical techniques
7. Ovomacroglobulin	0.5	900,000	4.5	—	—
8. Ovoglycoprotein	1.0	24,400	3.9	—	—
9. Ovoflavoprotein	0.8	32,000	4.0	Binding of riboflavin	Microbiological assay
10. Avidin	0.05	68,300	10	Binding of biotin	Microbiological assay

about the information available on the properties and characteristics of the egg white proteins. The same cannot be said about their function in the egg. The Editor's invitation to speculate has not been neglected in the discussion of novel roles of the egg white proteins. If this provokes studies of biological function, then the speculation will not have been in vain. The choice of references is another feature of the second part of the chapter which needs comment. Their selection was dictated by two main considerations: (1) to provide, as far as possible, references to recent papers so that adequate guidance to the literature was provided, and (2) to emphasize that the avian egg is more often than not the source of materials rather than a subject of study in its own right.

II. Plasma Proteins

A. Prealbumin

Marshall and Deutsch (1950), Heim and Schechtman (1954), and Vanstone *et al.* (1955) all reported the presence of prealbumin electrophoretic bands in chick embryo serums up to day 1 posthatch, with maximum concentrations of 0.14 gm% appearing at day 18 of incubation. Christov and Rashev (1961) confirmed their presence by day 6 but their disappearance by day 15 of incubation. Electrophoretic analysis is open to these aberrational differences, for the stated presence or absence of protein peaks is quite often dependent upon the buffer and pH employed (Table I) (Moore *et al.*, 1945; Marshall and Deutsch, 1950).

The prealbumin bands are rich in phospholipids (Marshall and Deutsch, 1950) and may play some part in the transport of yolk lipid reserves necessary for the developing embryo (Nobel and Moore, 1966). During egg production, 18-month-old laying hens show an electrophoretically fast moving component (Brandt *et al.*, 1952) which was correlated to egg formation when there is a mobilization of reserves. Prealbumin fractions have been confirmed in laying hens (Lush, 1963) and shown to be at maximum concentration when the shell is produced in the uterus (Kristjansson *et al.*, 1963). Several prealbumin bands have been detected in adult peafowl serum (Kimura *et al.*, 1970), pigeon serum (Baxendale *et al.*, 1971), and other species (Table II). Gel filtration chromatography indicates that these are probably low molecular weight (10,000 daltons) proteins (Baxendale *et al.*, 1971).

Although synthesis of the prealbumins is genetically determined by one autosomal locus with two alleles designated Pa^A and Pa^B (Stratil,

TABLE II

ELECTROPHORETIC MOBILITY DATA FOR AVIAN AND HUMAN SERUM COMPONENTS IN THREE BUFFERS[a]

Animal	Measurement[a]	Prealbumin	Albumin	Globulin					Fibrinogen
				α_1	α_2	β_1	β_2	γ	
Veronal–citrate buffer, pH 8.6., $\mu = 0.1$									
Man	m		6.6	5.4	4.3	3.1		1.3	2.3
	%		59.6	6.7	8.8	11.0		9.1	4.8
Chicken	m	8.1	7.3	6.1	4.6			2.9	
	%	0.5	38.2	15.8	7.7			37.5	
Duck	m	7.6	6.7	5.8	4.9			2.5	3.7
	%	2.6	47.8	21.9	6.1			6.0	15.5
Pheasant	m	6.1	5.2	4.2	3.6			1.7	2.9
	%	0.4	58.2	14.0	6.5			4.3	16.3
Pigeon	m	7.8	6.4	5.2	4.5			1.7	3.3
	%	3.1	64.1	7.2	4.5			7.7	17.4
Turkey	m	6.7	5.9	5.0	4.1			1.7	2.9
	%	1.0	51.5	13.4	4.3			8.1	21.6
Veronal buffer, pH 8.6, $\mu = 0.1$									
Man	m		5.9	5.1	4.1	2.8		1.0	
	%		63.0	5.0	7.0	13.0		12.0	
Chicken	m		5.7	4.5		3.5		1.9	
	%		58.0	16.0		9.0		17.0	
Pigeon	m		5.8	4.4	2.8		2.4	1.7	
	%		58.0	10.0	5.0		18.0	9.0	
Phosphate buffer, pH 7.4, $\mu = 0.2$									
Man	m		5.1		3.5	2.5		0.7	
	%		65.0		9.0	16.0		11.0	
Chicken (male)	m		5.2	4.2	3.2	2.3		1.5	
	%		44.0	15.0	10.0	14.0		14.0	
Chicken (female)	m	5.8	4.9	4.0	3.1	1.8		1.1	
	%	5.0	29.0	6.0	8.0	35.0		21.0	
Pigeon	m		4.1		2.2	1.6		0.3	
	%		57.0		19.0	16.0		8.0	

1970), Asofsky *et al.* (1962) suggested that they may not be synthesized in the embryo. It has generally been assumed (Butler, 1971) that prealbumins are transferred to the embryo via the yolk. Wise *et al.* (1964), noting the similarities of alkaline phosphatase reactions on proteins from sera and egg proteins, considered prealbumins as ovalbumins and thus components of egg white and not the yolk. Because prealbumins are present from at least day 11 of incubation and before the developing chick has swallowed egg white proteins, Wise *et al.* (1964) suggested that prealbumin synthesis does occur in the embryo. Such an observation would tend to indicate a positive function of these proteins in the embryo and not a "contaminant" from the egg.

B. Albumin

Although intensive biochemical and chemical investigations have been applied to the study of albumin, the compound still has no well-established biological function. In birds, its amino acid composition (Schjeide, 1963) and its molecular weight of 65,000 daltons show it to be similar to mammalian albumin. It probably consists of a single polypeptide chain. In common with mammals, the albumin from bird species has aspartic acid as the N-terminal amino acid (Peters *et al.*, 1958; Brown, 1975). By definition (Hughes, 1954) the molecule contains no carbohydrate, any being detected on crystallization usually being assigned to small amounts of α-globulins present as contaminants. The albumin molecule is stabilized by disulfide cross-linking (40 sulfurs per 65,000 gm) and, providing these links are unaltered, there may be a considerable amount of reversible configurational variation in the molecule. The C-terminal amino acid residues of three avian species (Table III) may be compared with some mammalian data (Peters *et al.*, 1958). Alanine is the C-terminal amino acid in two

TABLE III
C- and N-Terminal Amino Acids of Albumins in Three Avian and Two Mammalian Species

Species	Order	N-Terminal	C-Terminal					
			6	5	4	3	2	1
Species								
Duck	Anseriformes	Asp–	(Thr,	Ser,	Leu,	Val,	Gly)-	Ala
Chicken	Galliformes	Asp–	(Ser,	Thr,	Val,	Leu,	Gly)-	Ala
Turkey	Galliformes	Asp–		(Ala,	Thr,	Lęu,	Gly)-	Val
Mammalia								
Cow	Artiodactylia	Asp–	(Ala,	Ser,	Val,	Thr,	Leu)-	Ala
Man	Primata	Asp–			(Gly,	Val,	Ala)-	Leu

different orders of birds (Anseriforme and Galliforme) but in the turkey (a Galliforme) is replaced by valine.

Two or more serum albumins have been reported in fowl (McIndoe, 1962), in turkey (Quinteros *et al.*, 1964), and in ducklings under immobilization stress (Paulov, 1972). Such polymorphisms of the molecule are not reflected in any differences in their physicochemical properties (Fried and Chum, 1971).

Albumin is a highly charged molecule with a great affinity for ions, particularly anions. The binding of a large number of organic anions is probably explained by the reversible configurations within the molecule (Karush, 1950) and one of the functions of the molecule may be the transport of both desirable and nondesirable ions through the body (Foster, 1960). Fatty acid transport is considered to be one such important role (Fredrickson and Gordon, 1958). Although there are considerable structural similarities between mammalian and avian albumin, the binding patterns and sites may vary with different species. This is so for uric acid (Simkin, 1972) and for thyroxine (T_4) and triiodothyronine (T_3) (Tata and Shellaburger, 1959).

Up to the sixth day of incubation, embryo serum albumin is supplied from the yolk reservoir (Nace, 1953; Zaccheo and Grossi, 1967) and is manufactured by the yolk sac (Gitlin and Kitzes, 1967). This yolk-derived albumin may have a proliferative effect on liver cells (Konyshev, 1968) and, by day 6 of incubation, albumin is supplied by the liver (Nace, 1953), reaching a serum concentration of 0.21% (Christov and Rashev, 1961). By day 21 of incubation, concentrations are 0.62% and on day 3 of posthatch, 0.87% (Christov and Rashev, 1961). Thereafter, albumin levels remain constant (Brandt *et al.*, 1952).

C. Globulins

The conventional mammalian nomenclature of α- and β-globulin arises from electrophoretic studies. Such proteins are heterogeneous, and in the following discussion an attempt will be made to place them in a more homogeneous context.

1. *Lipoproteins*

Lipid transport systems of birds, in particular their lipoproteins, are in general less complicated than those in mammals (Hillyard *et al.*, 1972). In mammals exogenous fat (triglycerides) is transported in chylomicrons via the lymph. In chickens, chylomicrons are not formed and exogenous fat is transported as very low-density lipoproteins (VLDL) via the portal vein (Noyan *et al.*, 1964). Mammalian chylomicrons are the least dense of the plasma lipoproteins and are particulate (0.075–1 μm diameter) and by definition are found in chyle which is formed in the lymphatic system draining the intestine (Mayes, 1973). The chylomicrons reported in the rooster (Schjeide and Urist, 1956) are a convenience term by these authors referring to lipoproteins of density 1.003 gm cm^{-3} and may not be related to mammalian chylomicrons.

Three weeks prior to the onset of laying, serum VLDL is increased from trace levels to 2 gm% as a result of increased estrogen levels (Hillyard *et al.*, 1956) stimulating liver synthesis. This accumulation of serum VLDL serves as a reservoir for VLDL needed for egg yolk production (eyVLDL). The amino acid composition of apolipoproteins of serum VLDL and eyVLDL are similar, both having lysine at the N-terminus and tyrosine at the C-terminus (Hillyard *et al.*, 1972).

A further lipoprotein fraction in plasma is the low-density lipoprotein (LDL) which carries a larger proportion of plasma cholesterol than other lipoproteins (Hillyard *et al.*, 1955). Its amino acid composition, including C- and N-terminal amino acids, is similar to VLDL (Hillyard *et al.*, 1972) and it is precipitated by anti-VLDL antibody (Luskey *et al.*, 1974). Based on this evidence, Luskey and his colleagues suggest that plasma lipoproteins of the chicken be divided into two groups with densities above and below 1.063 gm cm^{-3}. Lipoproteins of density less than 1.063 gm cm^{-3} (VLDL and LDL) contain as their major protein component a single antigenic polypeptide with a molecular weight less than 25,000 daltons. Lipoproteins with a density greater than 1.063 gm cm^{-3} are referred to as high-density lipoprotein (HDL). HDL has aspartic acid at the N-terminus and alanine and leucine at the C-terminus (Hillyard *et al.*, 1972). The HDL contains more than 30% protein with the lipid portion held in pockets formed by a folded peptide chain (Butler, 1971).

When the VLDL levels increase in the laying hen and estrogenized rooster there are also dramatic rises in calcium levels associated with either shell formation (Winget and Smith, 1959) or calcium deposition in the egg yolk (Urist *et al.*, 1958). Associated with the high calcium levels is the increase in two specialized high-capacity binding proteins, X_1 (a phosphoprotein) and X_2 (a phospholipid lipoprotein) (Urist *et al.*, 1958). When calcium levels were raised artificially to 100 mg%, X_1 and X_2 (chiefly X_1) bound 79 mg%, the lipoproteins and albumin bound 15 mg% with the remaining 6 mg% of calcium remaining unbound (Urist *et al.*, 1958). The X_1 phosphoprotein may be the serum equivalent of phosvitin (Mecham and Olcott, 1949), a principal phosphoprotein of egg yolk. Phosvitin has a molecular weight of 40,000 daltons and contains 10% phosphorus bound as phosphoserine units (Mok *et al.*, 1961). It is produced by the liver and the ovary after estrogen stimulation (Greengard *et al.*, 1965; Gornall *et al.*, 1971).

The calcium-binding fraction of the laying hen's serum has been attributed to the presence in the serum of another typical egg yolk protein, vitellin (Greenberg *et al.*, 1936). Although there may be some doubt as to the absolute efficiency of chemical separation of vitellin and phosvitin (Mecham and Olcott, 1949), this cannot explain the assumptions made by Winget and Smith (1959) on the importance of vitellin, and not phosvitin, as the major calcium-binding protein of serum.

The X_2 lipoprotein of Urist *et al.* (1958) has been called lipovitellin (Schjeide, 1963) and is a high molecular weight protein (400,000 daltons) which migrates electrophoretically with the α_2-globulins. As pointed out previously it does not appear to be the major Ca^{2+} binding protein so the relation between vitellin and lipovitellin is doubtful. The lipovitellin of egg yolk has been fractionated into three components, referred to as α- and β-lipovitellin and lipovitellenin (Cook, 1961) which simply confuses the issue more. Clearly, there is much confusion about the terminology of lipoproteins in the early literature which was exacerbated by arguments over the purity and methods of analysis of individual components (see Cook, 1961; Schjeide and Urist, 1961). The grouping of lipoproteins according to density does not help this confused state and it will remain so until total fractionation of all components can be achieved.

2. *Iron-Binding Proteins*

Phosvitin binds ferric iron strongly (Taborsky, 1962) and may well contribute to half the organic bound iron found in the serum of the laying hen and the estrogenized nonlayers (Ali and Ramsay, 1968). The other iron-binding component is transferrin and *in vitro* the iron shows no exchange between the two organic forms (Ali and Ramsay, 1974). Ali and Ramsay (1974) suggested that in laying birds, transferrin iron is a precursor of phosphoprotein iron.

Fowl transferrin is a glycoprotein and differs from the egg white protein ovotransferrin (conalbumin) in its carbohydrate prosthetic group. The protein structures are identical (Williams, 1962). Transferrin has most of its carbohydrate in a single unit composed of two residues of mannose, two residues of galactose, three residues of *N*-acetylglucosamine and one or two residues of sialic acid (Williams, 1968). Ovotransferrin lacks galactose and sialic acid. Serum transferrin has a molecular size similar to serum albumin and a molecular weight similar to human transferrin (80,000 daltons) (Torres-Medina *et al.*, 1971). In moving boundary (Marshall and Deutsch, 1951) and starch gel (Williams, 1962) electrophoresis, transferrin was classed as a β-globulin but in cellulose acetate or paper electrophoresis (Torres-

Medina *et al.*, 1971) it moved as a γ-globulin. This illustrates the confusion that may arise when referring to plasma proteins according to their electrophetograms. Similarly, duck serum transferrins were electrophoretically heterogeneous (Richter *et al.*, 1969) when either the pH or buffer varied, 2–4 fractions being present whose mobility differed from the bird's ovotransferrins.

The liver and yolk sac are the major organs for transferrin synthesis (Gitlin and Kitzes, 1967) and the spleen may act as a reservoir (Giurgea, 1974). Transferrin functions to transport iron to the bone marrow (Katz and Jandl, 1964) and to the yolk (Williams, 1962) and plays an important role in the general control of iron metabolism (Fletcher and Huehns, 1968) and antimicrobial defense.

3. *Copper-Binding Proteins*

Ceruloplasmin, a copper-containing α_2-globulin, was discovered in mammals in 1948 (Holmberg and Laurell, 1948). When compared with the majority of mammals, the levels of ceruloplasm in poultry are low (Srivastava and Dwaraknath, 1971). Chick ceruloplasmin has a molecular weight of 158,000 daltons and contains 0.2% copper indicating 5 atoms of copper per mole. Amino acid composition and electrophoretic mobilities are similar to those reported for human ceruloplasmin (Starcher and Hill, 1966). It has the enzymatic properties of an oxidase but the evidence and the likely substrate are uncertain (Morrell *et al.*, 1962). *In vitro,* it catalyzes the oxidation of ferrous iron (Curzon and O'Reilly, 1960), and from this Osaki *et al.* (1966) suggested that it plays a biological role in the transfer of iron from cells to plasma transferrin. Such a role was demonstrated in copper-deficient swine (Ragan *et al.*, 1969) and in fowl stressed with *Escherichia* endotoxin (Butler *et al.*, 1973). The ferroxidase hypothesis has been refuted by Shokeir (1972) who has suggested that ceruloplasmin mediates copper transfer to copper-containing enzymes, notably cytochrome *c* oxidase and tyrosinase. It is possible that a minor function of ceruloplasmin functions as a haptoglobin in birds through which lyzed hemoglobin is transported and eliminated. Evidence is scanty, but an α_2-globulin does bind hemoglobin in the duck (Liang, 1957).

D. Blood Coagulation Proteins

Blood coagulation is a complex reaction resulting in the conversion of soluble fibrinogen to insoluble fibrin and catalyzed by thrombin. Thrombin is not present in the blood but it is produced from its precursor, prothrombin, by the action of thromboplastin. In mammals, thromboplastin is a component of platelets and many tissues which release thromboplastin when damaged. The currently known factors present in the mammalian serum that are involved in thromboplastin production are given in Table IV. Blood coagulation in fowl was reviewed recently (Archer, 1971) and only a brief survey will be presented here. The blood coagulation times for chickens are quoted as ranging from 10–300 seconds (Bigland and Triantaphyllopoulos, 1961) and it has always been noted that the shortest times were recorded when the bleeding was least efficiently done (Archer, 1970).

The intrinsic blood coagulation system of chickens is poor, probably because no discrete platelets occur in birds but analogous nucleated cells called thrombocytes are found (Didisheim *et al.*, 1959). Thrombocytes have a much slower thromboplastin generation time than mammalian platelets. When homologous brain extract is used to investigate avian coagulation times, consistent times as short as 10–15 seconds are normally recorded (Wartelle, 1957) but with heterologous systems times are longer (Dorn and Müller, 1965). Coagulation in birds obviously depends essentially on extrinsic thromboplastin production brought about by general tissue damage.

Fibrinogen is a high molecular weight (~340,000 daltons) glycoprotein (β-globulin) with a mean concentration in chicken plasma of 346 mg% compared to 250–400 mg% for humans (Bigland and Triantaphyllopoulos, 1961). Chicken fibrinogen is only clotted by chicken thrombin (Didisheim *et al.*, 1959). Much of the chemical analysis of fibrinogen has been conducted on mammalian species, particularly human (Blombäck, 1970). Doolittle *et al.* (1962) demonstrated close structural similarities in the fibrinogen molecule which can be traced from cyclostomes to mammals. The evolutionary changes in the fi-

TABLE IV
BLOOD COAGULATION FACTORS IN MAN

I	Fibrinogen
II	Prothrombin
III	Thromboplastin
IV	Calcium
V	Labile factor
VII	Stable factor
VIII	Antihemophilic factor
IX	Christmas factor
X	Stuart-Prower factor
XI	Plasma Thromboplastin Antecedent
XII	Hageman factor
XIII	Fibrin-Stabilizing factor

brinogen molecule have also been discussed (Blombäck, 1970; Doolittle, 1976). Certainly chicken and human fibrinogens share common antigenic determinants (Menache *et al.*, 1973).

The significance of the carbohydrate moieties of most glycoproteins is unknown and it has been suggested that the "nonessential" carbohydrate plays a supportive role by helping the polypeptide chain to assume the correct three-dimensional structure (Schmid, 1972). Removal of sialic acid from some of the components of the clotting mechanism, although not preventing clotting, changes the kinetics (Schmid, 1972). Differences in the hexosamine content as well as the tryptophan/tyrosine ratio between chicken and human fibrinogen have been reported by Guimbault *et al.* (1972) and in the latter case the differences were attributed to variable coagulability. The implication is that differences in avian and mammalian fibrinogen may relate to both peptide as well as carbohydrate units. Fibrinogen is produced by liver cells, probably under the influence of steroid hormones (Pindyck *et al.*, 1975).

Among the other factors concerned with avian blood coagulation (Table IV), prothrombin and thrombin are probably similar to the mammalian forms because chicken thrombin will clot mammalian fibrinogens (Didisheim *et al.*, 1959). Factor IX, the plasma thromboplastin component, (Griminger, 1965) and factor XII (Didisheim *et al.*, 1959) have been reported as absent in fowl, factor XI was assumed absent (Archer, 1971), factor VII present in low concentration (Stopforth, 1970), and factor V present only in low levels (Didisheim *et al.*, 1959). According to Wartelle (1957) all factors reported absent are present in very small quantities.

E. Fibrinolysis in Birds

The physiological converse of coagulation is fibrinolysis. In mammals the proteolytic enzyme (plasmin) which breaks down the fibrin clot is triggered by factor XII (Iatridis and Ferguson, 1961). Such a mechanism protects the animal from a buildup of fibrin deposits in the blood vessels. Because plasmin is an anticoagulant, it is present in the blood as an inactive precursor plasminogen. Fibrinolytic activity has not been produced experimentally in birds by any accepted plasminogen activators, but spontaneous lysis has been observed in the blood of some birds (Hawkey, 1970). The saliva of a bird-feeding bat *(Diaemus youngi)* will activate clot lysis, whereas the saliva of the common Vampire Bat *(Desmodus rotundus)* did not (Hawkey, 1970). Such results indicate the presence in avian blood of a plasminogen and low levels of activators.

F. The Immune Plasma Proteins

A complete discussion of the avian immune response is not possible in this chapter not only because of its complexity but also because immunity in an animal relies on both cell-mediated and humoral responses and as such the cell-mediated response is outside the scope of this review.

Immunity may be divided into innate or natural immunity (not acquired through contact with the antigen or infectious agent) and acquired immunity. Acquired immunity may similarly be divided into passive immunity (administration of antibodies manufactured in another individual) and active immunity (antibodies manufactured by the individual itself on being challenged with an antigen). The identification of whether a particular process is innate or passively acquired presents problems mainly because during an embryo's development antimicrobial agents pass into it from the maternal supply. Although the following section is titled passive immunity, it will also discuss natural agents.

1. Passive Immunity

During the development of the avian embryo an extra embryonic membrane encloses the yolk to become the yolk sac. Antibodies are secreted into this sac by specialized cells of the epithelial lining of the oviduct. The levels of these antibodies transmitted to the embryo increases from day 11 of incubation (Buxton, 1952). At the time of hatching, yolk sac and residual yolk is ingested and γ (IgM) antibodies appear in the circulation (Solomon, 1968a).

Newly hatched chickens are resistant to fowl plague and Newcastle disease because of the presence of high levels of maternal antibody (Solomon, 1971). The presence of such antibodies appears to have the secondary effect of suppressing active immunity (Hallaner, 1936). In the case of "fowl typhoid" produced by *Salmonella gallinarium*, there are agglutinins to the bacterium in the embryonic circulation from day 11 of incubation, but they do not appear to be protective and infected 1-day-old chicks die (Solomon, 1971). The presence of natural opsonins in the embryonic circulation plays only a minor role in combating this infection (Solomon, 1968b) and it is only during the first few days posthatch that there is an acquired resistance to the bacterium (Shaffer *et al.*, 1964). The presence of opsonins facilitating phagocytosis of *S. gallinarium* and not being effective implies either a low population of phagocytes or some deficiency in functional ability. With diseases such as pneumococcal infection, where protection is associated with humoral antibody, opsonins play an important role (Wright, 1927). Opsonins are generally specific antibodies of the IgM or IgG type.

Innate agglutinins for xenogenetic erythrocytes of different species appear in chick serum at different times after hatching. Erythrocytes of rat and rabbit were agglutinated with serum from 16-day-old chicks and guinea pig with serum from 30-day-old chicks (Bailey, 1923).

2. *Complement*

As with other work discussed in this section, mammalian complement has been investigated in considerable detail, particularly human (Müller-Eberhard, 1975). Complement is the general term, in humans, for eleven serum proteins that participate with antigen–antibody complexes in cytolysis, release of histamine, enhancement of phagocytosis, and capillary permeability changes. Complements are glycoproteins of molecular weight 79,000–400,000 daltons, which migrate electrophoretically with all classes of globulins and are designated C1 to C9. C1 is subdivided into C1q, C1r, and C1s. Avian complement and complement components have been little studied because of titration difficulties. Like the mammalian form, avian complement is calcium- and magnesium-dependent for activation (Wirtz, 1967). Sherman (1919) detected complement in 17-day-old embryos, but it was uncertain whether their presence was autogenic or of maternal origin (Solomon, 1971). Chicken C1 has been detected in the serum of normal and hypogammaglobulinemic animals and, in the latter case, correlated with susceptibility to disease (Gabrielson *et al.*, 1974).

3. *Acquired Immunity*

a. Interferon. These cellular proteins are produced in response to invading viruses, circulate in the blood and are taken up by other body cells to offer general body protection. They are species-specific but lack viral specificity, are nondialyzable, nonsedimentable at 105,000 *g* for 2 hours, and are destroyed by proteolytic enzymes, but are stable to acid (pH 2.0) (Kleinschmidt, 1972). They are produced in chick embryos as early as day 6 of incubation (Isaacs and Barm, 1960) and increase by a factor of 20 times by day 11 of incubation.

Attempts have been made to purify chick interferon (Fautes and Furminger, 1967) but there is still some doubt as to whether the carbohydrate found in the molecule is a functional requirement of antiviral activity or an impurity. Certainly, the amino, disulfide and the γ-S-methyl of methionine are necessary for activity (Kleinschmidt, 1972).

b. Immunoglobulins. The immune system of birds, in common with all vertebrate species, is characterized by the capacity to synthesize humoral antibodies in response to a challenge with antigen. There are five classes of immunoglobulins in humans IgG, IgM, IgA, IgE, and IgD; IgG is the major component. Whether the same components are present in birds is still speculative. Using starch gel electrophoresis and Sephadex gel filtration, the patterns obtained for IgG for several mammalian species were similar, but chicken IgG showed slower mobilities (Mehta and Tomasi, 1969). The IgM of all species examined were extremely variable.

When challenged with infectious bronchitis, chickens show an increase in serum complement fixation titers at 6 days postinoculation and high titers for 12–42 days (Marquardt, 1974). The antibodies were generally associated with the γ-globulin fraction in electrophoretic analysis and are referred to as immunoglobulins (Ig). Chicken, challenged with the parasite *Eimeria tenella* (Mukkur, 1969) or suffering from spirochetosis (Perk and Hart, 1966), showed significant drops in albumin fractions and increased levels of globulins, in particular β-globulin.

The response to antigen challenge is an inherited dominant trait in the case of inbred lines of chickens (Balcarova *et al.*, 1974) and is age dependent (Wolfe *et al.*, 1957). It is poorest soon after hatching and increases through the first 4 weeks, with a sudden increase thereafter.

Based on ultracentrifugal evidence, two immunoglobulins have been separated in chickens (Benedict, 1967), a high molecular weight form (HMW) with a sedimentation constant of 19 S and a low molecular weight form (LMW) of 7.8 S. In duck serum, an immunoglobulin of 5.7 S was found which was not a degradation product of the 7.8 S form and was reported present in a number of divergent vertebrates (Zimmerman *et al.*, 1971). The LMW Ig of chicken and the IgG of man possess several characteristics in common, both representing the predominant Ig class in serum (Walsh *et al.*, 1968). Before discussing their characteristics it will be necessary to outline briefly the chemical structure of the immunoglobulin molecule.

The molecular organization of Ig has been discussed in many excellent reviews (e.g., Edelman and Gall, 1969; Fleischman, 1966). The basic molecule (Fig. 1) is composed of two light (L) chains of molecular weight 22,500 and two heavy (H) chains of molecular weight 50,000–75,000 linked by covalent and disulfide bridges to form a T- or Y-shaped structure.

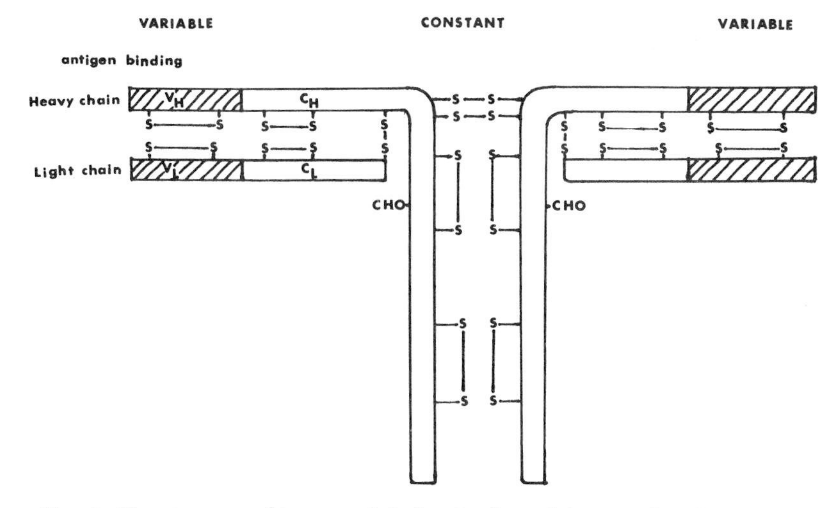

FIG. 1. The structure of immunoglobulin G, after Edelman and Gally (1971).

The L chains contain approximately 217 amino acids and are divided into two groups, κ and λ, distinguished by their amino acid sequences. The relative proportions of the two varies among species. Each light chain is divided into two approximately equal parts, the V chain with an amino terminal unit and the C chain with a carboxyl terminal unit.

The H chains contain 450–576 amino acids depending on Ig class, and are divided into V regions that are variant and C regions that are invariant within a class. In mammals (humans) there are five classes of heavy chains corresponding to the five Ig classes α (IgA), γ (IgG), μ (IgM), δ (IgD), ϵ (IgE). It is the C regions of the heavy chain that determine the Ig class.

Peptide mapping analysis of chicken and human heavy chains demonstrates nine peptides in the two species with identical mobilities (Walsh *et al.*, 1968). Amino acid studies of the peptides, however, show that structurally they are different (Sanders *et al.*, 1973) (Table V). The terms euglobulin and pseudoglobulin are not electrophoretic terms but are based upon sodium sulfate precipitation techniques (Cohn, 1945). The electrophoretic pattern usually remains unchanged after precipitation of the euglobulins (13.5% sodium sulfate). In removal of both the euglobulins and pseudoglobulins (17.4% sodium sulfate) approximately 50% of α-globulins and 25% of the β-globulins are eliminated but none of the γ-globulins.

TABLE V
COMPARISON OF LOW MOLECULAR WEIGHT IMMUNOGLOBULIN HEAVY CHAIN OF CHICKEN WITH γ-CHAIN (IgG) OF MAN

	Chicken	Man
Molecular weight	58,000	48,000
Carbohydrate content	3%, 12%[a]	3%
Number of cysteines	13	10–11
Lysine/arginine ratio	0.6	2.0

[a] Values for pseudoglobulin and euglobulin, respectively (see text).

It would appear that the structural homology between the IMW Ig and IgG first considered by Walsh *et al.* (1968) is not apparent (Table V). Leslie and Clem (1969) have suggested that the avian LMW Ig be referred to as IgY, signifying the fact that it does not resemble any human Ig classes. Sedimentation equilibria studies on LMW Ig indicate a molecular weight of approximately 170,000 daltons, suggesting a structure composed of two H and two L chains. Based on physicochemical, immunochemical, and proteolysis investigations, Tenenhouse and Deutsch (1966) suggested that LMW Ig was homologous to mammalian IgA, a point of view challenged by Lebacq-Verheyden *et al.* (1972) who found a third Ig of γ mobility not related to LMW Ig and referred to as IgA to suggest an IgA homology. This IgA homology was confirmed by Orlans and Rose (1972) and Bienenstock *et al.* (1973) who found that IgA exists in serum in polymeric form (molecular weight of 350,000–360,000 daltons) and monomeric form (170,000 daltons). It was found in and isolated from bile (Lebacq-Verheyden *et al.*, 1972). Because the secretion versus serum ratio for IgA was higher than that for IgG (Lebacq-Verheyden *et al.*, 1974) it confirms the fact that chickens have a secretory immunologic system similar to that found in man.

The HMW Ig of chicken, in terms of gross architecture, resembles the mammalian IgM class (Leslie and Clem, 1969). It has a molecular weight of approximately 900,000 daltons. The H and L chains account for 75 and 25%, respectively, of the mass of the molecule. It is now generally accepted that there are three definite immunoglobulins in avian blood designated IgY, IgA, and IgM.

The secretory immunologic system of birds is probably under the control of the Bursa of Fabricius (Leslie and Martin, 1973) and the IgA-forming cells are derived from cells formerly producing IgM. Certainly the Bursa is the first site of IgM synthesis which occurs in 18-day-old embryos (Thorbecke *et al.*, 1968) when the organ provides cells which give rise to antibody-producing clones. Animals lacking a Bursa cannot respond to antigens by antibody production even though their thymus is morphologically normal. Thymus cells are evidently involved in the homograph rejection mechanism and hence function similarly to the mammalian thymus (Warner *et al.*, 1962). The mammalian equivalent of the Bursa is uncertain although the human tonsils and Peyer's patches fit the necessary criteria (Cooper *et al.*, 1966). The thymus gland does not produce IgM but after 2 weeks IgY is found in the gland, but probably not in cells of thymic origin. Thorbecke *et al.* (1968) suggested that stem cells from either the bone marrow or the yolk enter the embryonic Bursa as well as other lymphoid tissue on day 14–16 of incubation. Bursal cells then synthesize a hormone that initiates early immunoglobulin synthesis in the organ. The other lymphoid tissues are stimulated to synthesize antibody later by either release of the free hormone or cell types.

III. Egg White Proteins

A. General Comment

Although the components of the avian egg have attracted the attention of workers in all branches of science, there has been a trend since the initial, descriptive investigations (Romanoff and Romanoff, 1949) for the studies to be concerned with fundamental problems of protein chemistry, molecular biology, genetics, etc. Thus, we have a paltry understanding of the overall contributions of the various protein components to the function of the embryo, but a large literature on their occurrence (Sibley, 1960, 1970; Sibley and Ahlquist, 1972), chemical and biological properties (Feeney and Allinson, 1969; Osuga and Feeney, 1974), and phylogenetic traits (Baker, 1970; Manwell and Baker, 1970). There are many admirable reviews and texts dealing with the findings of the protein chemist and their application to phylogenetic studies. This section will be concerned with a discussion of the possible functional roles of the proteins in the embryo.

B. The Role of the Cleidoic Egg

The pivotal role (Freeman and Vince, 1974) of the cleidoic egg in the breeding biology of birds was taken for granted until the demonstration (Ratcliffe, 1970) that the agricultural use of certain pesticides was associated with a reduction in hatchability. Through being weakened, the shell, particularly those of the eggs of raptors, could no longer resist the stresses to which it was exposed in the next. Those whose concern it is to hatch thousands of domestic chicks per day have long been aware of the contribution of egg traits (size and shape of egg, shell characters, albumen quality) to hatchability (Landauer, 1967) even when the performance of the incubator had been optimized by trial and error (Lundy, 1969). In studies of the natural history of incubation (Drent, 1975; White and Kenney, 1974) the emphasis was largely on behavioral adaptations whereby the parent(s) impose an environment within the nest so that the needs of the developing embryo were met. In an ecological context Lack (1968) sought an interpretation of the means whereby birds achieve population stability. He recognized "problems concerned with eggs" even though the discussion was limited to cryptic markings of the shell and the probable need for the eggshells of waterfowl to be waterproofed.

In discussing the egg and the environment, Needham (1950) noted a complex interplay of factors between the embryo and the environment—a concept that, as stressed by Vince (1973), has been largely ignored by those studying the embryo's behavior. Thus with the avian egg, its sole demand on the abiotic components of the environment is for oxygen, the parent(s) providing heat—or shade, movement and, perhaps, some control over relative humidity. In accepting that the egg's release from a requirement for exogenous water is the last fundamental step in the evolution toward minimum dependence on the external environment, zoologists accepted that the embryo's physiology would have had to have adapted to life in a closed system, and much effort has been directed at topics such as nitrogen metabolism leading to the end product, uric acid, and the acid–base relationship within the egg (Dawes, 1975; see also Freeman, Chapter 3, this volume).

Although achieving independence from exogenous water, the avian egg has become vulnerable to that water. With the high metabolic rate obtained with incubation at 35°–40°C, the embryo, especially in the few days preceding pipping, has a large requirement for O_2 (Freeman and Vince, 1974) that can be satisfied only by diffusion (Wangensteen and Rahn, 1970–1971) across pores (Board *et al.*, 1977) in the calcitic shell. The complement of pores provides a diffusive capacity which matches the demand of the embryo in the last day or so of incubation (Tullett and Board, 1976). It is recognized that the pores provide portals also for the outward diffusion of water vapor (Rahn and Ar, 1974). It is generally accepted that the egg contains at oviposition sufficient water to compensate for loss of up to 18% of egg weight (Freeman and Vince, 1974). In theory the diameter of the pores is such that free water could be drawn in by capillarity and it was surmised

(Board and Halls, 1973a,b) that this could lead to asphyxiation of the embryo, a situation analogous to that following the flooding of an insect's plastron (Hinton, 1968). Board (1974, 1975) demonstrated that the eggshell was water repellent, waterproofed, or both and that work needs to be done to overcome these resistances. When work is done, there is a marked tendency for the eggs to rot during storage (Haines and Moran, 1940).

The problems which confront the egg tend to be ignored in many of the discussions of strategies of egg production (Price, 1974) even though it was noted often that egg production was adapted to counter the relative hostile nature of the environment colonized by the parents. Actually both the parents and the eggs must counter such hostility. When the proposed strategy of adaptation had as its basis the allocation of resources then the concept is germane to this section. Given that there is a limited amount of energy available for reproduction, then this must be allocated optimally to eggs, avoidance of predation, and competitive ability. Predation tends to be associated popularly with food chains; it ought to be considered also in the sense of "energy wastage" by heterotrophic bacteria. If this is accepted, then two questions may be posed. First, have the selective pressures which favor the evolution of a few large eggs—and thus a relatively large allocation of energy—favored also a defense whereby the energy and developing embryo are protected from infection with heterotrophic bacteria? Second, have these pressures favored also a defense which operates in the neonate during both the time that it is acquiring immunological competence and during the period that it takes for the gut and mucous membranes to be colonized by nonpathogenic microorganisms? Those trained in classical immunology (Solomon, 1971) would brook no argument about the neonate's complement of maternal antibodies and they cite examples which provide a clear indication of a relationship between the immunobiological status of parent and offspring. In this context the avian embryo is analogous to those of guinea pigs, man, etc. (Brambell, 1958) in being dependent upon the transmission of antibodies within the parent's body. Nevertheless, human milk is rich in antibodies and other antimicrobial substances that create an environment in the gut which holds the enteropathogens in check while the commensal flora becomes established. Apart from its large content of antibodies, mammalian milk contains many components of the defense system which Tokin (1964) and Board and Fuller (1974) have advanced to account for the defense of the avian embryo and its food reserves. Although having teleological overtones, their hypothesis does permit a novel discussion of the egg white proteins without requiring excessive cataloging of the achievement of the protein chemists and molecular biologists. It also directs attention to a form of defense (Glynn, 1972) that has been often neglected by those trained in classical immunology. It may also provide another vantage point for discussion of the apparent evolutionary conservatism in birds in general (Prager *et al.*, 1974). Overall, it may reflect the outcome of a situation where selective pressures operate on two independent but interdependent units, the parent and its egg.

C. The Albumen

The albumen (white) occupies the space between the yolk (Fig. 2), bounded by the chalaziferous layer and four-layered vitelline membrane (Bellairs *et al.*, 1963), and the shell membranes which consist of anastomizing fibers with a protein core and a mantle rich in polysaccharides (Cooke and Balch, 1970). In the eggs of ten altricial species discussed by Romanoff and Romanoff (1949), the albumen makes a relatively larger contribution than in precocial species (Table VI). The albumen is synthesized and deposited in the oviduct (Gilbert, 1971) by mechanisms controlled by steroids (Palmiter and Smith, 1973; Rosen *et al.*, 1975; Tuohimaa and Söderling, 1976; Sharma *et al.*,

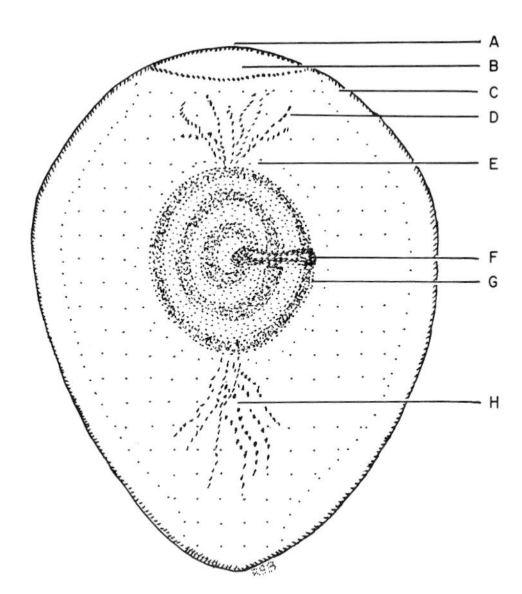

Fig. 2. Structure of egg as seen in longitudinal section. A, shell and shell membranes; B, air cell; C, outer thin white; D, albuminous sac; E, inner thin white; F, blastoderm; G, yolk; and H, chalaza.

TABLE VI
GROSS COMPOSITION OF EGGS

Component (wt %)	Precocial (domestic hen)	Altricial
Albumen[a]	56	61–78.6
Yolk	33	—
Shell/membranes	11	11

[a] Albumen of typical hen's egg (wt %): water (88.5), protein, (10.5), carbohydrate, (0.5), inorganic, [0.5–Na^+, 1.42 (1.27–1.83); K^+, 1.40 (0.94–1.77); Ca^{2+}, 0.13 (0.07–0.29; Mg^{2+}, 0.1; Fe^{3+}, 0.0001; S, 1.99; Cl^-, 1.27; PO_4^-, 0.11]; vitamins (trace).

1976). Amino acids from the blood are extracted in the magnam at a rate corresponding to the rate of albumen synthesis (Edwards *et al.*, 1976). The general composition of the albumen of the domestic hen is given in Table VI.

There are three layers of albumen in freshly laid eggs, each containing different amounts of dissolved material. The approximate percentage of total egg white : dry matter for hen's egg are 23 : 21% in the outer thin white, 57 : 12% in the albuminous sac, and 17 : 13% in inner thin white (Brooks and Taylor, 1955). The protein content of thin and thick white is similar but thick white contains about four times more ovomucin. The thick white in the hen's egg is a weak gel interpenetrated by a system of microscopic elastic fibers (Brooks and Hale, 1959). With storage at ambient temperature several changes occur in the hen's egg. The pH drifts from about 7.4 to 9.5 as CO_2 diffuses across the shell. The amount of outer thin white increases at the expense of the albuminous sac and eventually the inner thin white is transferred across the albuminous sac. The albuminous sac loses its gel state and the density of the white increases as it loses water by evaporation and absorption by the yolk. Finally, the freedom of movement of the yolk increases, due to decay of the albuminous sac, and it rises to make contact with the shell membranes. During the first 5–8 days of incubation much of the water in the albumen is transferred to the subembryonic fluid (Romanoff, 1967; New, 1956). This is associated with a reduction of the albumen's content of Na^+, and Tullett and Board (1976) concluded that a Na^+ pump was involved. There is a close association also between the albumen water loss and a diminution in the resistance of the shell membranes to gaseous diffusion (Tullett and Board, 1976), possibly as a result of the membrane becoming "dry" (Kutchai and Steen, 1971). On about the twelfth day of embryo development of the domestic hen, the albumen begins to pass via the seroamniotic connection to the amnion and eventually to the gut and lungs of the embryo (Witschi, 1956). There is a migration of ovomucoid from the albumen to the amnion from which it is swallowed and transported via the gut to the yolk sac where it is catabolized by proteases and glycosidases present in the yolk sac membrane (Oegema and Jourdian, 1974a,b). Ovomucoid is present in the embryo's serum on day 12 and in that of the neonate on the seventh day following hatching.

D. The Role of the Albumen and Antimicrobial Defense

The contributions made by the major components of the egg to the maintenance of the embryo (Table VII) emphasize the dependence on the concerted action of many functions of each component. Thus the vulnerable food store of the egg, the yolk, is protected from infection with microorganisms of extragenital origin by the chemical defenses of the albumen and the physical defenses of the shell. Moreover, as noted later, the central location of the yolk, and thus the biological structure of the egg, contributes to its defense.

The egg's vulnerability to addling by heterotrophic microorganisms has attracted insufficient attention. This is due partly to ornithologists assigning infertile eggs, dead embryos, and rotten eggs to the category, addled, and partly to the freedom from contamination which experimental embryologists have enjoyed with their rather casual approach to asepsis during experiments on incubating eggs. However, addling has long been recognized by those who deal with commercial eggs or by those who operate incubators which contain many thousands of eggs. The literature generated from the latter sources has been reviewed elsewhere (Board, 1966, 1968, 1969) and the principal findings can be summarized briefly. First, the majority of eggs of the domestic hen are free from microbial infection at oviposition (Brooks and Taylor, 1955; Harry, 1963a,b) but certain pathogens (*Salmonella*, mycoplasma and viruses) can be transmitted via the yolk and white and cause disease in the newly hatched chick (Payne, 1968; Harry and McClintock, 1972). Second, rot-producing bacteria which contaminate the shell have to be translocated along water-filled pores and be lodged in the shell membranes before the egg contents are at risk (Williams and Whittemore, 1967; Board and Halls, 1973a). Following translocation, there are three possible outcomes: (1) the organisms migrate through the albumen and colonize the surface of the yolk (Sharp and Whitaker, 1927); (2) the infection is confined to the shell membranes until the yolk contacts the shell membranes (Board, 1964); or (3) the bacteria

TABLE VII
THE CONTRIBUTIONS OF COMPONENTS OF THE EGG TO THE MAINTENANCE AND DEVELOPMENT OF THE EMBRYO

Component	Physiological function	Physical protection	Antimicrobial defense
The integument (preening oils, cuticle/cover, true shell and shell membranes[a])	Exchange of respiratory gases Conservation of water Reservoir of Ca^{2+} Depot of carbonate Insulation Conductance of sound waves and photostimuli	Protection from crushing Prevention of water logging Camouflage	Barrier to microbial invasion
Air cell	Air reservoir and induction of breathing Water conservation—a "cold nose" phenomenon	Compensation for changes in pressure	—
Albumen	Reservoir of water Depot of cations (Na^+, Mg^{2+}, K^+) Exchange of O_2, CO_2, H^+ with young embryo Source of protein	Cushioning against damage due to sudden movement Lag against temperature change Scaffolding for yolk and embryo	Viscosity an impediment to bacterial movement Controls rate and extent of microbial growth Passive immunity of chick
Yolk	Principal depot of all major and minor nutrients	Location of young embryo at least distance from heat of brood patches	Passive immunity of chick

[a] Board *et al.* (1977).

pass across the albumen, infect the yolk, and through modifying the food reserves, cause intoxication or starvation of the neonate (Harry, 1957). Of these, point (2) is a common feature of infertile eggs or those stored at subincubation temperatures (Board, 1969) and is probably the one which operates in nature during the laying of a clutch of eggs. Option (3) is a feature of commercial hatcheries if the hygiene is poor. The apparent failure of microorganisms to colonize and grow in the albumen is a feature common to all three. The inadequacy of the albumen as a medium for microbial growth was implicit in the observations of Gayon (1873), and the tendency in the past century has been to attempt to account for this in terms of the albumen's content of proteins having specific biological properties (Table I). The uniqueness of these properties relative to those of the components of the cellular and humoral defense systems has attracted little comment even though they may represent an extremely old (primitive?) form of defense. It operates without energy consumption, "training," neural or hormonal control in a system lacking a vascular network and it can act against one cell type (the prokaryotic heterotroph) while providing the physiochemical environment required by the eukaryotic cell of the avian embryo (Tokin, 1964; Board and Fuller, 1974).

E. Egg White Proteins

Information about the complement of proteins in the albumen of avian eggs has been a product of innovation in the methods used to isolate, purify, and characterize proteins. Thus gel electrophoresis demonstrates upward of 50 or so "bands" and purification can be achieved by affinity or ion-exchange chromatography, ultrafiltration, etc. (Neurath and Hill, 1975). In practice, most information pertains to the ten major proteins of the albumen of the hen's egg (Osuga and Feeney, 1974) many of which occur widely in bird eggs (Feeney and Allinson, 1969). Since the studies of McCabe and Deutsch (1952) and their assertion that "an egg contains more of its incipient phylogeny than the more superficial aspects of the bird's adult morphology," there have been many phylogenetic studies (Sibley, 1960, 1970; Sibley and Ahlquist, 1972), some of which have established clearly the relatedness of particular birds (Baker and Manwell, 1976) and the recognition that protein polymorphism is widespread (Ferguson, 1971). Support for the assertion "the selection for factors affecting egg white proteins is probably slow, indirect and less drastic" (McCabe and Deutsch, 1952) has come from immunological studies. Prager *et al.* (1974) surmise that slow evolution in some egg white proteins may be associated with the birds' high body temperature or low DNA content may need to be considered alongside our speculation (Section III,B) that selection pressures operate on two independent but interdependent units, the chicken and egg. Detailed studies of some of the highly purified proteins have shown differences in amino acid composition and biological properties (discussed in the following sections).

1. *Lysozymes and Ovomucins*

The lysozymes (muramidases) cleave the β(1–4) glycosidic linkages between *N*-acetylglucosamine and *N*-acetylmuramic acid in the peptidoglycan heteropolymers which confer geometry and resistance to osmotic lysis to prokaryotes. Their contribution to the total egg white ranges from 3–4% in the chicken, other galliformes, and anseriformes (Feeney and Allinson, 1969) to only trace amounts in penguins (Manwell and Baker, 1973).

The amino acid sequence of several avian lysozymes has been elucidated (summarized in Jollès *et al.*, 1976). Differences also exist in their enzymatic properties (Dianoux and Jollès, 1967; Arnheim *et al.*, 1973; Bailey and Geoffrey, 1975), crystallography (Bott and Sarma, 1975), and immunogenicity (Arnheim and Steller, 1970; Jollès *et al.*, 1976). The latter has been used in a survey, based on immunodiffusion techniques, of the distribution of the chicken (lysozyme *c*) and goose (lysozyme *g*) muramidases in the orders of birds (Prager and Wilson, 1974). Lysozyme *c* was demonstrated in two orders (Galliformes and Anseriformes), whereas lysozyme *g* was found in nine (Anseriformes, Struthioniformes, Rheiformes, Apterygiformes, Tinamiformes, Podicipediformes, Sphenisciformes, Casuriformes, Charadriiformes) and both *c* and *g* in some Anseriformes (Black Swan and Canada Goose). The domestic chicken produces both lysozymes *c* and *g*, the former occurring in egg white and both in the polymorphonuclear leukocytes (Hindenburg *et al.*, 1974). The genetic loci that code for both forms of the enzyme may be common to many species but species-specific regulatory mechanisms control whether one or both are expressed during the production of egg white proteins. Jollès *et al.* (1976) compared sequence differences with immunological differences and derived a cladogram of selected taxa based on lysozyme differences.

The lysis of bacteria by egg white was noted by Laschtschenko (1909) but this property did not become widely recognized until similar observations led Fleming (1922) to appreciate the enzymatic nature of the process and to propose the name, lysozyme, for the lytic agent. Since his account "on a remarkable bacteriolytic element found in tissues and secretions," the enzyme has been accorded a cardinal role in the antimicrobial defense of the egg. Manwell and Baker (1973) discussed the concept that large amounts of lysozyme in egg white may reflect selective pressures associated with eggs being challenged with a large load of microorganisms. Arnheim *et al.* (1973) pondered a possible correlation between the type of muramidase and the dominant prokaryotic type to which eggs are exposed. However, nothing is known about the level and type of bacterial infection of nests, so neither hypothesis can be considered further at the present. In fact, it is not possible to present evidence to support the general view that lysozymes, as enzymes, are important in the egg's defense. Fleming and those who have followed him (e.g., Garibaldi, 1960) have used a sensitive organism such as *Micrococcus lysodeikticus* in whose cell wall the peptidoglycans are a major component and, as a consequence, many β(1–4) glycosidic bonds are exposed. The addition of lysozyme c to the mixed bacterial flora from eggshells did not reduce the viable count by a significant amount (Board, 1968) and it has been established (Board, 1966, 1969) that gram-negative bacteria are the dominant contaminants of rotten eggs. Such organisms have a relatively small amount of peptidoglycan in their cell envelopes, and additional materials, such as the lipoproteins and lipopolysaccharides, impede the diffusion of lysozyme to its substrate. This may be disrupted experimentally with EDTA at pH 8–7, for example, but whether such disruption occurs in the egg has not been demonstrated. It is notable, even if at the present time anecdotal, that yet another diffusion barrier, a capsule, was a common feature of the gram-negative bacteria isolated from the rotten eggs of waterfowl (Seviour *et al.*, 1972). This may reflect the selectivity imposed by lysozyme g which attacks peptidoglycans from gram-negative bacteria more avidly than those from gram-positive ones (Arnheim *et al.*, 1973). If a cardinal role for lysozyme is accepted, particularly with respect to gram-negative bacteria, then it is pertinent to consider the consequences of bacterial lysis. Many of the lysates show great pharmacological activity and may enter the gut, etc. of the embryo. In this case could not the action of lysozyme exacerbate rather than resolve problems? Another, and possibly more fruitful, line of enquiry as to the role of lysozyme had its genesis in commercial problems associated with the breakdown of the albuminous sac and the loss of "quality" of eggs intended for human consumption (Almquist and Lorenz, 1932). Many investigators have concluded that the gel structure of the albuminous sac resulted from a protein–protein interaction and Brooks and Hale (1959) postulated that the mechanical properties of the sac were the product of a network formed from cross-linking of an ovomucin–lysozyme complex. Further, lysozyme will polymerize with itself when the ionic environment is appropriate and reduced soluble ovomucin derivatives can interact with stoichiometric amounts of lysozyme through salt linkages (Robinson and Monsey, 1969a,b). Subsquently, Robinson and Monsey (1971, 1972a,b, 1975) showed that ovomucin of the hen's egg consists of two components: α-ovomucin (MW 210,000) with about 1% (w/w) sialic acid and β-ovomucin [an aggregate (MW 720,000) formed from globular subunits (MW 112, 300)] containing more than 10% (w/w) sialic acid. With incubation at 37°C, thinning of the thick white was associated with degradation of β-ovomucin. Kato *et al.* (1975) endorsed the concept of an electrostatic attraction between the negative charges of the terminal sialic acid in ovomucin and the positive charges of lysyl ε-amino groups in lysozyme. They demonstrated that the extent of interaction of F-ovomucin (their ovomucin fraction rich in sialic acid) and lysozyme was reduced when sialic acid was removed by neuraminidase. It is noteworthy, also, that at the body temperature of the hen, lysozyme has a conformation different from that at 20°C and a diminished affinity for certain substrates (Jollès *et al.*, 1975).

These observations also provide the basis for comparative studies of the rheological properties of egg albumen in general and contribute to another feature of the avian egg, the role of the albuminous sac. It is suggested that the sac performs a scaffolding role which is equal in functional significance to the contribution of the extra embryonic membranes. Its contribution is considered important in two phases of the egg's existence: (1) initially by ensuring that the yolk is retained centrally and thereby protected from colonization by microorganisms (Board, 1964); (2) by ensuring that the embryo which has yet to complete its chorioallantoic membrane is prevented from adhering to the shell membranes, a phenomenon which results in malformation (a "stuck germ"). Both roles are transient. They are performed in the absence of neural or hormonal control and, on completion of their function, they must not be an impediment to the continued development of the embryo. The protein complexes must acquire physical properties which permit their passage across the seroamniotic raphe, and they presumably have to be adapted to the environment existing before the onset of brooding. Thus if the postulated lysozyme–ovomucin complex of the hen's egg were present in the eggs of the ostrich it could be anticipated that, because of damage by exposure for a few minutes to temperatures near 60°C (Garibaldi *et al.*, 1968), it might well be denatured in the time that the Ostrich egg spends on sand at 56°C (Sauer and Sauer, 1966) before the onset of brooding. Dr. S. G. Tullett and I (unpublished) have noted in determinations of the freezing points of albumen that some separate into two phases on thawing whereas others return to a state resembling untreated samples. Such observations imply that some environmental components, especially temperature, may apply strong selective pressures via the egg to the synthetic attributes of the oviduct. Thus a broader biological setting might well aid our understanding of the reasons underlying observations such as the resistance to decay in the albumen of duck and goose eggs (Rhodes and Feeney, 1957); that the albumen of penguin eggs, which contains little lysozyme but large amounts of sialic acid (Feeney *et al.*, 1968; Manwell and Baker, 1973), deteriorates at about the same rate as that of domestic hens (Feeney and Allinson, 1969); and the correlation between the quality of the thick white and its content of sialic acid (Sauveur, 1975).

2. *Avidin and Ovoflavoprotein*

The basic, tetrameric protein, avidin (Green, 1975), is secreted in the oviducts of reptilia, amphibia and birds, but not mammals, when they are supplied with progestagenic steroids (Tuohimaa, 1975; Tuohimaa and Söderling, 1976). Avidin binds with four moles of biotin (vitamin H), one per subunit, with a dissociation constant of 10^{-5} M (Green, 1963a,b, 1964; Green *et al.*, 1971; Chignell *et al.*, 1975). The avidin content in egg white (Feeney and Allinson, 1969) ranges from 0.1 units/gm dried weight of the Herring Gull *(Larus argentatus)* to 16.2 units for the Turkey *(Meleagris gallopavo)*. The ease with which this potent binder can be studied by microbiological assay (Wright and Skeggs, 1944) is probably sufficient evidence to justify the common belief that it can contribute to the defense of the albumen against colonization by microorganisms requiring this vitamin.

Ovoflavoprotein, which can be isolated from the hen's egg white by cellulose ion exchange (Rhodes *et al.*, 1958), and purified by affinity chromatography (Blackenhorn *et al.*, 1975), binds riboflavin (Rhodes *et al.*, 1959) more avidly ($K_a = 7.9 \times 10^8\ M^{-1}$) than the more commonly occurring cofactors of flavoenzymes, FMN ($K_a = 7.3 \times 10^5\ M^{-1}$) or FAD ($K_a < 7 \times 10^4\ M^{-1}$). It is a glycoprotein (32,000 daltons) containing 16 half cystine residues (8 Cyst–Cyst bridges) and consists of two subunits the larger of which contains five of the dithio linkages (Clagett, 1971). The flavoproteins are fairly widespread in birds (Feeney and Allinson, 1969). Homogeneity of precipitin lines was demonstrated by immunodiffusion of the flavoproteins isolated from the yolk, white, and blood serum of domestic hens (Clagett, 1971). By selective breeding of certain strains of domestic hens, the genotypes (RdRd, Rdrd and rdrd) were obtained with the ratio of riboflavin : carrier : protein in the blood stream of 2 : 1 : 0. Hens with the double recessive rapidly excrete riboflavin in the urine and lay eggs which, with little ovoflavoprotein in the albumen, will not support embryo development beyond 10–14 days unless the vitamin (200 mg) is injected through the shell (Buss, 1969; Clagett, 1971).

Although the inhibition of the growth of *Streptococcus pyogenes* and *Lactobacillus casei* has been demonstrated *in vitro* when 10 moles of apoprotein was present for every mole of riboflavin (Rhodes *et al.*, 1959), the half-saturation of the apoprotein in the hen's egg white would give this protein only a minor role in the egg's antimicrobial defense (Baker, 1968). This suggestion was based on a consideration of apoprotein in isolation. In practice a riboflavin-requiring microorganism may not be able to glean sufficient vitamin in an environment having many other extremes of environmental parameters (high pH, unavailability of combined nitrogen, Fe^{3+}, avidin, etc).

3. *Ovotransferrin*

Although the inadequacy of the albumen of the hen's egg as a medium for microbial growth had been recognized for about 100 years, this property was not explained satisfactorily until the studies of Schade and Caroline (1944). They noted that several species of bacteria and a yeast would grow in albumen if Fe^{3+} was added. This led to the demonstration (Alderton *et al.*, 1946) of the iron-binding (2 binding sites per mole; Evans and Holbrook, 1975) properties of conalbumin (ovotransferrin), a protein which is a member of a class of ligands studied extensively in the past 20 years (Crichton, 1975). The transferrins, glycoproteins of ca. 80,000 daltons (Tsao *et al.*, 1974a,b,c), are a common component of bird eggs (Sibley, 1960, 1970), often in polymorphic form (Ferguson, 1971), and occur in a wide range of biological fluids (Feeney and Allinson, 1969). Their evolution can be ascribed presumably to the need for an iron-transport system to operate in situations of pH and oxygen tension where the tendency is for Fe^{3+} to form complex, insoluble hydroxides. In such situations, those cells which require Fe^{3+} have a receptor site whereby transfer of the metal ion from the transferrin is achieved (Crichton, 1975).

In studies with bacteriological media containing known amounts of Fe^{3+} and conalbumin, Theodore and Schade (1965a) noted that the initiation of growth, rate of growth, and total crop of *Staphylococcus aureus* were dependent on the concentration of free iron, which was a function of the percentage iron saturation of the ligand. With albumin *in vitro* deprivation of Fe^{3+} did not cause the death of those organisms which are commonly associated with rotten eggs (Board and Halls, 1973c). It can be inferred, therefore, that a fermentative prokaryote suspended in albumen does not have its metabolism switched from an oxidative to a fermentative type, a phenomenon noted by Theodore and Schade (1965b) when *St. aureus* was deprived of iron by the addition of ovotransferrin to a bacteriological medium which had been depleted of Fe^{3+}. Microorganisms do not retain their viability when suspended in vertebrate serum *in vitro*. Their death was attributed to an interplay of antibody, transferrin, and complement (Fletcher, 1971). From studies of *Pasteurella septica*, Griffiths (1974) concluded that iron deprivation initiated changes which resulted in the rapid degradation of ribosomal RNA.

The bacteriostatic action of ovotransferrin can be negated if another chelate, 8-hydroxyquinoline, is added to a bacteriological medium (Feeney and Nagy, 1952). It is of interest to note that monoferric ovotransferrins are formed in the presence of the chelates, ATP, and nitrilotriacetate (Donovan *et al.*, 1976), whereas Fe_2-ovotransferrin is formed in their absence (Aisen and Leibman, 1968). Owing presumably to the formation of insoluble hydroxides in systems having O_2 in solution at neutral pH, a wide range of prokaryotes have evolved compounds associated with the scavenging of iron (Snow, 1970). Although such compounds can negate the action of ovotransferrin in albumin *in vitro* (Garibaldi, 1970), there is as yet no evidence that they are produced when a microorganism invades the albumen of the egg. As some members of the enterobacterioceae (Wilkins and Lankford, 1970) can grow even when the transferrin of vertebrate serum is unsaturated, the possibility exists that iron-scavenging compounds, enterobactins, may aid the bacterial colonization of the gut of avian neonates.

The conclusion (Board and Fuller, 1974) that ovotransferrin is the principal component of the antimicrobial defense of the hen's egg is supported by studies on the pigeon, *Columba livia* (Frelinger, 1972). The egg albumen produced by females heterozygous for ovotransferrin was more inhibitory to the growth of *Saccharomyces cerevisiae* than was that produced by homozygous birds. Moreover, the hatching success of eggs containing the polymorphs of ovotransferrin was greater than with those laid by homozygous females. This was taken to be an important factor in the maintenance of polymorphism in pigeons at Hardy–Weinberg equilibrium. This equilibrium was found also with *C. livia* in Belfast (Ferguson, 1971) but not with the recently introduced *Streptopelia decaocto*. The egg data imply that the heterozygous birds are better pioneers than homozygous individuals.

4. Ovomucoid and Ovoinhibitor

The inhibition of proteases is a general property of the ovomucoids and ovoinhibitors of the egg whites of many species of birds (Spiro, 1973; Osuga and Feeney, 1974; Osuga *et al.*, 1974; Donovan and Beardslee, 1975). The ovomucoids tend to have a narrow spectrum of activity. Those of chicken, cassowary, ostrich, emu and rhea inhibit mainly bovine trypsin; that of the quail inhibits both human and bovine trypsin, and that of the pheasant inhibits bovine α-chymotrypsin and subtilisin but not trypsin of human or bovine origin. The ovoinhibitors have a broader spectrum of activity. With chicken ovoinhibitor, for example, bovine trypsin, bovine α-trypsin, subtilisin, and fungal proteinase are inhibited. Although a role in the antimicrobial defense of the egg has been implied (Ayres and Taylor, 1956), there is no direct evidence to support such a view. It was suggested (Board and Fuller, 1974) that these compounds may act as inhibitors of proteases released by cells which lyse within the egg, for example, those spermatozoa that did not take part in fertilization. The suggestion that bovine colostrum inhibitor may protect immunoglobulins from digestion in the alimentary tract directs attention to another role for the ovoinhibitors and ovomucoids of egg white.

5. *Ovalbumin, Ovomacroglobulin and Ovoglycoproteins*

Ovalbumin, the major protein of the albumin and one having no biological property of note, and two derivatives, Plakalbumin (Linderstrom-Lang and Otteson, 1947; Otteson, 1958) and S-ovalbumin (Smith and Back, 1964), are separated electrophoretically from the prealbumins (Baker, 1968; Feeney and Allinson, 1969). Polymorphism occurs in the ovalbumin of some birds (Lush, 1961).

The 2–3 bands obtained by electrophoretic separation of hens' egg white reflect differences in the proteins' content of phosphate (Perlman, 1952). Ovalbumin A_1 has two phosphates per mole, A_2 has one, and A_3 none. This does not represent a protein polymorphism.

Ovomacroglobulin, a little studied protein, is notable for its marked antigenicity. It contains immunodeterminants which are common to the proteins present in a range of egg whites (Baker, 1968; Feeney and Allinson, 1969). No particular biological property has been ascribed to the minor component of egg white, the ovoglycoprotein (Ketterer, 1965; Baker, 1968; Osuga and Feeney, 1974).

6. *Miscellaneous Components*

In addition to the ten major proteins of the albumin, the secretory immunoglobulins, IgA and IgM, are present in the albumen of unembryonated chicken eggs and in washings of the 19-day-old embryo gut (Rose *et al.*, 1974). The occurrence of enzymes other than lysozyme was summarized by Baker (1968).

7. *Changes with Incubation*

Although the resolution of separation by electrophoresis is impaired by incubation (Csuka *et al.*, 1973), the biological properties of ovotransferrin (Annau and Cochrane, 1962) and lysozyme (Cunningham, 1974) can be demonstrated up to days 12–16 of incubation of hens' eggs. It was noted above that the albumen passes via the seroamniotic raphe to the amnion, and it has been established that proteins of egg white origin, viz., ovotransferrin (Frelinger, 1970), are included in the embryo's blood. Does this give a passive immunity analagous to that provided via the placenta or colostrum in mammals (Brambell, 1958)?

REFERENCES

Aisen, P., and Leibman, A. (1968). *Biochem. Biophys. Res. Commun.* **30**, 407.

Alderton, G., Ward, W. H., and Fevold, H. L. (1946). *Arch. Biochem.* **11**, 9.

Ali, K. E., and Ramsay, W. N. M. (1968). *Biochem. J.* **110**, 36P.

Ali, K. E., and Ramsay, W. N. M. (1974). *J. Exp. Physiol. Cogn. Med. Sci.* **59**, 159.

Almquist, H. J., and Lorenz, F. W. (1932). *U.S. Egg Poult. Mag.* **38**, 20.

Annau, E., and Cochrane, D. (1962). *Nature (London)* **193**, 879.

Archer, R. K. (1970). *Symp. Zool. Soc. (London)* **27**, 121.

Archer, R. K. (1971). *In* "Physiology and Biochemistry of Domestic Fowl" (D. J. Bell and B. M. Freeman, eds.), Vol. 2, p. 897. Academic Press, New York.

Arnheim, N., and Steller, R. (1970). *Arch. Biochem. Biophys.* **141**, 656.

Arnheim, N., Inouye, M., Law, L., and Laudin, A. (1973). *J. Biol. Chem.* **248**, 233.

Asofsky, R., Trnka, Z., and Thorbecke, G. J. (1962). *Proc. Soc. Exp. Biol. Med.* **111**, 497.

Ayres, J. C., and Taylor, B. (1956). *Appl. Microbiol.* **4**, 356.

Bailey, C. E. (1923). *Am. J. Hyg.* **3**, 370.

Bailey, P., and Geoffrey, C. J. (1975). *Biochem. Soc. Trans.* **3**, 1212.

Baker, C. M. A. (1968). *In* "Egg Quality: A Study of the Hen's Egg" (T. C. Carter, ed.), p. 67–108. Oliver & Boyd, Edinburgh.

Baker, C. M. A. (1970). *Adv. Genet.* **15**, 147.

Baker, C. M. A., and Manwell, C. (1976). *Philos. Trans. R. Soc. London.* **275**, 109.

Balcarova, J., Hala, K., and Hraba, T. (1974). *Folia Biol. (Prague)* **19**, 329.

Baxendale, W., Courtenay, J. S., Phillips, A. W., and Zola, H. (1971). *Int. J. Biochem.* **2**, 419.

Bellairs, R., Harkness, W., and Harkness, E. D. (1963). *J. Ultrastruct. Res.* **8**, 339.

Benedict, A. A. (1967). *Methods Immunol. Immunochem.* **1**, 229-232.

Bienenstock, J., Perey, D. Y. E., Gauldie, J., and Underdown, B. J. (1973). *J. Immunol.* **110**, 524.

Bigland, C. H., and Triantaphyllopoulos, D. C. (1961). *Am. J. Physiol.* **200**, 1013.

Blackenhorn, G., Osuga, D. T., Lee, H. S., and Feeney, R. E. (1975). *Biochim. Biophys. Acta* **386**, 470.

Blombäck, B. (1970). *Symp. Zool. Soc. London.* **27**, 167.

Board, R. G. (1964). *J. Appl. Bacteriol.* **27**, 350.

Board, R. G. (1966). *J. Appl. Bacteriol.* **29**, 319.

Board, R. G. (1968). *In* "Egg Quality: A Study of the Hen's Egg" (T. C. Carter, ed.), p. 133. Oliver & Boyd, Edinburgh.

Board, R. G. (1969). *Adv. Appl. Microbiol.* **11**, 245.

Board, R. G. (1974). *Br. Poult. Sci.* **15**, 415.

Board, R. G. (1975). *Br. Poult. Sci.* **16**, 89.

Board, R. G., and Fuller, F. (1974). *Biol. Rev. Cambridge Philos. Soc.* **49**, 15.

Board, R. G., and Halls, N. A. (1973a). *Br. Poult. Sci.* **14**, 69.

Board, R. G., and Halls, N. A. (1973b). *Br. Poult. Sci.* **14**, 311.

Board, R. G., and Halls, N. A. (1973c). *Br. Poult. Sci.* **14**, 357.

Board, R. G., Tullett, S. G., and Perrott, H. R. (1977). *J. Zool. (London)* **182**, 251–265.

Bott, R., and Sarma, R. (1975). *Acta Crystallogr., Sec. A* **31**, 525.

Brambell, F. W. R. (1958). *Biol. Rev. Cambridge Philos. Soc.* **33**, 488.

Brandt, L. W., Smith, H. D., Andrews, A. C., and Clegg, R. E. (1952). *Arch. Biochem. Biophys.* **36,** 11.
Brooks, J., and Hale, H. P. (1959). *Biochim. Biophys. Acta* **32,** 237.
Brooks, J., and Taylor, D. I. (1955). *Rep. Food Invest. Board, London* No. 6a.
Brown, J. R. (1975). *Fed Proc., Fed. Am. Soc. Exp. Biol.* **34,** 591.
Buss, E. G. (1969). *In* "The Fertility and Hatchability of the Hen's Egg" (T. C. Carter and B. M. Freeman, eds.), p. 109. Oliver & Boyd, Edinburgh.
Butler, E. J. (1971). *In* "Physiology and Biochemistry of the Domestic Fowl" (D. J. Bell and B. M. Freeman, eds.), Vol. 2, pp. 934–961. Academic Press, New York.
Butler, E. J., Curtis, M. S., and Watford, M. (1973). *Res. Vet. Sci.* **15,** 267.
Buxton, A. (1952). *J. Gen. Microbiol.* **7,** 268.
Chignell, C. F., Starkweather, D. K., and Sinha, B. K. (1975). *J. Biol. Chem.* **250,** 5622.
Christov, D., and Rashev, Z., (1961). *Sofii Univ. Biol. Geol. Geogr. Fak. God Kni Biol. (Zool).* **56,** 29.
Clagett, C. O. (1971). *Fed. Proc., Fed. Am. Soc. Exp. Biol.* **30,** 127.
Cohn, E. J. (1945). *Science* **101,** 51.
Cook, W. H. (1961). *Nature (London)* **190,** 1173.
Cooke, A. S., and Balch, D. A. (1970). *Br. Poult. Sci.* **11,** 345.
Cooper, M. D., Peterson, R. D. A., and Good, R. A. (1966). *In* "Phylogeny of Immunity" (R. T. Smith, P. A. Miescher, and R. A. Good, eds.), pp. 243–252. Univ. of Florida Press, Gainesville.
Crichton, R. R., ed. (1975). "Proteins of Iron Storage and Transport in Biochemistry and Medicine." North-Holland Publ., Amsterdam.
Csuka, J., Novŷi, J., and Jirošovâ, Z. (1973). *Br. Poult. Sci.* **14,** 203.
Cunningham, F. E. (1974). *Poult. Sci.* **53,** 1561.
Curzon, G. and O'Reilly, S. (1960). *Biochem. Biophys. Res. Commun.* **2,** 284.
Dawes, C. M. (1975). *Biol. Rev. Cambridge Philos. Soc.* **50,** 351.
Dianoux, A. C., and Jollès, P. (1967). *Biochim. Biophys. Acta* **133,** 472.
Didisheim, P., Hattori, K., and Lewis J. H. (1959). *J. Lab. Clin. Med.* **53,** 866.
Donovan, J. W., and Beardslee, R. A. (1975). *J. Biol. Chem.* **250,** 1966.
Donovan, J. W., Beardslee, R. A., and Ross, K. D. (1976). *Biochem. J.* **153,** 631.
Doolittle, R. F. (1976). *Fed. Proc., Fed. Am. Soc. Exp. Biol.* **35,** 2145.
Doolittle, R. F., Oncley, J. C., and Surgenor, D. M. (1962). *J. Biol. Chem.* **237,** 3123.
Dorn, P., and Müller, F. (1965). *Zentralbl. Veterinaer Med., Reihe A* **12,** 380.
Drent, R. (1975). "Avian Biology" (D. S. Farner and J. R. King, eds.), Vol. V. p. 333. Academic Press, New York.
Edelman, G. M., and Gall, W. E. (1969). *Annu. Rev. Biochem.* **38,** 415.
Edelman, G. M., and Gally, J. (1971). *New Sci.* **52,** 213.
Edwards, N. A., Luttrell, V., and Nir, I. (1976). *Comp. Biochem. Physiol. B* **53,** 183.
Evans, R. W., and Holbrook, J. J. (1975). *Biochem. J.* **145,** 201.
Fautes, K. H., and Furminger, E. G. S. (1967). *Nature (London)* **216,** 71.
Feeney, R. B., and Nagy, D. A. (1952). *J. Bacteriol.* **64,** 629.
Feeney, R. E., and Allison, R. G. (1969). "Evolutionary Biochemistry of Proteins." Wiley (Interscience), New York.
Feeney, R. E., Allison, R. G., Osuga, D. T., Bigler, J. C., and Miller, H. T. (1968). *Antarct. Res. Ser.* **12,** 151.
Ferguson, A. (1971). *Comp. Biochem. Physiol. B* **38,** 477.
Fleming, A. (1922). *Proc. R. Soc. London, Ser. B* **93,** 306.
Fletcher, F. (1971). *Immunology* **20,** 493.
Fletcher, J. C., and Huehns, E. R. (1968). *Nature (London)* **218,** 1211.
Fleischman, J. B. (1966). *Annu. Rev. Biochem.* **35,** 835.

Foster, J. F. (1960). *In* "The Plasma Proteins" (F. W. Putham, ed.), Vol. 1, p. 179. Academic Press, New York.

Fredrickson, D. S., and Gordon, R. S., Jr. (1958). *Physiol. Rev.* **38**, 585.

Freeman, B. M., and Vince, M. A. (1974). "Development of the Avian Embryo." Chapman & Hall, London.

Frelinger, J. A. (1970). *Science* **171**, 1260.

Frelinger, J. A. (1972). *Proc. Natl. Acad. Sci. U.S.A.* **69**, 326.

Fried, M., and Chum, P. W. (1971). *Comp. Biochem. Physiol. B* **39**, 523.

Gabrielson, A. E., Linna, T. J., Weite-Kamp, D. P., and Pickering, R. J. (1974). *Immunology* **27**, 463.

Garibaldi, J. A. (1960). *Food Res.* **25**, 337.

Garibaldi, J. A. (1970). *Appl. Microbiol.* **20**, 558.

Garibaldi, J. A., Donovan, J. W., Davis, J. G., and Cimino, S. L. (1968). *J. Food Sci.* **33**, 514.

Gayon, M. U. (1873). *C. R. Hebd. Seances Acad. Sci.* **76**, 232.

Gilbert, A. B. (1971). *In* "Physiology and Biochemistry of the Domestic Fowl" (D. J. Bell and B. M. Freeman, eds.), Vol. 3, p. 1163. Academic Press, New York.

Gitlin, D., and Kitzes, J. (1967). *Biochim. Biophys. Acta* **145**, 334.

Giurgea, R. (1974). *Stud. Univ. Babes-Bolyai, Ser. Biol.* **19**, 137.

Glynn, A. A. (1972). *In* "Microbial Pathogenicity in Man and Animals" (H. Smith and J. H. Pearce, eds.), p. 75. Cambridge Univ. Press, London and New York.

Gornall, D. A., Kuksis, A., Pinteric, L., and Mookerjea, S. (1971). *Can. J. Biochem. Physiol.* **49**, 671.

Green, N. M. (1963a). *Biochem. J.* **89**, 585.

Green, N. M. (1963b). *Biochem. J.* **89**, 609.

Green, N. M. (1964). *Biochem. J.* **92**, 16C–17C.

Green, N. M. (1975). *Adv. Protein Chem.* **29**, 85.

Green, N. M., Konieczyny, L., Toms, E. J., and Valentine, R. C. (1971). *Biochem. J.* **125**, 781.

Greenberg, D. M., Larsan, C. E., Pearson, P. B., and Bumester, B. M. (1936). *Poult. Sci.* **15**, 483.

Greengard, O., Sentenac, A., and Acs, G. (1965). *J. Biol. Chem.* **240**, 1687.

Griffiths, E. (1974). *Biochim. Biophys. Acta* **340**, 400.

Griminger, P. (1965). *In* "Avian Physiology" (P. D. Sturkie, ed.), 2nd ed., pp. 21–25. Cornell Univ. Press (Comstock), Ithaca, New York.

Guimbault, P. R., Szabados, L., and Mester, L. (1972). *C. R. Hebd. Seances Acad. Sci., Ser. D.* **275**, 3013.

Haines, R. B., and Moran, T. (1940). *J. Agric. Sci.* **40**, 453.

Hallaner, C. (1936). *Z. Hyg. Infektionskr.* **118**, 605.

Harry, E. G. (1957). *Vet. Rec.* **69**, 1.

Harry, E. G. (1963a). *Br. Poult. Sci.* **4**, 63.

Harry, E. G. (1963b). *Br. Poult. Sci.* **4**, 91.

Harry, E. G., and McClintock, M. (1972). *In* "Safety in Microbiology" (D. A. Shapton and R. G. Board, eds.), p. 121. Academic Press, New York.

Hawkey, C. M. (1970). *Symp. Zool. Soc. London.* **27**, 133.

Heim, W. G., and Schechtman, A. M. (1954). *J. Biol. Chem.* **209**, 241.

Hillyard, L. A., Entenman, C., Feinberg, H., and Chaikoff, I. L. (1955). *J. Biol. Chem.* **214**, 79.

Hillyard, L. A., Entenman, C., and Chaikoff, I. L. (1956). *J. Biol. Chem.* **223**, 359.

Hillyard, L. A., White, H. M., and Pangburn, S. A. (1972). *Biochemistry.* **11**, 511.

Hindenburg, A., Spitzragel, J., and Aynheim, N. (1974). *Proc. Natl. Acad. Sci. U.S.A.* **71**, 1974.
Hinton, H. E. (1968). *Adv. Insect Physiol.* **5**, 65.
Holmberg, C. G., and Laurell, C. B. (1948). *Acta Chem. Scand.* **2**, 550.
Hughes, L. (1954). *In* "The Proteins" (H. Neurath and K. Bailey, eds.), 1st ed., Vol. 2, Part B, pp. 663–754. Academic Press, New York.
Iatridis, S. G., and Ferguson, J. H. (1961). *Thromb. Diath. Haemorrh.* **6**, 411.
Isaacs, A., and Barm, S. (1960). *Lancet* **2**, 946.
Jollès, J. F., Schoentgen, F., Jollès, P., Prager, E. M., and Wilson, A. C. (1976). *J. Mol. Evol.* **8**, 59.
Jollès, P., Saint-Blanchard, M., Allary, M., Perrin, J. P., and Cozzone, P. (1975). *FEBS Lett.* **55**, 165.
Karush, F. (1950). *J. Am. Chem. Soc.* **72**, 2705.
Kato, A., Imoto, T., and Tagishita, K. (1975). *Agric. Biol. Chem.* **39**, 541.
Katz, J. H., and Jandl, J. H. (1964). *In* "Iron Metabolism" (F. Gross, ed.), pp. 103–117. Springer-Verlag, Berlin and New York.
Ketterer, B. (1965). *Biochem. J.* **96**, 372.
Kimura, M., Yokoyama, Y., Hashimoto, I., Arakig, T., Sakurazaw, M., Isogai, I., and Mahita, N. (1970). *Jpn. Poult. Sci.* **7**, 144.
Kleinschmidt, W. J. (1972). *Annu. Rev. Biochem.* **41**, 517.
Konyshev, V. A. (1968). *Dokl. Akad. Nauk SSSR* **182**, 477.
Kristjansson, P. K., Taneja, G. C., and Gowe, R. S. (1963). *Br. Poult. Sci.* **4**, 239.
Kutchai, H., and Steen, J. B. (1971). *Respir. Physiol.* **11**, 265.
Lack, D. (1968). "Ecological Adaptations for Breeding in Birds." Methuen, London.
Landauer, W. (1967). *Storrs, Agric. Exp. Stn. Bull* **1** (rev.).
Laschtschenko, P. (1909). *Z. Hyg. Infektions.* **64**, 419.
Lebacq-Verheyden, A. M., Vaerman, J. P., and Heremans, J. F. (1972). *Immunology.* **22**, 165.
Lebacq-Verheyden, A. M., Vaerman, J. P., and Heremans, J. F. (1974). *Immunology.* **27**, 683.
Leslie, G. A., and Clem, L. W. (1969). *J. Exp. Med.* **130**, 1337.
Leslie, G. A., and Martin, L. N. (1973). *J. Immunol.* **110**, 959.
Liang, C. C. (1957). *Biochem. J.* **66**, 552.
Linderstrom-Lang, K., and Ottesen, M. (1947). *Nature, (London)* **159**, 807.
Lundy, H. (1969). *In* "The Fertility and Hatchability of the Hen's Egg" (R. C. Carter and B. M. Freeman, eds.), p. 143. Oliver & Boyd, Edinburgh.
Lush, I. E. (1961). *Nature (London)* **189**, 981.
Lush, I. E. (1963). *Br. Poult. Sci.* **4**, 255.
Luskey, K. L., Brown, M. S., and Goldstein, J. C. (1974). *J. Biol. Chem.* **249**, 5939.
McCabe, R. A., and Deutsch, H. F. (1952). *Auk* **69**, 1.
McIndoe, W. M. (1962). *Nature, (London)* **195**, 353.
Manwell, C., and Baker, C. M. A. (1970). "Molecular Biology and the Origin of Species: Heterosis, Protein Polymorphism and Animal Breeding." Sidgwick & Jackson, London.
Manwell, C., and Baker, C. M. A. (1973). *Ibis* **115**, 586.
Marquardt, W. W. (1974). *Avian Dis.* **18**, 105.
Marshall, M. E., and Deutsch, H. F. (1950). *J. Biol. Chem.* **185**, 155.
Marshall, M. E., and Deutsch, H. F. (1951). *J. Biol. Chem.* **189**, 1.
Mayes, P. (1973). *In* "Review of Physiological Chemistry" (H. A. Harper, ed.), p. 14. Lange Med. Rubl, Los Altos, California.

Mecham, D. K., and Olcott, H. S. (1949). *J. Am. Chem. Soc.* **71,** 3670.
Mehta, P. D., and Tomasi, T. B. (1969). *Fed. Proc., Fed. Am. Soc. Exp. Biol.* **28,** 820. Abstr. 3131
Menache, D., Cesbron, N., Guillon, M. C., and Shlegel, N. (1973). *Thromb. Diath. Haemorrh.* **30,** 72.
Mok, C. C., Martin, W. G., and Common, R. H. (1961). *Can. J. Biochem. Physiol.* **39,** 109.
Moore, D. H., Shen, S. C., and Alexander, C. S. (1945). *Proc. Soc. Exp. Biol. Med.* **58,** 307.
Morrell, A. G. Aisen, P., and Scheinberg, I. H. (1962). *J. Biol. Chem.* **237,** 3455.
Mukkur, T. K. S. (1969). *Exp. Parasitol.* **26,** 1.
Müller-Eberhard, H. J. (1975). *Annu. Rev. Biochem.* **38,** 697.
Nace, G. W. (1953). *J. Exp. Zool.* **122,** 423.
Needham, J. (1950). "Biochemistry and Morphogenesis." Cambridge Univ. Press, London and New York.
Neurath, H., and Hill, R. L., eds. (1975). "The Proteins," 3rd ed., Vol. 1. Academic Press, New York.
New, D. A. T. (1956). *J. Embryol. Exp. Morphol.* **4,** 221.
Noble, R. C., and Moore, J. H. (1966). *In* "Physiology of the Domestic Fowl" (C. Horton-Smith and E. C. Amoroso, eds.), p. 87. Oliver & Boyd, Edinburgh.
Noyan, A., Lossow, W. V., Brot, N., and Chaikoff, I. L. (1964). *J. Lipid Res.* **5,** 538.
Oegema, T. R., and Jourdian, G. W. (1974a). *Arch. Biochem. Biophys.* **160,** 26.
Oegema, T. R., and Jourdian, G. W. (1974b). *J. Exp. Zool.* **189,** 147.
Orlans, E., and Rose, M. E. (1972). *Immunochemistry* **9,** 833.
Osaki, S., Johnson, D. A., and Frieden, E. (1966). *J. Biol. Chem.* **241,** 2746.
Osuga, D. T., and Feeney, R. E. (1974). *In* "Toxic Constituents of Animal Foodstuffs" (I. E. Liener, ed.), 2nd., p. 39. Academic Press, New York.
Osuga, D. T., Bigler, J. C., Uy, R. L., Sjöberg, L., and Feeney, R. E. (1974). *Comp. Biochem. Physiol. B* **43,** 519.
Ottesen, M. (1958). *C. R. Trav. Lab. Carlsberg* **30,** 211.
Palmiter, R. D., and Smith, L. T. (1973). *Nature (London), New Biol.* **246,** 74.
Paulov, S. (1972). *Biologia (Bratislava).* **27,** 887.
Payne, L. N. (1968). *In* "Egg Quality: A Study of the Hen's Egg" (T. C. Carter, ed.), p. 181. Oliver & Boyd, Edinburgh.
Perk, K., and Hart, I. (1966). *Avian Dis.* **10,** 208.
Perlman, G. E. (1952). *J. Gen. Physiol.* **25,** 711.
Peters, T., Logan, A. C., and Sanford, C. A. (1958). *Biochim. Biophys. Acta* **30,** 88.
Phelps, C. F., and Antonini, E. (1975). *Biochem. J.* **147,** 385.
Pindyck, J., Mosesson, M. W., Rooni, M. W., and Levere, R. D. (1975). *Biochem. Med.* **21,** 22.
Prager, E. M., and Wilson, A. C. (1974). *J. Biol. Chem.* **249,** 7295.
Prager, E. M., Brush, A. H., Nolan, R. G., Nakanishi, M., and Wilson, A. C. (1974). *J. Mol. Evol.* **3,** 243.
Price, P. W. (1974). *Evolution* **28,** 76.
Quinteros, I.R., Stevens, R. W., Stomont, C., and Asmundson, U.S. (1964). *Genetics* **50,** 579.
Ragan, H. A., Nacht, S., Lee, G. R., Bishop, C. R., and Cartwright, G. E. (1969). *Am. J. Physiol.* **217,** 1320.
Rahn, H., and Ar, A. (1974). *Condor* **76,** 147.
Ratcliffe, D. A. (1970). *J. Appl. Ecol.* **7,** 67.
Rhodes, M. B., and Feeney, R. E. (1957). *Poult. Sci.* **36,** 891.
Rhodes, M. B., Azari, P. R., and Feeney, R. E. (1958). *J. Biol. Chem.* **230,** 399.

Rhodes, M. B., Bennett, N., and Feeney, R. E. (1959). *J. Biol. Chem.* **234**, 2054.
Richter, H., Gurtler, H., Hoffman, G., and Diettrich, M. (1969). *Acta Biol. Med. Ger.* **23**, 863.
Robinson, D. S., and Monsey, J. B. (1969a). *Biochem. J.* **115**, 64P.
Robinson, D. S., and Monsey, J. B. (1969b). *Biochem. J.* **115**, 65P.
Robinson, D. S., and Monsey, J. B. (1971). *Biochem. J.* **121**, 537.
Robinson, D. S., and Monsey, J. B. (1972a). *J. Food Sci. Agric.* **23**, 29.
Robinson, D. S., and Monsey, J. B. (1972b). *J. Food Sci. Agric.* **23**, 893.
Robinson, D. S., and Monsey, J. B. (1975). *Biochem. J.* **147**, 55.
Romanoff, A. L. (1967). "Biochemistry of the Avian Embryo." Wiley, New York.
Romanoff, A. L., and Romanoff, A. J. (1949). "The Avian Egg." Wiley, New York.
Rose, M. E., Orlans, E., and Buttress, N. (1974). *Eur. J. Immunol.* **4**, 521.
Rosen, J. M., Woo, S. L. C., Holder, J. W., Means, A. R., and O'Malley, B. W. (1975). *Biochemistry* **14**, 69.
Sanders, B. G., Travis, J. C., and Wiley, K. L. (1973). *Comp. Biochem. Physiol. B* **45**, 189.
Sauer, E. G. F., and Sauer, E. M. (1966). *Living Birds* **5**, 45.
Sauveur, B. (1975). *Ann. Zootech.* **24**, 170.
Schade, A. L., and Caroline, L. (1944). *Science* **100**, 14.
Schjeide, O. A. (1963). *Prog. Chem. Fats Other Lipids* **6**, 253.
Schjeide, O. A., and Urist, M. R. (1956). *Science* **124**, 1242.
Schjeide, O. A., and Urist, M. R. (1961). *Nature (London)* **90**, 1175.
Schmid, K. (1972). *Chimia* **26**, 405.
Seviour, E. M., Sykes, F. R., and Board, R. G. (1972). *Br. Poult. Sci.* **13**, 549.
Shaffer, M. F., Bridges, J. F., Clemmer, D. E., and Pontobbidan, K. C. (1964). *Am. J. Hyg.* **80**, 377.
Shamra, O. P., Borek, E., and Martinez-Hernandez, A. (1976). *Nature (Lond.)* **259**, 588.
Sharp, P. F., and Whitaker, R. (1927). *J. Bacteriol.* **14**, 17.
Sherman, H. W. (1919). *J. Infect. Dis.* **25**, 256.
Shokeir, M. H. K. (1972). *Clin. Biochem.* **5**, 115.
Sibley, C. G. (1960). *Ibis* **102**, 215.
Sibley, C. G. (1970). *Bull. Peabody Mus.* No. 32.
Sibley, C. G., and Ahlquist, J. E. (1972). *Bull. Peabody Mus.* No. 39.
Simkin, P. A. (1972). *Proc. Soc. Exp. Biol. Med.* **139**, 604.
Smith, A. B., and Back, J. F. (1964). *Aust. J. Biol. Sci.* **18**, 365.
Snow, G. A. (1970). *Bacteriol Rev.* **34**, 99.
Solomon, J. B. (1968a). *Immunology* **15**, 197.
Solomon, J. B. (1968b). *Immunology* **15**, 219.
Solomon, J. B. (1971). "Foetal and Neonate Immunity." North-Holland Pub., Amsterdam.
Spiro, R. G. (1973). *Adv. Protein Chem.* **27**, 350.
Starcher, B., and Hill, C. H. (1966). *Biochim. Biophys. Acta* **127**, 400.
Stopforth, A. (1970). *J. Comp. Pathol.* **80**, 525.
Stratil, A. (1970). *Anim. Blood Groups Biochem. Genet.* **1**, 15.
Srivastava, K. B., and Dwaraknath, P. K. (1971). *Indian J. Anim. Sci.* **41**, 1044.
Taborsky, G. (1962). *Biochemistry.* **2**, 266.
Tan, A. T., and Woodworth, R. C. (1968). *Fed. Proc., Fed. Am. Soc. Exp. Biol.* **27**, 780.
Tata, J. R., and Shellaburger, C. J. (1959). *Biochem. J.* **72**, 608.
Tenenhouse, H. S., and Deutsch, H. F. (1966). *Immunochemistry* **3**, 11.
Theodore, T. S., and Schade, A. L. (1965a). *J. Gen. Microbiol.* **39**, 75.
Theodore, T. S., and Schade, A. L. (1965). *J. Gen. Microbiol.* **40**, 385.

Thorbecke, G. J., Warner, N. L., Hochwald, G. M., and Ohanian, S. H. (1968). *Immunology.* **15,** 123.
Tokin, B. P. (1964). *Folia Biol. (Prague)* **10,** 61.
Torres-Medina, A., Rhodes, M. B., and Mussman, H. C. (1971). *Poult. Sci.* **50,** 1115.
Tsao, D., Azari, P., and Phillips, J. L. (1974a). *Biochemistry.* **13,** 397.
Tsao, D., Morris, D. H., Azari, P., Tengerdy, R. P., and Phillips, J. L. (1974b). *Biochemistry.* **13,** 403.
Tsao, D., Azari, P. R., and Phillips, J. L. (1974c). *Biochemistry.* **13,** 408.
Tullett, S. G., and Board, R. G. (1976). *Br. Poult. Sci.* **17,** 441.
Tuohimaa, P. (1975). *Histochemie* **44,** 95.
Tuohimaa, P., and Söderling, E. (1976). *Acta Endocrinol. (Copenhagen)* **81,** 593.
Urist, M. R., Schjeide, C. A., and McLean, F. C. (1958). *Endocrinology* **63,** 570.
Vanstone, W. E., Maw, W. A., and Common, R. H. (1955). *Can. J. Biochem. Physiol.* **33,** 891.
Vince, M. A. (1973). *In* "Studies on the Development of Behavior and the Nervous System" (G. Gottlieb, ed.), Vol. 1. Academic Press, New York.
Walsh, J. A., Lisowska-Berstein, B., and Lamm, M. F. (1968). *Biochim. Biophys. Acta* **168,** 67.
Wangensteen, O. D., and Rahn, H. (1970–1971). *Respir. Physiol.* **11,** 31.
Warner, N. L., Szenberg, A., and Burnet, F. A. (1962). *Aust. J. Exp. Biol.* **40,** 373.
Wartelle, O. (1957). *Rev. Hematol.* **12,** 350.
White, F. N., and Kenney, J. L. (1974). *Science* **186,** 107.
Wilkins, T. D., and Lankford, C. E. (1970). *J. Infect. Dis.* **121,** 129.
Williams, J. (1962). *Biochem. J.* **83,** 355.
Williams, J. (1968). *Biochem. J.* **108,** 57.
Williams, J. E., and Whittemore, A. D. (1967). *Avian Dis.* **11,** 467.
Winget, C. M., and Smith, A. H. (1959). *Am. J. Physiol.* **196,** 371.
Wirtz, G. H. (1967). *Immunochemistry.* **4,** 118.
Wise, R. W., Ketterer, B., and Hansen, I. A. (1964). *Comp. Biochem. Physiol.* **12,** 439.
Witschi, E. (1956). "Development of Vertebrates." Saunders, Philadelphia, Pennsylvania.
Wolfe, H. R., Mueller, A., Neess, J., and Tempelis, C. (1957). *J. Immunol.* 9, 142.
Wright, H. D. (1927). *J. Pathol. Bacteriol.* **30,** 185.
Wright, L. D., and Skeggs, H. R. (1944). *Proc. Soc. Exp. Biol. Med.* **56,** 95.
Zaccheo, D., and Grossi, C. E. (1967). *J. Embryol. Exp. Morphol.* **18,** 289.
Zimmerman, B., Shalatin, N., and Grey, H. M. (1971). *Biochemistry* **10,** 482.

CHAPTER 3

Chemical Embryology

B. M. Freeman

I. Introduction

In about 21 days, given a suitable physical environment (see Lundy, 1969), the fertilized ovum of the domestic fowl is transformed into a viable chick. The incubation period for some birds is shorter, for others longer, and a direct relationship between period and egg weight has been established (Rahn and Ar, 1974). Nevertheless the relative rate of development (measured as growth rate) for all species examined is similar (Laird, 1966).

Space does not allow a full discussion of chemical embryology: the subject is grounded in Needham's two monumental works (1931, 1942), while more recent treatments may be found in Weber (1965, 1967) and Romanoff (1967). This chapter will be restricted to those areas of biochemical development which have excited most interest in recent years and will be dominated by data for the domestic fowl since this is the species of choice for most research workers. Occasionally reference will be made to a particular stage of development and refers to the morphogenic state of the embryo as classified by Hamburger and Hamilton (1951).

II. Gaseous Metabolism

A. Oxygen

The oxygen requirements of the embryo are met, ultimately, by diffusion of the gas through the porous shell from the environment. From an early point of development (stage 9) the oxygen diffusing into the egg is actively transported within a circulatory system with hemoglobins as the carrier molecules (see Section VII,A).

Because the embryo is poikilothermic its metabolism varies directly with the environmental temperature. As a general rule, however, development of the bird proceeds best at an environmental temperature of 37° or 38°C. The values for oxygen uptake shown in Table I may be regarded as typical for the domestic fowl. Calculations based on measurements of the partial pressure of oxygen in the air space suggests that the average values computed by Romanoff (1967) are too low (see Freeman and Vince, 1974).

Three features of oxygen uptake during development call for comment. First there is a marked (62%) increase between days 10 and 11 which is a response to the initiation of thyroid hormone secretion. The second feature is the constancy of uptake between days 15 and 19. The embryo continues to grow during this period with a resulting fall in metabolic rate (Table I). It is likely that the constancy of uptake re-

TABLE I
GASEOUS EXCHANGE (μL/MINUTE) OF THE EMBRYONIC DOMESTIC FOWL[a]

Age (days)	$\dot{V}O_2$	$\dot{V}CO_2$	RQ	Metabolic rate[b] (μl O_2/gm minute)
1	1	—[c]	—	5000
2	2	2[c]	1.00	667
3	3	3	1.00	143
4	7	7	1.00	116
5	13	12	0.92	81
6	20	17	0.85	59
7	33	26	0.79	52
8	40	37	0.92	37
9	53	47	0.89	34
10	80	68	0.85	33
11	130	107	0.82	37
12	195	143	0.73	39
13	276	192	0.70	39
14	343	244	0.71	35
15	398	275	0.69	32
16	417	288	0.69	28
17	396	280	0.71	22
18	420	290	0.69	19
19	373	265	0.71	14
20	400	280	0.70	14
21 (just hatched)	650	462	0.71	21
22 (1-day-old chick)	1050	735	0.70	32

[a] Values calculated from Romijn and Lokhorst (1960), Visschedijk (1962), and Freeman (1962, and unpublished data).

[b] Wet weight of embryo taken from Romanoff (1967).

[c] Data are unreliable in this period due to the loss of CO_2 from the shell itself, i.e., not produced by the embryo.

flects the physiological maximum flux of oxygen across the shell. Finally there is another large increase at hatching: part of which is due to greater activity and part to the assumption of homeothermy (Freeman, 1965a).

B. CARBON DIOXIDE

Accurate measurements of the production of carbon dioxide by the embryo *in ovo* during the first days of incubation are not possible because much carbon dioxide bound in the shell and albumen during their formation is released. Since the respiratory quotient is relatively constant during the second half of development (Table I), the patterns of CO_2 production and O_2 uptake are similar.

C. Permeability of the Shell

The initial permeability of the shell and its membranes is insufficient to meet the eventual requirements of the embryo. Permeability increases by a factor of between 10 and 20 during the first week of incubation following a partial dehydration of the inner shell membrane (Kutchai and Steen, 1971; Tullett and Board, 1976) which is the main barrier to gaseous exchange by diffusion. This change is probably brought about by a translocation of water from the albumen and the inner shell membrane to the subembryonic fluid by the secretory activity of the blastodermal endoderm (Howard, 1957).

D. Energy Sources

The values of the respiratory quotient (Table I) indicate clearly that carbohydrate forms the exclusive energy source at the beginning but that lipid becomes progressively more important. Overall it is estimated (Murray, 1925) that the catabolism of lipid provides about 94% of the energy requirements for embryonic development.

III. Carbohydrate Metabolism

A. Pentose Phosphate Pathway

Glycolysis is vital at the beginning of development and the glycolytic flow rate increases progressively during incubation (Arese *et al.*, 1967). The pentose phosphate pathway is active also, providing pentose phosphate for nucleic acid synthesis and reduced nicotinamide adenine dinucleotide phosphate (NADPH) for the many reductive syntheses associated with development. Indeed this pathway is preferred from 2½ to 3 days and from 4 to 8 days (Coffey *et al.*, 1964; Wenger *et al.*, 1967) and is used preferentially by the brain from 12 to 15 days (Liuzzi and Angeletti, 1964; Wang, 1968).

B. Glycogen Metabolism

1. Glycogen Stores

The stores amount to only 4 mg initially but at the peak (19 days) reach 70–80 mg. Glycogen synthesis has been demonstrated in the stage 8 blastoderm (Allen, 1919). Discrete stores can be found in striated and cardiac muscle and in the yolk sac membrane by the third day (Lee, 1951; Thommes and Just, 1964) and in the liver, perhaps as early as the fourth (Houssaint *et al.*, 1970), and certainly by the sixth day (Thommes and Firling, 1964; Daugèras, 1971).

2. *Changes in Enzymatic Activities during Development*

The enzymes concerned in glycogen synthesis and mobilization (for reviews of the metabolic pathways, see Smith *et al.*, 1968; Ryman and Whelan, 1971) have been detected in the fowl at about the same time discrete particles of glycogen can be detected histochemically (Grillo *et al.*, 1964; Grillo and Baxter-Grillo, 1966; Rinaudo, 1966; Rinaudo *et al.*, 1968).

a. Glycogenesis. Generally there is a marked increase in the activity of UDPG–glycogen synthetase (EC 2.4.1.11) from the time of its appearance in each tissue but there is little agreement as to the timing of the major changes. Thus synthetase activity in the liver has been found to reach a peak at about 11 days (Grillo *et al.*, 1964; Benzo and de la Haba, 1972) while others have observed increasing activity to about 18 days (Ballard and Oliver, 1963; Rinaudo *et al.*, 1968). Similar uncertainties exist with the activity of UDPG–pyrophosphorylase (EC 2.7.7.9) and phosphoglucomutase (EC 2.7.5.1) (Ballard and Oliver, 1963; Rinaudo *et al.*, 1968). Insufficient information is available to determine whether these enzymes have different patterns of activity in different tissues.

Benzo and de la Haba (1972) and Benzo and de Gennaro (1974) found that the D-form of glycogen synthetase predominates in the three tissues examined—glycogen body, skeletal muscle, and liver.

b. Glycogenolysis. Glycogen phosphorylase (EC 2.4.1.1) activity is present in the yolk sac membrane on the third day (Grillo and Baxter-Grillo, 1966), in striated muscle by the fourth (Rinaudo and Bruno, 1968) and in the liver by the fifth day (Benzo and de la Haba, 1972). Activity increases in all tissues examined to day 18 (Ballard and Oliver, 1963; Grillo and Baxter-Grillo, 1966; Goris and Merlevede, 1969) although Tezuka *et al.* (1974) reported a decline in activity in the yolk sac membrane after 12 days while Benzo and de la Haba (1972) found a relatively constant activity from 8 to 18 days.

The activities of phosphorylase kinase (EC 2.7.1.38) increase during the third week but phosphorylase phosphatase (EC 3.1.3.17) does not (Goris and Merlevede, 1969).

3. *Hormonal Regulation of Glycogen Metabolism*

Research into the regulation of glycogen metabolism by the avian embryo has been somewhat haphazard in that the known differences in sensitivity of the different organs to various metabolic modifiers have not been exploited fully or systematically. At the same time one needs to proceed with caution in interpreting some observations of the kind that demonstrate a certain reaction *in vitro* when similar conditions do not pertain *in vivo.*

Insulin is secreted by the fifth day (Benzo and Green, 1974) and may be concerned in regulating glycogenesis since it stimulates that pathway in the yolk sac membrane (Zwilling, 1951; Thommes and Mathew, 1969) and the liver (Benzo and de la Haba, 1972). As in mammals the action of insulin is directed toward stimulating glycogen synthetase activity (Benzo and de la Haba, 1972). Glucagon, also secreted by the fifth day (Przybylski, 1967; Benzo and Stearns, 1975) mobilizes glycogen from the liver (Thommes and Firling, 1964; Freeman and Manning, 1971) and yolk sac membrane (Thommes and Just, 1964) by activating glycogen phosphorylase ($b \rightarrow a$) (Grillo and Baxter-Grillo, 1966). In the liver at least, adrenaline is a potent glycogenolytic agent (Gill, 1938; Freeman, 1969) while glucocorticoids stimulate glycogenesis (Thommes and Just, 1966).

The contribution of hormones in regulating glycogen metabolism remains somewhat circumstantial though the work of Konigsberg (1954) and Thommes and Aglinskas (1966) has shown that pituitary and adrenocortical hormones are necessary and that thyroxine may have a permissive role (Thommes *et al.*, 1968; Daugèras, 1971). A transient fall in hepatic stores on day 11 has been linked with the concomitant increase in metabolic rate (Table I) initiated by the onset of thyroid hormone (Thommes and Pall, 1974) though the maintenance of the stores in the yolk sac membrane is problematic in this context. The mobilization of the stores of the yolk sac membrane (Thommes and Just, 1964; Freeman, 1965b, 1969) and liver prior to the commencement of hatching has been variously explained in terms of adrenaline or glucagon stimulation (see Freeman and Manning, 1971; Freeman and Vince, 1974). It is perhaps significant that this mobilization is not general, the stores of the heart (Freeman, 1965b) and the brain (Edwards and Rogers, 1972) remaining intact.

C. Gluconeogenesis

The carbohydrate reserves of the unincubated egg do not exceed 500 mg and are virtually exhausted by the tenth day (Yarnell *et al.*, 1966). It is not unexpected therefore that the pathways of gluconeogenesis are active during incubation. The reader is referred to Pontremoli and Grazi (1968) for a general account of gluconeogenesis.

Phosphoenolpyruvate carboxykinase (EC 4.1.1.32) probably exists in four forms in the liver, two in the mitochondrial fraction and two in the cytosol (Jo *et al.*, 1974). The authors have shown that the developmental changes of these forms differ significantly although their activities are similar. The cytosol forms have little or no activity initially but there is a rapid increase from about day 16. The molecular weight of all forms is about 24,000.

Pyruvate carboxylase (EC 6.4.1.1) is present, according to Nelson *et al.* (1966) and Rinaudo and Giunta (1967) in both cytosol and mitochondria of the liver and shows peak activity on the eleventh and sixteenth days respectively, while Hendrick and Moller (1973) suggest it is bound exclusively to the mitochondria and maintains a relatively constant activity from 9 to 21 days. Fructose-1,6-diphosphatase (EC 3.1.3.11) and glucose-6-phosphatase (EC 3.1.3.9) increase in activity in the liver to a peak at about 16 days (Kilsheimer *et al.*, 1960; Ballard and Oliver, 1963; Nelson *et al.*, 1966; Wang, 1968) though Kuroda and Nagatani (1965) found the peak of fructose-1,6-diphosphatase on the sixth day.

In the heart the changes are less certain. Wang (1968) and Rinaudo and Ponzetto (1972) found the activity of glucose-6-phosphatase steady and low until day 18 when there was a slight increase, while the activity of fructose-1,6-diphosphatase rose sevenfold to a peak at 18 days. Kuroda and Nagatani (1965), however, reported low activity throughout incubation.. Glucose-6-phosphatase is confined mainly to the endoplasmic reticulum (Pollak and Shorey, 1968).

Studies on pyruvate kinase (EC 2.7.1.40) isozymes by Strandholm *et al.* (1975) show that only type K isozyme was present until shortly before hatching when type M appeared in skeletal muscle and brain. It is thought that type K pyruvate kinase is used in the gluconeogenic effort of the developing bird.

A role for fructose-1,6-diphosphate aldolase B in gluconeogenesis also was proposed (Rutter *et al.*, 1963) and a transition from synthesizing aldolase C to B was observed in the liver from 5 to 8 days which coincided with the onset of gluconeogenesis in that organ (Lebherz, 1972). There is, moreover, aldolase B activity in the yolk sac membrane from 3 to 10 days (Lebherz, 1972).

Gluconeogenic enzymes appear in the blastoderm after as little as 24 hours incubation (Meyerhof and Haley, 1975) and in the liver from the third day (Kilsheimer *et al.*, 1960). Presumably they are also present in the yolk sac from a similar time. There is evidence that the induction of fructose, 1,6-diphosphatase utilizes a different mechanism from that used in glucose-6-phosphatase induction (Meyerhof and Haley, 1975).

While [2-^{14}C] glutamate and [U-^{14}C] alanine (Yarnell *et al.*, 1966) and [2-^{14}C] pyruvate (Kilsheimer *et al.*, 1960) can be incorporated into glucose it seems likely that lactate and malate are important glycogenic precursors (Solomon, 1958; Ballard and Oliver, 1963).

Serine, available in large amounts from phosvitin, may be an important glycogenic amino acid (Willier, 1968).

That gluconeogenesis should be particularly active in the third week of incubation, when lipid is the main energy source, may be more than fortuitous. The production of acetyl-CoA inhibits oxidation of pyruvate and favors its carboxylation while the oxidation of fatty acids supplies reducing equivalents for the reduction of oxaloacetate to malate.

IV. Lipid Metabolism

A. Absorption and Assimilation

The mechanisms of absorption are poorly understood. Mono- and diglycerides and free fatty acids are never present in either the yolk or yolk sac membrane in more than trace amounts (Noble and Moore, 1967b). Claims for the presence of a lipase in the membrane (Buño and Mariño, 1952) have not been confirmed (Juurlink and Gibson, 1973). The positional distribution of fatty acids in the triglycerides and phospholipids found in the membrane and in the yolk itself is similar, indicating that for the most part both classes of lipid are absorbed intact (Noble and Moore, 1967b,c).

During the first two weeks of incubation, about 350 mg lipid are absorbed from the yolk. The rate of uptake increases rapidly thereafter to reach a value in excess of 1 gm/day (Noble and Moore, 1964). However, there is evidence of a preferential absorption of the fractions of the phospholipid phosphatidyl ethanolamine which are rich in stearic acid in the α-position and arachidonic or docosahexanoic acids in the β-position (Noble and Moore, 1967c). Cholesterol is actively esterified, mostly with oleic acid, in the yolk sac membrane (Noble and Moore, 1967b) and there is some transport of this ester back to the yolk. The purpose of this esterification is not understood but the abundance of cholesterol oleate in the chylomicrons and lipoproteins of the plasma (Schjeide, 1963) suggests that it may have an important role in the assembly of lipoprotein for transport to embryo.

The lipid stores of the embryo increase markedly in the last week of incubation to about 2.2–2.4 gm/embryo. In the liver, cholesterol esters from the main storage product (64%) while in the other tissues triglycerides account for 75% of the total (Noble and Moore, 1964; Wood, 1972; Gómez-Capilla *et al.*, 1975). About 80% of the cholesterol esters are accounted for by cholesterol oleate.

The differences in fatty acid composition of phosphatidylcholine, phosphatidylethanolamine, phosphatidylserine, and diphosphatidylglycerol from the liver of the embryo as compared with the compositions of the corresponding phospholipids in the yolk suggest that there is extensive breakdown and resynthesis of these substances in the liver. The fatty acid compositions of the sphingomyelins of yolk and liver are sufficiently similar to indicate that there is little modification to the molecule (Noble and Moore, 1967a).

B. Fatty Acid Metabolism

1. *Fatty Acid Oxidation*

Palmityl-CoA synthetase activity of the liver and heart is at a peak at 15 days while acetyl-CoA synthetase is active in the heart, but not in the liver, with a peak at 15 days (Warshaw, 1972).

Both long- and short-chain fatty acids are oxidized by the mitochondria of both heart and liver (Koerker and Fritz, 1970). The ability generally increases with development though Pugh and Sidbury (1971) reported a decline in the activity of the liver. The addition of carnitine leads to a two- to tenfold increase in oxidation indicating the increase in oxidative ability may be the result of increased carnitine palmitoyltransferase activity (Rose and Shrago, 1975). The oxidation of fatty acids proceeds by the β-oxidative pathway (Pugh and Sidbury, 1971).

There is an accumulation of ketone bodies in the plasma during fatty acid oxidation rising from 120 μg/ml at 14 days to 177 μg/ml at 17 days and then falling to 100 μg/ml until hatching (Rinaudo and Passano, 1972). β-Hydroxybutyrase activity is reported to be high in the mesonephros (Pearse *et al.*, 1963).

2. *Fatty Acid Synthesis*

Fatty acid synthesis proceeds at a very low rate in the embryo liver, perhaps as little as 1000th the activity seen in the 4-week-old bird (Goodridge, 1968a). A similarly low activity was observed in adipose tissue (Goodridge, 1968b). The activity of enzymes involved in lipogenesis is very low in both tissues during development (Goodridge, 1968c).

The limited synthesis of fatty acids that does occur in the liver is mostly by mitochondrial elongation with very little microsomal elongation or fatty acid synthetase activity (Donaldson *et al.*, 1971). Both fatty acid synthetase and acetyl-CoA carboxylase (EC 6.4.1.2) activities in the liver are enhanced by exogenous glucose (Donaldson *et al.*, 1971) though there is a concomitant fall in the concentration of plasma free fatty acids (Koerker and Fritz, 1970). Glucose loading of adipose tissue *in vitro* results in only a 2% incorporation into fatty acids (Goodridge, 1968b). Two factors might be implicated in the low rate of lipogenesis in the liver: the abundance of fatty acids continuously being made available from the yolk lipids and the very low rate of glucose oxidation in the liver (Goodridge, 1968a).

In contrast the synthetase activity of the heart is high from an early stage but disappears by hatching while mitochondrial chain elongation, initially low, increases during incubation (Joshi and Sidbury, 1975). Insulin stimulates lipogenesis from day 5 (Foà *et al.*, 1965), particularly the production of phospholipids (Berger and Foà, 1971). Koerker and Fritz (1970) and Langslow (1975) have shown that concentrations of plasma FFA and glucose are inversely related. Thus as plasma glucose rises from about 140 mg/100 ml on day 14 to 220 mg/100 ml on day 20 there is a decline in plasma FFA from 1200 μEq/liter to 400 μEq/liter.

3. *Lipolysis*

Triglycerides are laid down in the subcutaneous fat pads during the last week of incubation (Langslow and Lewis, 1972). In view of the abundance of yolk-derived fatty acids it is not surprising that the lipolytic mechanism is poorly developed (Goodridge, 1968d; Freeman and Manning, 1971; Langslow, 1972). Glucagon is a potent lipolytic stimulus in the newly hatched bird but has only slight activity in the embryo (Freeman and Manning, 1971; Langslow, 1972). It appears likely that this is due to late maturation of the adenyl cyclase system in the tissue (Langslow, 1972). Insulin is without effect on lipolysis (Goodridge, 1968d).

V. Protein Metabolism

A. Absorption of Yolk Proteins

The debate on whether proteins are absorbed intact continues (see Williams, 1967, for a review of earlier work). Certainly the cells of the area opaca absorb intact proteins (Hassell and Klein, 1971) by phagocytosis (Litke and Low, 1975) but this may be preceded by enzymatic emulsification (Bellairs, 1964). There is also evidence that the yolk sac membrane, which becomes functional on the third day of incubation, is able to absorb proteins intact. Thus livetins are taken up to form, without alteration, plasma proteins (Nace, 1953).

Recent work has strengthened the view that most protein is absorbed intact and then modified intracellularly. Indeed Lambson (1970) concluded on the basis of an ultrastructural survey that the yolk sac membrane is "structurally adapted for . . . macromolecular absorption and intracellular alteration." Juurlink and Gibson (1973) support Lambson's conclusion for they were unable to confirm the presence of extracellular proteinases which was described previously (Emanuelsson, 1955; Ito, 1957). They provide evidence that the absorbed proteins are hydrolyzed in the lysosomes and resynthesized in the apical region of the endoderm where there is a high activity of alkaline phosphatase. However the existence of an active amino acid transport system in the yolk sac membrane (Holdsworth and Wilson, 1967) indicates that some extracellular digestion of yolk proteins cannot be ruled out.

B. Amino Acid Transport

1. *Yolk Sac Membrane*

Active transport mechanisms are fully established by the sixth day and the activity remains high until the fifteenth day (Holdsworth and Wilson, 1967).

2. *Cardiac Muscle*

The heart is able to accumulate amino acids by the fifth day or earlier (Guidotti *et al.*, 1968b). Gazzola *et al.* (1972) have shown that the uptake of neutral amino acids is mediated by the A system (Christensen, 1969). The activity of the transport system is regulated by a repression–derepression mechanism by the substrate amino acids for which it is competent. Franchi-Gazzola *et al.* (1973) presented evidence that the mRNA specific for the synthesis of protein(s) needed for amino acid transport by the A mediation accumulated in cardiac cells under conditions of inhibited translation and that transcription of this mRNA can be repressed by amino acids pertaining to the A transport system. Insulin stimulates cardiac uptake of glycine and α-aminoisobutyric acid but not leucine (Guidotti *et al.*, 1968a,b).

3. *Nervous Tissue*

The pattern of amino acid uptake differs between the brain and the spinal ganglia. Thus spinal ganglia accumulate three to four times more α-amino-isobutyrate than D-glutamate per unit time while in the brain the situation is practically reversed (Levi and Lattes, 1969).

Nerve growth factor also stimulates the uptake of acidic amino acids by the spinal ganglia while insulin stimulates both acidic amino acid and neutral amino acid transport (Levi-Montalcini and Cohen, 1960).

The accumulation rates of α-aminobutyrate, glycine, α-aminoisobutyrate, leucine, D-glutamate, and lysine by the brain increase during development (Levi, 1972). The increases for the latter four are the result of an increase in the availability or turnover rates of the carrier molecules whereas the former two probably have two modes of entry into the cell. Levi (1972) has shown also that there is a progressive overlapping of the α-aminoisobutyrate and leucine transport systems during development.

4. *Cartilage and Bone*

Adamson and Anast (1966) and Adamson *et al.* (1966a) found that the active transport mechanism in embryonic cartilage was stimulated by the presence of serum, the active substance probably being identical to the serum sulfation factor described for rats. This transport system is dependent upon concomitant protein synthesis (Adamson *et al.*, 1966b).

In embryonic bone there are at least two transport sites for amino acids (Adamson and Ingbar, 1967a)—the A and the L (Christensen, 1969). Triiodothyronine stimulates the L-site specifically (Adamson and Ingbar, 1967b) and, in contrast to cartilage, concomitant protein synthesis is not a requirement for the continued transport of amino acids (Adamson and Ingbar, 1967c).

5. *Small Intestine*

Intestinal transport mechanisms for amino acids become functional about day 17 (Holdsworth and Wilson, 1967; Pratt and Terner, 1971) though there is evidence that the time at which the intestine can transport different amino acids varies (Pratt and Terner, 1971). $Na^{+}K^{+}$–ATPase (EC 3.6.1.3) has been implicated in the mechanism but the relationship is uncertain (Pratt and Terner, 1971).

C. Amino Acid Metabolism

1. *Glutamic Acid*

Glutamic acid plays a central role in intermediary metabolism since it is related directly or indirectly with protein synthesis, nucleic acid synthesis, and the Krebs' cycle.

a. α-Ketoglutaric Acid. Glutamic dehydrogenase (EC 1.4.1.2) catalyzes the formation of α-ketoglutaric acid from glutamate. Activity is located in the mitochondria in the liver and increases during development (Mason and Hooper, 1969). An inverse relationship between the activities of glutamic dehydrogenase and glutamic-oxaloacetic transaminase and alanine transaminases (EC 2.6.1.1 and 2.6.1.2, respectively) has been shown (Sheid and Hirschberg, 1967).

b. γ-Aminobutyric Acid (GABA). GABA is an inhibitory neurotransmitter mainly associated with various interneurons and Purkinje cells and is formed from glutamate by glutamate decarboxylase (EC 4.1.1.15), an enzyme which is restricted to nervous tissue. There are two forms of the enzyme, one requiring pyridoxal-5′-phosphate as a coenzyme (type I), the other not (type II). Both forms are present in the brain of the chick embryo, type I having primarily a synaptosomal location, type II being mainly localized in the mitochondria (Haber *et al.*, 1970a,b).

The activity of glutamate decarboxylase and the concentration of GABA in the brain rise rapidly from the seventh day (Vos *et al.*, 1967; McGeer *et al.*, 1974). Kuriyama *et al.* (1968) correlated the development of the GABA system with that of the synaptic structures of the brain. Both forms of the decarboxylase are inhibited by a high intracellular concentration of GABA and this appears to be the main method of control (Sze *et al.*, 1971).

c. Glutamine. The synthesis of glutamine from glutamic acid is catalyzed by glutamine synthetase (EC 6.3.1.2) (but see Herzfeld and Raper, 1975). This enzyme is prominent in nervous tissue but its activity is particularly marked in that part of the brain concerned with vision, i.e., in the neural retina (Rudnick and Waelsch, 1955; Piddington, 1967), in the optic tectum (Shimada *et al.*, 1967), in the large lateral nucleus of the ectostriatum of the cerebrum (Piddington, 1971), and in the nucleus rotundus of the diencephalon (Piddington, 1973). The activity of glutamine synthetase rises rapidly from day 17 and coincides with the development of visual activity. This strongly suggests that the enzyme is concerned in the processing of visual information. There is some evidence that glutamine synthetase acts as an inactivating enzyme at synapses where glutamate has been implicated as a neurotransmitter (Krnjevic, 1970; Johnson, 1972). At the same time the occurrence of the enzyme in other nervous tissue indicates it may have another role. The requirement for ammonia in the synthesis of glutamine suggests that the removal of this toxic product might be an alternate role.

Glutamine synthetase activity can be enhanced or induced prematurally by certain steroid hormones including hydrocortisone (Piddington, 1967, 1970, 1973; Moscona, 1971). It is likely therefore that the activity of this enzyme is controlled by the adrenal cortical secretions.

d. Aspartate. Glutamic-oxaloacetic transaminase (EC 2.6.1.1) catalyzes the conversion of glutamate to aspartate. In the liver its activity reaches a transient peak at the tenth day and then declines thereafter (Sheid and Hirschberg, 1967). Activity in the chick brain, however, increases during incubation. In the hemisphere it increases from a very low level on day 14, by a factor of 4 by day 20 while in the optic lobe the rise is initiated on day 12 (Vos *et al.*, 1967).

Asparagine is synthesized in the liver from aspartate and glutamine. The activity of asparagine synthetase (EC 6.3.1.1) is relatively constant to 17 days and then declines rapidly (Arfin, 1967).

2. *Tryptophan*

The synthesis of nicotinamide from tryptophan is initiated by tryptophan oxygenase (EC 1.13.1.12). Its activity can be detected in homogenates of embryo by about the sixth day (Peterkofsky, 1968) and in the liver by the eighth day (Knox and Eppenberger, 1966; Boucek *et al.*, 1967; Peterkofsky, 1968; Wagner *et al.*, 1969; Leibenguth, 1969). The intermediates kynurenic and xanthurenic acids have been detected in the allantoic fluid from the seventh day (Boucek *et al.*, 1967). Activity in the liver remains low during incubation though a doubling in activity has been noted between 8 and 13 days (Peterkofsky, 1968). In contrast with the adult, exogenous hydrocortisone is without effect on the activity of hepatic tryptophan oxygenase (Wagner *et al.*, 1969).

Some evidence has been adduced showing a relationship between tryptophan metabolism and pyridoxine (see Boucek *et al.*, 1967).

3. *Proline and Lysine*

These two amino acids will be considered together since they are important precursors of collagen (see Section VII,D). The general steps in the synthetic pathway are (1) formation, on ribosomal complexes, of a proline-rich and lysine-rich polypeptide, procollagen; (2) hydroxylation of the appropriate proline and lysine of the procollagen, and; (3) glycosylation of some of the hydroxylysine to galactosylhydroxylysine and glucosylgalactosylhydroxyproline.

The hydroxylation enzymes, procollagen proline hydroxylase and procollagen lysine hydroxylase both require O_2, Fe^{2+}, α-ketoglutarate, and probably ascorbate as a reducing agent. There is no cross reactivity between the enzymes. Both enzymes have been isolated, partially purified, and characterized (Halme *et al.*, 1970; Popenoe and Aronson, 1972; Kivirikko and Prockop, 1972). The location of procollagen proline hydroxylase in tendon cells has been shown to be in discrete areas of the cytoplasm (Berg *et al.*, 1972) probably within the cisternae of the rough endoplasmic reticulum (Guzman *et al.*, 1976). The activity of the enzyme in liver has been measured and shows high activity at day 10 and a minor peak at day 18 (Helfre *et al.*, 1974).

Some of the characteristics of collagen galactosyltransferase and collagen glucosyltransferase were determined by Myllylä *et al.* (1975, 1976) and Risteli *et al.* (1976). Blumenkrantz and Prockop (1970) report that the ratio of glucosylgalactosylhydroxylysine to galactosylhydroxylysine increases from 7 to 10 days but that this ratio is susceptible to hormonal influence.

D. Embryo-Specific Proteins

Three embryo-specific proteins have been identified in the chick embryo (Lindgren *et al.*, 1974). One is an ovalbumin and the more abundant of the remaining two is an α_2-globulin with similar characteristics to mammalian fetoprotein. The avian "fetoprotein" is synthesized exclusively by the yolk sac membrane (Gitlin and Kitzes, 1967). The pheasant embryo also synthesizes a fetoprotein analogous to that of the fowl and also a species-specific α_3-globulin (Weller, 1966).

E. Elimination of Nitrogen

Most of the nitrogen is eliminated in the form of uric acid. An amount of urea is produced but in the absence of a complete urea cycle this is assumed to derive from the action of arginase on arginine.

The oxidation of hypoxanthine to uric acid proceeds via xanthine and is catalyzed by xanthine dehydrogenase. Activity of this enzyme can be detected on about the fifth day in the mesonephros and remains high until nitrogen elimination is taken over by the metanephros. Activity in the latter increases suddenly about one day before the mesonephros begins to regress. Throughout incubation xanthine dehydrogenase activity in the liver, intestine, and pancreas is very low but increases markedly at hatching (Lee and Fisher, 1971; Croisille, 1972) with liver activity predominating over all other tissues. The characteristics of the enzyme are similar between sites (Croisille, 1972). Attempts to induce xanthine dehydrogenase activity have failed (Hwang, 1968; Croisille, 1972).

VI. Calcium Metabolism

A. Calcium Stores

About 5 gm of calcium are found in the shell in the form of calcium carbonate while the yolk and albumen contain about 22 mg. Since the newly hatched chick contains about 125 mg Ca, clearly about 100 mg calcium are absorbed from the shell during development. This occurs between day 13 and hatching.

B. Calcium Absorption

1. *Ionic and Bound Calcium*

The removal of calcium from the shell poses serious problems of maintaining the intracellular calcium concentration within normal limits if a variety of processes are not to be jeopardized. Absorption proceeds at a rate of 7.3 μmoles/hour (Crooks and Simkiss, 1974) and if this were to be transported in its ionic form the chorioallantoic cells would have to withstand intracellular concentrations several thousand times greater than found in any known cell. Experiments using ^{45}Ca have shown that the transport of calcium across the chorioallantoic membrane to the capillary blood vessels is achieved with no intracellular mixing (Terepka *et al.*, 1969; Coleman and Terepka, 1972b,c). The suggestion of Coleman and Terepka (1972b,c) that the calcium is compartmentalized as it enters the cells of the chorioallantois by endocytosis and after transport across the cells is released into the blood by exocytosis has been shown to be unlikely. Simkiss (1974) calculates no more than 10% of the observed flow rate could be accounted for in this way. It seems more likely, therefore, that calcium is bound immediately upon entry into the cell. The mechanisms of transport across the cell and the nature of the protein are not known (see Chapter 8).

2. *Solubilization of Calcium*

Recent work has shown that a cell type found in the chorioallantois secretes carbonic acid which is responsible for mobilizing the shell stores of calcium. Earlier claims that hydrochloric acid was responsible now seem unlikely (see Narbaitz, 1974). The cell is variously described as the intercalating cell (Leeson and Leeson, 1963), the calcium absorbing cell (Owczarzak, 1971), or the villus cavity cell (Coleman and Terepka, 1972a). Its morphology is similar to the gastric parietal cell which also secretes acid. Carbonic anhydrase (EC 4.2.1.1) activity has been shown to be high in the cells from the eleventh day when active absorption begins (Heckey and Owczarzak, 1972). Carbon dioxide necessary is available in abundance from the blood where its partial pressure is around 50 mm Hg during this period (Dawes and Simkiss, 1969; Freeman and Misson, 1970; Tazawa *et al.*, 1971).

3. *Transport of Calcium*

The uptake of calcium by the chorioallantoic ectoderm is an active process (Garrison and Terepka, 1972a). It is a unidirectional transport process and is specific for Ca^{2+}. Calcium receptors are probably located on the surface of the ectoderm and contain sulfhydryl (–SH) groups (Garrison and Terepka, 1972a). Ectodermal transport of Ca^{2+} into the cells is a sodium-dependant process (Garrison and Terepka, 1972b; Armbrecht and Terepka, 1975). Ca^{2+}–ATPase has not been implicated in the mechanism (Saleuddin *et al.*, 1976).

It is not yet certain whether the transport mechanism is common to the whole chorioallantoic ectoderm or whether only certain cells are concerned. Coleman *et al.* (1970) and Coleman and Terepka (1972a,b,c) have shown that the "capillary covering" cells takes up Ca^{2+} but it was pointed out above that the pinocytotic mechanism proposed for these cells was insufficient to meet the observed efflux of calcium (113 nmole/cm^{2}hour—Crooks and Simkiss, 1975). Another objection to this interpretation rests on the nonuniform distribution of these cells (Narbaitz, 1972). It is possible therefore that Owczarzak's (1971) suggestion that the acid-secreting cells also transport Ca^{2+} may be correct.

C. Plasma Calcium

1. *Normal Values*

Early claims that the concentration of calcium in the plasma rises progressively from the fourteenth day (Taylor, 1963) have not been substantiated (Stewart and Terepka, 1969; Taylor *et al.*, 1975a). Improved methodology has shown that between days 13 and 16 the concentration is constant at 2.42–2.45 mmole/liter. It then increases to 2.61 mmole/liter by day 18, remains constant to day 20 and then falls slightly (Taylor *et al.*, 1975a).

2. *Control of Plasma Calcium*

As pointed out above (Section VI,B,1) calcium concentrations need to be controlled within fine limits if the animal is to function normally. However the mechanisms of control are poorly understood both in the embryo and adult. Calcitonin and parathormone are generally considered to have antagonistic activity and the transient low plasma concentration of calcium on the twentieth day coincides with a peak concentration in plasma calcitonin (Taylor and Lewis, 1972; Taylor *et al.*, 1975b). Similarly parathormone may stimulate the increase in plasma calcium occurring from day 16 (Narbaitz, 1975) and also bone resorption by inducing osteoclastic activity (Semba *et al.*, 1966) or by inhibiting ossification (Jones, 1970).

The early embryo develops in a medium (subembryonic fluid) which is deficient in Ca^{2+} and it appears that this may be a prerequisite for normal development (Grau *et al.*, 1963). There is, moreover, relatively little calcium available to the embryo before the thirteenth day when the chorioallantoic transport system matures (Section VI,B). The limited bone mineralization that takes place before this time (Section VI,D) relies on the small yolk stores. Parathormone secretion probably begins at about $10\frac{1}{2}$ days (Gaillard, 1959) but calcitonin secretion does not occur before 17 days (Taylor and Lewis, 1972). How the calcium concentration is controlled, therefore, remains uncertain.

D. Bone Mineralization

Mineralization begins on about the eighth day. For a description of the essential cellular features of this process the reader is referred to Crisman and Low (1974), and to Waddell (1972) for a consideration of chemicophysical aspects.

Initially calcium is deposited as amorphous neutral or acidic calcium phosphate. Thereafter there is a progressive replacement of HPO_4^{2-} by CO_3^{2-} which transforms the bone material to crystalline hydroxyapatite (Pellegrino and Biltz, 1972).

VII. Biochemical Aspects of Certain Macromolecules

A. Hemoglobins

1. *Multiple Hemoglobins*

The literature concerning avian hemoglobins and their nomenclature is confused. There is general agreement, however, that embryo-specific hemoglobins exist though their number is disputed. In the domestic fowl it seems likely that there are two major embryonic hemoglobins (Denmark and Washburn, 1969; Fraser *et al.*, 1972; Schalekamp *et al.*, 1972; Bruns and Ingram, 1973; Brown and Ingram, 1974; Cirotto *et al.*, 1975; Irving *et al.*, 1976) and perhaps up to four minor components (Schalekamp *et al.*, 1972; Shimizu, 1972). The duck also has two embryonic hemoglobins (Borgese and Bertles, 1965).

Adult hemoglobins coexist with the embryonic forms during much of the incubation period with the result that up to eight species may be found at one time (Fraser *et al.*, 1972).

The composition of the globins of the embryonic hemoglobins differs from that of the adult forms (D'Amelio, 1966; D'Amelio and Costantino, 1968; Schalekamp *et al.*, 1972; D'Amelio *et al.*, 1973). Irving *et al.* (1976) have isolated a polyadenylic acid-rich species of RNA which has a sedimentation coefficient of 9 S–10 S and which codes specifically for the polypeptide chains of the early hemoglobins.

2. *Hemoglobin Synthesis*

a. Blood Islands. In the domestic fowl hemoglobin is synthesized from about stage 8 and is visible by stage 9. Values for total hemoglobin and mean corpuscular hemoglobin to stage 18 are given by Drupt *et al.* (1975).

Heme and globin syntheses are normally closely coordinated but globin synthesis can proceed independently in the very young (stage 6) embryo (Ingram, 1974; Wilt, 1974). Indeed the cells of the presumptive blood islands are probably determined by this stage since mRNA's for globin synthesis can be detected at around stage 4 (Wainwright and Wainwright, 1966; Irving *et al.*, 1976). It seems likely, therefore, that the onset of hemoglobin synthesis is regulated at the translation level of protein synthesis (Wainwright and Wainwright, 1967a,c; Incefy and Kappas, 1971).

δ-Aminolevulinate synthase (EC 2.3.1.37) catalyzes the formation, from succinyl-CoA and glycine, of δ-aminolevulinate, the rate-limiting step in the biosynthetic pathway for porphyrin synthesis (Levere and Granick, 1967). Its activity is high in the blastoderm in the 12 hours before hemoglobin can be detected (Irving *et al.*, 1976). Notwithstanding this and the demonstration that exogenous δ-aminolevulinate stimulates the earlier appearance of hemoglobin (Wainwright and Wainwright, 1966, 1967b; Levere and Granick, 1967; Battikh, 1971), Wilt (1968) has presented evidence to indicate that δ-aminolevulinate is not the unique controlling factor in hemoglobin synthesis.

b. Yolk Sac, Liver, and Bone Marrow. Erythropoiesis begins in the yolk sac of the domestic fowl on the fifth day and is at its maximum activity from the tenth to the fifteenth day. Bone marrow begins forming erythrocytes on the fourteenth day but is probably not fully active until the seventeenth (Godet, 1974). A limited amount of erythropoiesis occurs in the liver of the fowl though this is an important site in other species.

Activity of δ-aminolevulinate synthase in the liver is low compared with that of the heart but it is very sensitive to the actions of inducing agents whereas that of the heart is not (Incefy and Kappas, 1971; Israels *et al.*, 1974; Granick *et al.*, 1975). Changes in the hepatic glycine pool also affect the rate of porphyrin synthesis (Cowtan *et al.*, 1973).

Syntheses of adult and embryonic hemoglobins go on in the yolk sac and the bone marrow contemporaneoulsy though the amount of the embryonic form declines rapidly in the latter from the seventeenth day (Godet, 1974).

c. Regulation of Hemoglobin Synthesis. The stimulus for synthesis remains uncertain. A component of the wax fraction of the yolk has stimulatory activity (Wainwright and Wainwright, 1972) but further characterization of this substance is awaited. Steroids related to 5β-androstane have a highly specific inductive action on δ-aminolevulate synthase (Levere and Granick, 1967; Levere *et al.*, 1967; Irving *et al.*, 1976) and hence the synthesis of heme may be important in regulating hemoglobin synthesis. The blastoderm shows a high capacity for reducing testosterone to 5β-reduced androgens (Parsons, 1970) and the presence of receptors for such metabolites has been demonstrated (Spooner and Mainwaring, 1973). This mechanism has been implicated in hemoglobin synthesis (Irving *et al.*, 1976) though the source of the testosterone or the 5β-androstanes remains problematic since the gonads do not become active until the seventh day (Galli and Wassermann, 1973).

Erythropoietin stimulates erythropoiesis by a different mechanism but it has not been identified in the embryo with certainty. Salvatorelli (1967) has shown that liver cells stimulate erythropoiesis in bone marrow, although Malpoix (1967) and Rifkind (1974) found an indifferent response to the pure hormone.

During development the various hemoglobins are synthesized or not, according to a reasonably precise sequence reflecting the changes in gene expression. It is not known how these changes are controlled but it is known that the sequence can be modified—e.g., the appearance of an adult hemoglobin delayed and the synthesis of an embryonic one extended by changing the partial pressure of the oxygen in the atmosphere surrounding the egg (Atherton and Timiras, 1970; Ackerman and Ramm, 1971; Pagram and Parson, 1974).

Mean corpuscular hemoglobin concentration is also susceptible to changes in the partial pressure of oxygen (Ackerman, 1970). Thyroid activity does not influence hemoglobin synthesis (Atherton, 1969).

3. *Oxygen Affinities of Hemoglobins*

The importance of adenosine triphosphate (ATP) and inositol hexaphosphate in modifying the oxygen affinity of avian hemoglobin is now recognized (Benesch and Benesch, 1969; Vandecasserie *et al.*, 1973). The higher affinity of embryonic hemoglobins for oxygen is due to a difference in the modifying effects of these organic phosphates. "Stripped" embryonic and adult hemoglobins have a similar affinity for oxygen (Cirotto and Geraci, 1975). Early claims that oxygen affinity of blood declines during incubation as a result of the progressive replacement of embryonic hemoglobin by adult hemoglobin were questioned by Misson and Freeman (1972) who found that affinity increased from 14 to 17 days as a result of a marked decline in the intra-erythrocytic concentration of ATP.

B. Myoglobin

Myoglobin synthesis begins on about day 10 in cardiac muscle (Kagen *et al.*, 1969) but not until day 16 in striated muscle (Low and Rich, 1973). It is synthesized on small polysomes, containing between five and nine ribosomes (Low and Rich, 1973) and its mRNA is found in the 8–12 S RNA fraction (Thompson *et al.*, 1973).

The primary sequence of chicken myoglobin was elucidated by Deconinck *et al.* (1975).

C. Muscle Proteins

For a general review of the contractile proteins of muscle, with particular reference to the heart the reader is referred to Katz (1970). Discussion here will be limited to the four contractile proteins, myosin, actin, tropomysin, and troponin. It should be remembered that these proteins are not necessarily exclusive to muscle (see, for instance, Garnett *et al.*, 1973; Masaki, 1975).

The synthesis of muscle proteins before the appearance, at the ultrastructural level, of myofibrils has been demonstrated from a number of sites (Orkin, 1973) suggesting that the ability to synthesize these molecules is general in the very young embryo and that during ontogeny there is a progressive restriction of this synthetic capability.

1. *Myosin*

Myosin appears as thick filaments (100 Å), and in striated muscle is found in the A band. Early work (see Katz, 1970, for references) suggested that the myosins from voluntary muscle differed from that obtained from cardiac muscle, the former probably being a dimer of the latter. It is now known, however, that they have a similar molecular weight though differences in amino acid composition have been noted.

Although Baril and Herrmann (1967) could detect no differences between embryonic and adult myosin, Obinata (1969) reported the presence of two myosins in the young (8 days) embryo both of which differ from adult myosin. One has a sedimentation coefficient of 3 S and the other 6 S. Both have ATPase activity (see below). In the later stages of embryonic development the 3 S component could no longer be isolated. Masaki (1974) and Masaki and Yoshizaki (1974) have since confirmed the existence of different components of myosin. They identified three types of H-meromyosin coexisting in muscle, only one of which persists after hatching.

Recent research has indicated that the heavy and light chains of myosin (H- and L-meromyosin) are translated independently. The H-meromyosin chain is controlled by a 26 S mRNA (Heywood and Nwagwu, 1969) some properties of which have been determined by Mondal *et al.* (1974). Heywood *et al.* (1975) have further shown that much of the myosin messenger is stored in a 70–90 S ribonuclear protein particle and that accumulation of this material precedes by a considerable time period its transcription. Synthesis of myosin in the presumptive limb bud can be detected by about stage 24 (Medoff and Zwilling, 1972), that is, some time before myoblasts become differentiated, and the material can be visualized by stage 28 (Thorogood, 1973).

The ATPase activity of skeletal muscle myosin is low compared with that of the adult form (Trayer and Perry, 1966) and does not change significantly during incubation (Dow and Stracher, 1971). The activity of the 6 S component is higher than that of the 3 S component (Obinata, 1969). Activity is also relatively unstable and may reflect differences in conformation compared with the adult form. Actin may help stabilize the activity in the embryo (Dow and Stracher, 1971).

2. *Actin*

The synthesis of actin filaments, which are usually described as thin (60–80 Å) to distinguish them from the thick myosin filaments, begins slightly earlier than that of myosin (Allen and Pepe, 1965; Obinata *et al.*, 1966). Work on other vertebrates suggests that there is only one species of actin (see Katz, 1970). Synthesis begins at about stage 13 (Masaki and Yoshizaki, 1972).

3. *Tropomyosin and Troponin*

The synthesis of these molecules begins at stage 13–14 though the presumptive wing bud does not show activity until stage 24 (Llewellyn-Smith, 1974). Partial characterization of tropomyosin has been carried out (Potter and Herrmann, 1970).

D. Collagen

Collagen is a ubiquitous molecule and the method of its synthesis in the embryo has excited much research. Some aspects of the amino acid precursors have been discussed (Section V,C,3) and for extensive reviews of the subject the reader is referred to Bornstein (1974) and Gallup and Paz (1975).

The "basic" collagen molecule consists of three α-chains though there are specific variations not only in their sequencing but also in their length. In the collagen from skin and bone from older embryos two of the α-chains are identical and designated $\alpha1$ type I while the third has a different amino acid composition and is designated $\alpha2$. In collagen from cartilage the three chains are identical but a second variant of the $\alpha1$—$\alpha1$ type II. There are two more genetic variants of the $\alpha1$ chain: type III has been found from several sites including young skin (Vinson and Seyer, 1974) while type IV is restricted to the basement membrane. The latter is distinguished by having a longer chain length and contains more of the 3-isomer of hydroxyproline (Grant *et al.*, 1972a,b). The various forms of collagen identified thus far have the following molecular formulas: $[\alpha1(I)]_2\alpha2$, $[\alpha1(II)]_3$, $[\alpha1(III)]_3$, and $[\alpha1(IV)]_3$. Despite the marked molecular difference in the collagens their physicochemical properties are often similar (Igarashi *et al.*, 1973) though Rosenbloom *et al.* (1973) have shown that the denaturation temperature of collagen is determined by its hydroxyproline content.

Procollagen (see Section V,C,3) is synthesized on polysomes bound to the membranes of the endoplasmic reticulum (Burns *et al.*, 1973; Berman *et al.*, 1975). Two classes of ribosome have been identified, those which are bound tightly to the membrane and those which are bound loosely (Harwood *et al.*, 1975a) and while the significance of this arrangement has not been fully elucidated it seems that the former class are more active in synthesizing procollagen (Harwood *et al.*, 1975a). The enzymes concerned with hydroxylation of proline and lysine are probably located in the cisternae of the rough endoplasmic reticulum (see Guzman *et al.*, 1976) and the reader is referred to Harwood *et al.* (1976) for a consideration of the general secretory process. Details of the methods used for isolating and characterizing collagen-synthesizing polysomes are given in Wang *et al.* (1975) while those for optimum synthesis may be found in Traut and Petruska (1976).

The procollagen messenger RNA's differ from the majority of mRNA's isolated to date in that they have a high degree of secondary structure. That isolated from embryonic tendon cells has an apparent molecular weight of 1.65×10^6 (Harwood *et al.*, 1975b). The mRNA's are probably present in the "correct" ratios for the synthesis of the appropriate collagen (Vuust, 1975). It also seems likely that there is a characteristic complement of tRNA's associated with collagen synthesis (Christner and Rosenbloom, (1976).

Preliminary data on the carbohydrate moieties of procollagen may be found in Clark and Kefalides (1976).

The conversion of procollagen to collagen is probably extracellular and usually takes place as the molecule is synthesized but in at least the tendon procollagen persists for some time (Jimenez *et al.*, 1971) and may in some cases, e.g., basement membrane, escape complete conversion. The mechanism of transporting procollagen out of the cell is not well understood (see Bornstein, 1974).

The synthesis of collagen begins early in the embryo, with type $\alpha 1(\text{II})$ appearing in cartilage collagen from the notochord by stage 15 and type $\alpha 1(\text{I})$ being synthesized from stage 17 in the dermatomal myotomal plate (von der Mark *et al.*, 1976). When the bone matrix begins to develop, type $[\alpha 1(\text{I})]_2\ \alpha 2$ is produced (Linsenmayer *et al.*, 1973). The variant $[\alpha 1(\text{III})]_3$ is produced in the skin of embryos aged less than 13 days (Vinson and Seyer, 1974). The rate of collagen synthesis in the limb bud increases sixfold between the sixth day, when synthesis begins (Mottet and Hall, 1966), and the twelfth day (Diegelmann and Peterkofsky, 1972).

E. Glycosaminoglycans

The chondroitin sulfates are probably the best known glycosaminoglycans since they are normal components of connective tissue and cartilage. There are three forms of chondroitin sulfates in the avian embryo: A (4 sulfate), B (4 sulfate, but with iduronic acid replacing glucuronic acid residues), and C (6 sulfate).

All three chondroitin sulfates are synthesized from as early as stage 4 (O'Hare, 1973) and are found in a wide range of tissues (Abrahamsohn *et al.*, 1975). The activities of some of the enzymes concerned in their synthesis was determined by Medoff (1967) while the rate-limiting reactions, particularly that of xylosyltransferase, were considered by Stoolmiller *et al.* (1972). The location of the enzymatic activities has not yet been resolved but the Golgi apparatus has been implicated. Jansen and Bornstein (1974) provided evidence that microtubules were involved.

The differentiating heart (around stage 10) is capable of synthesizing chondroitin but fully sulfated chondroitin sulfate is synthesized only after the muscle differentiates (Manasek *et al.*, 1973). There is a marked increase in the synthetic ability of the presumptive limb bud at stage 22, some time before the cells differentiate into chondrocytes (Searls, 1965a,b). The major glycosaminoglycan secreted by developing limb cartilage is chondroitin sulfate (Huffer, 1970). Insulin may stimuate glycosaminoglycan synthesis (Hajek and Solursh, 1975).

Keratan sulfate, another glycosaminoglycan, is found in cartilage from some sites, but it is present in only low concentrations (Huffer, 1970; O'Hare, 1973).

F. Elastin

Elastin and collagen are the main components of the vertebrate aorta and provide much of the tissue's strength and resilience. In the developing chick a soluble precursor of elastin, tropoelastin, is actively synthesized by the developing aorta (Murphy *et al.*, 1972). It has a molecular weight of approximately 68,000 and appears to be only a transient intermediate in elastin synthesis (Narayanan *et al.*, 1974), the cross linkages occurring either on lysine residues by the action of lysyl oxidase and subsequent spontaneous condensation reactions or by a totally spontaneous reaction.

Details of the amino acid composition of elastin from the aorta may be found in Keeley and Labella (1972). There is limited evidence that there is a change in its structure between 8 and 9 days (Newman and Low, 1973).

VIII. Biochemical Aspects of Development and Functional Differentiation

A. Alimentary Tract

The alimentary tract becomes functional before hatching. This may be seen as an adaptation to the embryonic existence since the amniotic fluid is actively imbibed by the embryo from the thirteenth day. This fluid contains much of the albumen protein since the seroamniotic connection separating the two compartments has failed by this time.

The amino acid transport system of the gut is functional (Section V,B,5) thus allowing the embryo to utilize these materials.

1. *Esophagus*

The esophagus becomes occluded at stage 26 (5 days) when the roof epithelium collapses and adheres to the floor; it reopens $2\frac{1}{2}$–3 days later (stage 33+) largely as a result of obligatory cell death (Allenspach, 1964). Wilson and Allenspach (1974) have shown that the lysosomal activities of acid phosphatase and β-glucuronidase increase independently from 7 to 12 days and may be implicated in the remodeling procedure (Allenspach, 1976). Thus the activity of acid phosphatase is highest in the anterior occluded segment initially but once the lumen is reformed it becomes uniformally active throughout the esophagus.

2. *Pancreas*

Between days 6 and 12 there is an increase in amylase (EC 3.2.1.1) specific activity but the greatest increase is seen between day 18 and hatching (Heller and Kulka, 1968a). Two amylase isozymes have been identified and these are produced in the same proportions during development. There are, however, three phenotypes (Heller and Kulka, 1968a). There are also increases in the specific activities of the other digestive enzymes chymotrypsinogen (EC 3.4.4.5), procarboxypeptidase (EC class 3.4.2), lipase, and Mg^{2+}-ribonuclease from day 18 but these do not parallel that in amylase activity (Marchaim and Kulka, 1967; Heller and Kulka, 1968b; Dieterlen-Lièvre and Hadorn, 1972). The latter is influenced by glucocorticoids (Yalovsky *et al.*, 1969; Cohen *et al.*, 1972), the hormone acting as an "amplifier" rather than an inducer. Three forms of chymotrypsin are recognized: type 3 predominates in the embryo to be replaced by types 1 and 2 after hatching (Cohen and Kulka, 1973). Type 3 chymotrypsin appears to have a high activity towards albumen proteins.

3. *Duodenum*

The rapid and sudden maturation of the duodenum is a well-known phenomenon though most attention has been given to the increase in alkaline phosphatase (EC 3.1.3.1) activity (see Moog, 1965). It is controlled by the pituitary–adrenal axis (see Bellware and Betz, 1970) and probably the pituitary–thyroidal axis (Hart and Betz, 1972). The role of somatotrophin is equivocal (Betz, 1971).

The 400 to 500-fold increase in alkaline phosphatase activity has been shown to be a result not of the initiation of synthesis but of the removal of an inhibitory protein thereby allowing activation of the enzyme (Moog and Grey, 1966). Three forms of phosphatase have been isolated: two (F) with low (200,000) molecular weight and one which is a series of polydisperse molecules (S) (Chang and Moog, 1972a,b). The proportion of F to S rises during hatching from 20% to 50%. The catalytic activities of F_1, F_2, and S are virtually indistinguishable (Chang and Moog, 1972a). Other enzymes also show marked increases in their activity at this time and all may be initiated by glucocorticoids (see Hijmans and McCarty, 1966; Strittmatter, 1972). They include glucose-6-phosphatase, fructose-1,6-diphosphatase, Mg^{2+}-acid phosphatase, Zn^{2+}-acid phosphatase, and sucrase (EC 3.2.1.26) (Brown and Moog, 1967; Strittmatter, 1972).

4. *Jejunum*

Maltase (EC 3.2.1.20) and sucrase show a marked increase in activity about the time of hatching (Dautlick and Strittmatter, 1970) though there may be a transient fall in the activity of the latter between 19 and 20 days (Brown and Moog, 1967). Lactase activity remains unchanged (Dautlick and Strittmatter, 1970). There is evidence that there are embryo and adult forms of maltase (Dautlick and Strittmatter, 1970).

Like the duodenal enzymes, those of the jejunum show an increased activity following treatment with glucocorticoids (Dautlick and Strittmatter, 1970), the increase possibly being due to the removal of an inhibitor (Hijmans and McCarty, 1966).

B. Cartilage and Bone

Bone may be synthesized either directly (membranous bone) or by the ossification of existing cartilage (endochondral bone). Both cartilage and bone contain collagen, 90% or more than organic matrix of the latter being collagen while the cartilage contains considerable amounts of chondroitin sulfate.

It has already been noted (Section VII,D) that different types of collagen are produced by different tissues. It should also be emphasized that the same tissue may produce different collagen types during development. Thus the skin produces both $[\alpha 1(I)]_2\alpha 2$ and $[\alpha 1(III)]_2$ prior to 13 days but mainly the former thereafter (Vinson and Seyer, 1974) while cartilage from the keel produces less type 1 material during embryogenesis (Seyer and Vinson, 1974). The genetic control of these syntheses is not understood.

Insulin has been implicated in the regulation of cartilage synthesis though its action seems directed towards glycosaminoglycan synthesis alone (Hajek and Solursh, 1975). This is somewhat surprising in view of the close correlation that exists between glycosaminoglycans and collagen (Hall, 1973a). Endogenous insulin might be concerned with modeling endochondral bone (Rabinovitch and Gibson, 1972a,b). Chondroblastic proliferation and glycosaminoglycan synthesis is reduced in hypophysectomized embryos (Hall and Girouard, 1973) perhaps because of the lack of thyroxine (Hall, 1973b) and glucocorticoid (Hall and Kalliecharan, 1975) but these hormonal effects can only influence development after 10 (thyroid) or 12 days (adrenal), when the endocrine glands become active. The chondrocytes have an enzyme system capable of metabolizing sex hormones and cortisol to yield 3α-hydroxy 5β-steroids (Murota and Tamaoki, 1967).

Vertebral cartilage differs from that in the limbs in that the predominent glycosaminoglycan is chondroitin 4-sulfate (Shulman and Meyer, 1968). Ossification begins at about the tenth day (see Section VI,D). The cartilage shows significant acid hydrolase activity capable of breaking down both cells and extracellular matrix which is a necessary preliminary to bone formation (Arsenis *et al.*, 1971). Alkaline phosphatase activity rises rapidly in differentiating bone (McWhinnie and Thommes, 1973) and osteoblasts become evident, quite possibly as a result of a transformation of the chondrocytes (see Hall, 1972, for references) mediated by either an altered enzyme-secretory ability or more simply by altering the rates of synthesis of existing products (Fitton-Jackson, 1970). Calcitonin may stimulate bone development (McWhinnie, 1975).

C. Muscle

1. *Striated Muscle*

In addition to the usual three classes of muscle, striated, smooth, and cardiac, birds have two types of striated muscle fiber, fast and slow. Most skeletal muscles contain both types but there are some which have only one. Thus the posterior latissimus dorsi muscle contains only fast fibers and the anterior latissimus dorsi muscle contains only slow fibers. The structural differences of these two muscle types include little or no transverse tubule system in the slow fiber, a multiple "*en grappe*" type of innervation in slow fibers but a single innervation in the fast fiber. Further details of the slow fiber are found in Hess (1970).

The differentiation of these muscle types offers a considerable problem to the biochemist since striated and smooth muscle appear to develop from similar myoblastic cell lines. Furthermore evidence is beginning to accumulate which indicates that the differentiation of slow and fast fiber is influenced by the innervation (Purves and Vrbová, 1974; Gordon and Vrbová, 1975).

Details of the changes in the nucleotides and other macromolecular components of the differing muscle types have been given by Radha and Krishnamoorthy (1973) and Radha (1975).

In vitro studies (de la Haba *et al.*, 1966, 1968; de la Haba and Amundsen, 1972) have shown that insulin and at least two other substances or groups of substances are required for myogenesis. In a medium containing these substances aggregation and fusion of the myoblasts proceeds and the formation of myotubes is promoted. Ca^{2+} also appears necessary (Ozawa, 1972) while collagen may have a permissive role. Paterson and Strohman (1972), also from *in vitro* studies, suggest that the regulation of myogenesis, cell fusion, and myosin synthesis are developed sequentially rather than simultaneously.

The rate of myosin synthesis was determined by Baril and Herrmann (1967) and Nwagwu and Stevens (1970) who found a rapid increase from day 9 to a peak at 14 days and that accumulation showed a fivefold increase from 9 to 18 days. The increased rate is not brought about by an increase in the ribosome or polysome content of the muscle. Nwagwu and Nana (1974) suggest that it results from an increased efficiency of the synthetic pathways. During this period the activity of protein phosphokinase increased by a factor of 8 (Piras *et al.*, 1972a) and glycogen synthetase kinase by a factor of 3 (Piras *et al.*, 1972b).

2. *The Hatching Muscle*

The hatching muscle, musculus complexus, appears to be concerned in the hatching process and some aspects of its development suggest there are specific differences in its ontogeny compared with other striated muscles. Thus there is an imbibition of water and increased synthesis of mucopolysaccharide just before hatching which is accompanied by decreases in lipid and carbohydrate of the muscle and increases in the activity of certain hydrolytic enzymes (Ramachandran *et al.*, 1969; Klicka *et al.*, 1969; Hsiao and Ungar, 1969). These changes may be controlled by adrenal steroids (Brooks and Ungar, 1967; but see Oppenheim, 1973).

D. Central Nervous System

1. *Brain*

Generally there is an increase in the glycolytic activity of the brain from 10 days (Seltzer and McDougal, 1975) no doubt reflecting the increased energy demands of the organ. At the same time there is evidence of glycogen accumulation (Medda and Das, 1972). There is also a considerable increase in the synthesis of phospholipids (Peter *et al.*, 1975) and in the activities of acetylcholinesterase (EC 3.1.1.7) and the enzymes involved in glutamine and GABA metabolism (Vos *et al.*, 1967; McGeer *et al.*, 1974; McGeer and Maler, 1975; Herzfeld and Raper, 1975). Peter *et al.* (1975) have correlated the increases in activity of ($Na^+ + K^+$)-ATPase (EC 3.6.1.3) with the concentrations of phosphatidylserine and phosphatidylcholine and also in the increase in activity of Mg^{2+}-ATPase with the concentration of phosphatidylethanolamine. Thyroxine may be implicated in some of these changes since *in vitro* work has shown that the hormone stimulates enzymatic activities (Werner *et al.*, 1971). Day 17 appears to be critical in the rate of tryptophan metabolism in the brain. Suzuki *et al.* (1975) have shown marked increases in tryptophan, 5-hydroxytryptamine and 5-hydroxyindoleacetic acid accumulation and in monoamine oxidase activity.

The cerebral hemispheres begin to show spontaneous electrical activity from about day 12 although unexpectedly this is not accompanied by any apparent increases in energy utilization (Jongkind *et al.*, 1972). The local activity of ($Na^+ + K^+$)-ATPase also increases at this time and there is some evidence to suggest that this occurs in response to hydrocortisone (Šťastný, 1971, 1972). Tryptophan decarboxylase activity (EC 4.1.1.27) also appears but is delayed until day 18 in the cerebellum (Wainwright, 1974). Aspects of RNA metabolism of the cerebral hemispheres have been investigated by Judes *et al.* (1973) and Judes and Jacob (1973a,b).

The presence of estrogen target cells in the developing brain has been shown by Martinez-Vargas *et al.* (1975). It is presumed that these cells are concerned in such activities as sexual behavior, aggression, and controlling gonadotrophin secretion. The authors demonstrated the receptors in the median preoptic and ventral hypothalamic regions as early as 10 days. They appear in the telencephalon, diencephalon, and mesencephalon later in incubation.

2. *Spinal Cord*

There is a close relationship between morphological development of spinal cord synapses and the development of choline acetyltransferase (EC 2.3.7.6) and acetylcholinesterase activities (EC 3.1.1.7) (Burt, 1968; Marchisio and Consolo, 1968; Giacobini *et al.*, 1970; Kim *et al.*, 1975; Fairman *et al.*, 1976) and catecholamine transmitters (Tunnicliff and Kim, 1973; Fairman *et al.*, 1976). Aspects of RNA metabolism have been studied by Mezei and Hu (1972).

3. *Myelination of Nerve Fibers*

Myelination begins on about day 12 but becomes particularly active from day 15. The pattern of development is paralleled by changes in the activities of 2′,3′-cyclic nucleotide 3′-phosphohydrolase (Kurihara and Tsukada, 1968) and cholesterol esterase (EC 3.1.1.13) (Mezei *et al.*, 1971) which appear to be located in the myelin sheath. The proportions of phosphatidylserine and phosphatidylcholine in the sheath fall from day 18 to 21 while that of phosphatidylethanolamine increases (Oulton and Mezei, 1973).

4. *Cerebrospinal Fluid*

Details of the composition of cerebrospinal fluid during embryonic development may be found in Birge *et al.* (1974) and Sedláček (1972, 1975a,b,c).

E. Reproductive System

During the first week of incubation the genital tracts of the male and female cannot be distinguished morphologically. Thereafter the right gonad and Müllerian duct degenerate in the female while the left gonad and Müllerian duct become the ovary and the oviduct respectively. Both Wolffian ducts also degenerate. In the male both gonads develop into testes, the Müllerian ducts degenerate and the Wolffian ducts become the vasa deferentia. The genetic control of these changes is not understood.

While morphological differentiation cannot be made before day 7 the biosynthetic dichotomy of the gonads is already detectable. Thus only the presumptive ovary can synthesize estradiol-17β and estrone though both gonadal types can synthesize 20α- and 20β-dihydroprogesterone, 17α-hydroxyprogesterone, androstenedione, and testosterone (Haffen and Cèdard, 1968; Galli and Wassermann, 1973; Guichard *et al.*, 1973). These latter synthetic pathways are retained by both gonadal types to about day 15 (Galli and Wassermann, 1972) but by day 18 testosterone is synthesized by only the testis (Guichard *et al.*, 1973). Not surprisingly synthetic activity increases from day 7 as differentiation accelerates.

The regression of the Müllerian ducts in the male is probably controlled by a testicular secretion (Maraud *et al.*, 1969) but it is not an androgen (Stoll *et al.*, 1972; Weniger and Zeis, 1973). The hormone responsible has a molecular weight of not less than 1000 (Weniger *et al.*, 1975). The control of the regression of only the right Müllerian duct in the female is poorly understood but the mechanism differs from that in the male (Thiebold, 1973).

F. Skin

Between 12 and 19 days the three- to four-layered undifferentiated epidermis is transformed into mature, keratinized epidermis. The ultrastructural changes were detailed by Mottet and Jensen (1968) and biochemical changes by Pane *et al.* (1974). During this period collagen synthesis is markedly increased (Woessner *et al.*, 1967) and there are major changes in the proteins (Beckingham Smith, 1973). Kemp and Rogers (1972) have shown the proteins of the embryonic skin and feathers have a low molecular weight and that there are differences both between the keratins of these tissues and between those tissues from the embryo and adult. A mRNA with a sedimentation coefficient of 12 S has been identified with the keratinization process (Desveaux-Chabrol, 1974). Additional aspects of keratinization are discussed in Chapter 4.

The initiation of keratinization, at least in the skin, may be controlled hormonally. Sugimoto and Endo (1971a,b) and Kojima *et al.* (1976) have found the process may be induced with very low concentrations of hydrocortisone though it would appear that a sufficient level of thyroid hormones is necessary (Lawrenz and Johnson, 1970).

REFERENCES

Abrahamsohn, P. A., Lash, J. W., Kosher, R. A., and Minor, R. R. (1975). *J. Exp. Zool.* **194**, 511–518.

Ackerman, N. R. (1970). *Dev. Biol.* **23**, 310–323.

Ackerman, N. R., and Ramm, G. M. (1971). *Teratology* **4**, 445–452.

Adamson, L. F., and Anast, C. S. (1966). *Biochim. Biophys. Acta* **121**, 10–20.

Adamson, L. F., and Ingbar, S. H. (1967a). *J. Biol. Chem.* **242**, 2646–2652.

Adamson, L. F., and Ingbar, S. H. (1967b). *Endocrinology* **81**, 1362–1371.

Adamson, L. F., and Ingbar, S. H. (1967c). *Endocrinology* **81**, 1372–1378.

Adamson, L. F., Langeluttig, S. G., and Anast, C. S. (1966a). *Biochim. Biophys. Acta* **115**, 345–354.

Adamson, L. F., Langeluttig, S. G., and Anast, C. S. (1966b). *Biochim. Biophys. Acta* **115**, 355–360.

Allen, E. R., and Pepe, F. A. (1965). *Am. J. Anat.* **116**, 115–148.

Allen, H. J. (1919). *Biol. Bull.* **36**, 63–70.

Allenspach, A. L. (1964). *J. Morphol.* **114**, 287–302.

Allenspach, A. L. (1976). *Cytobiologie* **12**, 356–362.

Arese, P., Rinaudo, M. T., and Bosia, A. (1967). *Eur. J. Biochem.* **1**, 207–215.

Arfin, S. M. (1967). *Biochim. Biophys. Acta* **136**, 233–244.

Armbrecht, H. J., and Terepka, A. R. (1975). *Fed. Proc., Fed. Am. Soc. Exp. Biol.* **34**, 310.

Arsenis, C., Eisentein, R., Soble, L. W., and Kuettner, K. E. (1971). *J. Cell Biol.* **49**, 459–467.

Atherton, R. W. (1969). *J. Embryol. Exp. Morphol.* **22**, 99–105.

Atherton, R. W., and Timiras, P. S. (1970). *Am. J. Physiol.* **218**, 75–79.

Ballard, F. J., and Oliver, I. T. (1963). *Biochim. Biophys. Acta* **71**, 578–588.

Baril, E. F., and Herrmann, H. (1967). *Dev. Biol.* **15**, 318–333.

Battikh, H. K. (1971). *Arch. Anat., Histol. Embryol.* **54**, 113–122.

Beckingham Smith, K. (1973). *Dev. Biol.* **30**, 249–262.

Bellairs, R. (1964). *Adv. Morphog.* **4**, 217–272.

Bellware, F. T., and Betz, T. W. (1970). *J. Embryol. Exp. Morphol.* **24**, 335–355.

Benesch, R., and Benesch, R. E. (1969). *Nature (London)* **221**, 618–622.

Benzo, C. A., and De Gennaro, L. D. (1974). *J. Exp. Zool.* **188**, 375–380.

Benzo, C. A., and de la Haba, G. (1972). *J. Cell. Physiol.* **79**, 53–64.

Benzo, C. A., and Green, T. D. (1974). *Anat. Rec.* **180**, 491–496.

Benzo, C. A., and Stearns, S. B. (1975). *Am. J. Anat.* **142**, 515–518.

Berg, R. A., Olsen, B. R., and Prockop, D. J. (1972). *Biochim. Biophys. Acta* **285**, 167–175.

Berger, C. K., and Foà, P. P. (1971). *Horm. Metab. Res.* **3**, 98–102.

Berman, A. E., Oborotova, T. A., and Mazurov, V. I. (1975). *Biokhimiya* **40**, 364–371.

Betz, T. W. (1971). *In* "Hormones in Development" (M. Hamburgh and E. J. W. Barrington, eds.), pp. 75–94. Appleton, New York.

Birge, W. J., Rose, A. D., Haywood, J. R., and Doolin, P. F. (1974). *Dev. Biol.* **41**, 245–254.

Blumenkrantz, N., and Prockop, D. J. (1970). *Biochim. Biophys. Acta* **208**, 461–466.

Borgese, T. A., and Bertles, J. F. (1965). *Science* **148**, 509–511.

Bornstein, P. (1974). *Annu. Rev. Biochem.* **43**, 567–603.

Boucek, R. J., Boucek, R. J. Jr., Hlavackova, V., and Dietrich, L. S. (1967). *Biochim. Biophys. Acta* **141**, 473–482.

Brooks, W. S., and Ungar, F. (1967). *Proc. Soc. Exp. Biol. Med.* **125**, 488–492.

Brown, J. L., and Ingram, V. M. (1974). *J. Biol. Chem.* **249**, 3960–3972.

Brown, K. M., and Moog, F. (1967). *Biochim. Biophys. Acta* **132**, 185–187.

Bruns, G. A. P., and Ingram, V. M. (1973). *Dev. Biol.* **30**, 455–459.

Buño, W., and Mariño, R. G. (1952). *Acta Anat.* **16**, 85–92.

Burns, T. M., Spears, C. L., and Kerwar, S. S. (1973). *Arch. Biochem. Biophys.* **159**, 880–884.

Burt, A. M. (1968). *J. Exp. Zool.* **169**, 107–112.

Chang, C.-H., and Moog, F. (1972a). *Biochim. Biophys. Acta* **258**, 154–165.

Chang, C.-H., and Moog, F. (1972b). *Biochim. Biophys. Acta* **258**, 166–177.

Christensen, H. N. (1969). *Adv. Enzymol.* **32**, 1–20.

Christner, P. J., and Rosenbloom, J. (1976). *Arch. Biochem. Biophys.* **172**, 399–409.

Cirotto, C., and Geraci, G. (1975). *Comp. Biochem. Physiol. A* **51**, 159–163.

Cirotto, C., Scotto Di Telia, A., and Geraci, G. (1975). *Cell Differ.* **4**, 87–99.

Clark, C. C., and Kefalides, N. A. (1976). *Proc. Natl. Acad. Sci. U.S.A.* **73**, 34–38.

Coffey, R. G., Cheldelin, V. H., and Newburg, R. W. (1964). *J. Gen. Physiol.* **48**, 105–112.

Cohen, A., and Kulka, R. G. (1973). *Nature (London)* **244**, 97–99.

Cohen, A., Heller, H., and Kulka, R. G. (1972). *Dev. Biol.* **29**, 293–306.

Coleman, J. R., and Terepka, A. R. (1972a). *J. Membr. Biol.* **7**, 111–127.

Coleman, J. R., and Terepka, A. R. (1972b). *J. Histochem. Cytochem.* **20**, 401–413.

Coleman, J. R., and Terepka, A. R. (1972c). *J. Histochem. Cytochem.* **20**, 414–424.

Coleman, J. R., Dewitt, S. M., Batt, P., and Terepka, A. R. (1970). *Exp. Cell Res.* **63**, 216–220.

Cowtan, E. R., Yoda, B., and Israels, L. G. (1973). *Arch. Biochem. Biophys.* **155**, 194–202.

Crisman, R. S., and Low, F. N. (1974). *Am. J. Anat.* **140**, 451–470.

Croisille, Y. (1972). *J. Embryol. Exp. Morphol.* **27**, 261–275.
Crooks, J. R., and Simkiss, K. (1974). *J. Exp. Biol.* **61**, 197–202.
Crooks, R. J., and Simkiss, K. (1975). *Q. J. Exp. Physiol. Cogn. Med. Sci.* **60**, 55–63.
D'Amelio, V. (1966). *Biochim. Biophys. Acta* **127**, 59–65.
D'Amelio, V., and Costantino, E. (1968). *Biochim. Biophys. Acta* **155**, 614–615.
D'Amelio, V., Cirotto, C., and Costantino-Ceccarini, E. (1973). *Dev. Biol.* **32**, 446–452.
Daugèras, N. (1971). *J. Embryol. Exp. Morphol.* **25**, 377–384.
Dautlick, J., and Strittmatter, C. F. (1970). *Biochim. Biophys. Acta* **222**, 447–454.
Dawes, C. M., and Simkiss, K. (1969). *J. Exp. Biol.* **50**, 79–86.
Deconinck, M., Peiffer, S., Depreter, J., Paul, C., Schnek, A. G., and Leonis, J. (1975). *Biochim. Biophys. Acta* **386**, 567–575.
de la Haba, G., and Amundsen, R. (1972). *Proc. Natl. Acad Sci. U.S.A.* **69**, 1131–1135.
de la Haba, G., Cooper, G. W., and Elting, V. (1966). *Proc. Natl. Acad Sci. U.S.A.* **56**, 1719–1723.
de la Haba, G., Cooper, G. W., and Elting, V. (1968). *J. Cell. Physiol.* **72**, 21–28.
Denmark, C. R., and Washburn, K. W. (1969). *Poult. Sci.* **48**, 464–474.
Desveaux-Chabrol, J. (1974). *C. R. Hebd. Seances Acad. Sci., Ser. D* **278**, 2355–2358.
Diegelmann, R. F., and Peterkofsky, B. (1972). *Dev. Biol.* **28**, 443–453.
Dieterlen-Lièvre, F., and Hadorn, H. B. (1972). *Wilhelm Roux' Arch. Entwicklungsmech. Org.* **170**, 175–184.
Donaldson, W. E., Mueller, N. S., and Mason, J. V. (1971). *Biochim. Biophys. Acta* **248**, 34–40.
Dow, J., and Stracher, A. (1971). *Biochemistry* **10**, 1316–1321.
Drupt, F., Vagner, D., Lenicque, P., and Leclerc, M. (1975). *Ann. Pharm. Fr.* **33**, 85–92.
Edwards, C., and Rogers, K. J. (1972). *J. Neurochem.* **19**, 2759–2766.
Emanuelsson, H. (1955). *Acta Physiol. Scand.* **34**, 124–134.
Fairman, K., Giacobini, E., and Chiappinelli, V. (1976). *Brain Res.* **102**, 301–312.
Fitton-Jackson, S. (1970). *Proc. R. Soc. London, Ser. B* **175**, 405–453.
Foà, P. P., Melli, M., Berger, C. K., Billinger, D., and Guidotti, G. G. (1965). *Fed. Proc., Fed. Am. Soc. Exp. Biol.* **24**, 1046–1050.
Franchi-Gazzola, R., Gazzola, G. C., Ronchi, P., Saibene, V., and Guidotti, G. G. (1973). *Biochim. Biophys. Acta* **291**, 545–556.
Fraser, R., Horton, B., Dupourque, D., and Chernoff, A. (1972). *J. Cell. Physiol.* **80**, 79–88.
Freeman, B. M. (1962). *Br. Poult. Sci.* **3**, 63–72.
Freeman, B. M. (1965a). *Br. Poult. Sci.* **6**, 67–72.
Freeman, B. M. (1965b). *Comp. Biochem. Physiol.* **14**, 217–222.
Freeman, B. M. (1969). *Comp. Biochem. Physiol.* **28**, 1169–1176.
Freeman, B. M., and Manning, A. C. (1971). *Comp. Gen. Pharmacol.* **2**, 198–204.
Freeman, B. M., and Misson, B. H. (1970). *Comp. Biochem. Physiol.* **33**, 763–772.
Freeman, B. M., and Vince, M. A. (1974). "Development of the Avian Embryo." Chapman & Hall, London.
Gaillard, P. J. (1959). *Dev. Biol.* **1**, 152–181.
Galli, F. E., and Wassermann, G. F. (1972). *Gen. Comp. Endocrinol.* **19**, 509–514.
Galli, F. E., and Wassermann, G. F. (1973). *Gen. Comp. Endocrinol.* **21**, 77–83.
Gallop, P. M., and Paz, M. A. (1975). *Physiol. Rev.* **55**, 418–487.
Garnett, H., Gröschel-Stewart, V., Jones, B. M., and Kemp, R. B. (1973). *Cytobios* **7**, 163–169.
Garrison, J. C., and Terepka, A. R. (1972a). *J. Membr. Biol.* **7**, 128–145.
Garrison, J. C., and Terepka, A. R. (1972b). *J. Membr. Biol.* **7**, 146–163.

Gazzola, G. C., Franchi, R., Saibene, V., Ronchi, P., and Guidotti, G. G. (1972). *Biochim. Biophys. Acta* **266**, 407–421.

Giacobini, G., Marchisio, P. C., Giacobini, E., and Koslow, S. H. (1970). *J. Neurochem.* **17**, 1177–1185.

Gill, P. M. (1938). *Biochem. J.* **32**, 1792–1799.

Gitlin, D., and Kitżes, J. (1967). *Biochim. Biophys. Acta* **147**, 334–340.

Godet, J. (1974). *Dev. Biol.* **40**, 199–207.

Gómez-Capilla, J. A., Macarulla, J. M., Martín-Andrés, A., and Osorio, C. (1975). *Rev. Esp. Fisiol.* **31**, 177–182.

Goodridge, A. G. (1968a). *Biochem. J.* **108**, 655–661.

Goodridge, A. G. (1968b). *Am. J. Physiol.* **214**, 897–901.

Goodridge, A. G. (1968c). *Biochem. J.* **108**, 663–666.

Goodridge, A. G. (1968d). *Am. J. Phýsiol.* **214**, 902–907.

Gordon, T., and Vrbová, G. (1975). *Pfluegers Arch.* **360**, 199–218.

Goris, J., and Merlevede, W. (1969). *Arch. Int. Physiol. Biochim.* **77**, 44–56.

Granick, S., Sinclair, P., Sassa, S., and Grieninger, G. (1975). *J. Biol. Chem.* **250**, 9215–9225.

Grant, M. E., Kefalides, N. A., and Prockop, D. J. (1972a). *J. Biol. Chem.* **247**, 3539–3544.

Grant, M. E., Kefalides, N. A., and Prockop, D. J. (1972b). *J. Biol. Chem.* **247**, 3545–3551.

Grau, C. R., Walker, N. E., Fritz, H. I., and Peters, S. M. (1963). *Nature (London)* **197**, 257–259.

Grillo, T. A. I., and Baxter-Grillo, D. L. (1966). *Gen. Comp. Endocrinol.* **7**, 420–423.

Grillo, T. A. I., Okuno, G., Price, S., and Foà, P. P. (1964). *J. Histochem. Cytochem.* **12**, 275–280.

Guichard, A., Cèdard, L., and Haffen, K. (1973). *Gen. Comp. Endocrinol.* **20**, 16–28.

Guidotti, G. G., Borghetti, A. F., Gaja, G., Loreti, L., Ragnotti, G., and Foà, P. P. (1968a). *Biochem. J.* **107**, 565–574.

Guidotti, G. G., Gaja, G., Loreti, L., Ragnotti, G., Rottenberg, A., and Borghetti, A. F. (1968b). *Biochem. J.* **107**, 575–580.

Guzman, N. A., Rojas, F. J., and Cutroneo, K. R. (1976). *Arch. Biochem. Biophys.* **172**, 449–454.

Haber, B., Kuriyama, K., and Roberts, E. (1970a). *Biochem. Pharmacol.* **19**, 1119–1136.

Haber, B., Kuriyama, K., and Roberts, E. (1970b). *Brain Res.* **22**, 105–112.

Haffen, K., and Cèdard, L. (1968). *Gen. Comp. Endocrinol.* **11**, 220–234.

Hajek, A. S., and Solursh, M. (1975). *Gen. Comp. Endocrinol.* **25**, 432–446.

Hall, B. K. (1972). *Anat. Rec.* **173**, 391–404.

Hall, B. K. (1973a). *Can. J. Zool.* **51**, 771–776.

Hall, B. K. (1973b). *Anat. Rec.* **176**, 49–64.

Hall, B. K., and Girouard, R. J. (1973). *Anat. Rec.* **177**, 343–358.

Hall, B. K., and Kalliecharan, R. (1975). *Teratology* **12**, 111–120.

Halme, J., Kivirikko, K. I., and Simons, K. (1970). *Biochim. Biophys. Acta* **198**, 460–470.

Hamburger, V., and Hamilton, H. L. (1951). *J. Morphol.* **88**, 49–92.

Hart, D. E., and Betz, T. W. (1972). *Dev. Biol.* **27**, 84–99.

Harwood, R., Durrant, B., Grant, M. E., and Jackson, D. S. (1975a). *Biochem. Soc. Trans.* **3**, 914–915.

Harwood, R., Grant, M. E., and Jackson, D. S. (1975b). *Biochem. Soc. Trans.* **3**, 916–917.

Harwood, R., Grant, M. E., and Jackson, D. S. (1976). *Biochem. J.* **156**, 81–90.

Hassell, J., and Klein, N. W. (1971). *Dev. Biol.* **26**, 380–392.

Heckey, R. P., and Owczarzak, A. (1972). *J. Cell Biol.* **55**, 110a.

Helfre, G., Farjanel, J., Fonvielle, J., and Frey, J. (1974). *Experientia* **30**, 995.
Heller, H., and Kulka, R. G. (1968a). *Biochim. Biophys. Acta* **165**, 393–397.
Heller, H., and Kulka, R. G. (1968b). *Biochim. Biophys. Acta* **167**, 110–121.
Hendrick, D., and Moller, F. (1973). *Comp. Biochem. Physiol. B* **45**, 197–211.
Herzfeld, A., and Raper, S. M. (1975). *Biol. Neonate* **27**, 163–176.
Hess, A. (1970). *Physiol. Rev.* **50**, 40–62.
Heywood, S. M., and Nwagwu, M. (1969). *Biochemistry* **8**, 3839–3845.
Heywood, S. M., Kennedy, D. S., and Bester, A. J. (1975). *FEBS Lett.* **53**, 69–72.
Hijmans, J. E., and McCarty, K. (1966). *Proc. Soc. Exp. Biol. Med.* **133**, 633–637.
Holdsworth, C. D., and Wilson, T. H. (1967). *Am. J. Physiol.* **212**, 233–240.
Houssaint, E., LeDouarin, N. M., Le Douarin, G., and Weaver, G. (1970). *C. R. Hebd. Seances Acad. Sci.* **271**, 1315–1318.
Howard, E. (1957). *J. Cell. Comp. Physiol.* **50**, 451–470.
Hsiao, C. Y. Y., and Ungar, F. (1969). *Proc. Soc. Exp. Biol. Med.* **132**, 1047–1051.
Huffer, W. E. (1970). *Calcif. Tissue Res.* **6**, 55–69.
Hwang, U. K. (1968). *Experientia* **24**, 683–684.
Igarashi, S., Trelstad, R. L., and Kang, A. H. (1973). *Biochim. Biophys. Acta* **295**, 514–519.
Incefy, G. S., and Kappas, A. (1971). *FEBS Lett.* **15**, 153–155.
Ingram, V. M. (1974). *Ann. N.Y. Acad Sci.* **241**, 93–98.
Irving, R. A., Mainwaring, W. I. P., and Spooner, P. M. (1976). *Biochem. J.* **154**, 81–93.
Israels, L. G., Schacter, B. A., Yoda, B., and Goldenberg, G. J. (1974). *Biochim. Biophys. Acta* **372**, 32–38.
Ito, Y. (1957). *Acta Embryol. Morphol. Exp.* **1**, 118–130.
Jansen, H. W., and Bornstein, P. (1974). *Biochim. Biophys. Acta* **362**, 150–159.
Jimenez, S. A., Dehm, P., and Prockop, D. J. (1971). *FEBS Lett.* **17**, 245–248.
Jo, J.–S., Ishihara, N., and Kikuchi, G. (1974). *Arch. Biochem. Biophys.* **160**, 246–254.
Johnson, J. L. (1972). *Brain Res.* **37**, 1–19.
Jones, H. S. (1970). *Am. J. Anat.* **127**, 89–100.
Jongkind, J. F., Corner, M. A., and Bruntink, R. (1972). *J. Neurochem.* **19**, 389–394.
Joshi, V. C., and Sidbury, J. B. (1975). *Dev. Biol.* **42**, 282–291.
Judes, C., and Jacob, M. (1973a). *Brain Res.* **51**, 253–267.
Judes, C., and Jacob, M. (1973b). *Brain Res.* **52**, 333–344.
Judes, C., Sensenbrenner, M., Jacob, M., and Mandel, P. (1973). *Brain Res.* **51**, 241–251.
Juurlink, B. H. J., and Gibson, M. A. (1973). *Can. J. Zool.* **51**, 509–519.
Kagen, L., Linder, S., and Gurevich, R. (1969). *Am. J. Physiol.* **217**, 591–595.
Katz, A. (1970). *Physiol. Rev.* **50**, 63–158.
Keeley, F. W., and Labella, F. S. (1972). *Connect. Tissue Res.* **1**, 113–120.
Kemp, D. J., and Rogers, G. E. (1972). *Biochemistry* **11**, 969–975.
Kilsheimer, G. S., Weber, D. R., and Ashmore, J. (1960). *Proc. Soc. Exp. Biol. Med.* **104**, 515–518.
Kim, S. U., Oh, T. H., and Johnson, D. D. (1975). *Neurobiology* **5**, 119–127.
Kivirikko, K. I., and Prockop, D. J. (1972). *Biochim. Biophys. Acta* **258**, 366–379.
Klicka, J., Edstrom, R., and Ungar, F. (1969). *J. Exp. Zool.* **171**, 249–252.
Knox, W. E., and Eppenberger, H. M. (1966). *Dev. Biol.* **13**, 182–198.
Koerker, D. J., and Fritz, I. B. (1970). *Can. J. Biochem.* **48**, 418–424.
Kojima, A., Sugimoto, M., and Endo, H. (1976). *Dev. Biol.* **48**, 173–183.
Konigsberg, I. R. (1954). *J. Exp. Zool.* **125**, 151–169.
Krnjevic, K. (1970). *Nature (London)* **228**, 119–124.
Kurihara, T., and Tsukada, Y. (1968). *J. Neurochem.* **15**, 827–832.

Kuriyama, K., Sisken, B., Ito, J., Simonsen, D. G., Haber, B., and Roberts, E. (1968). *Brain Res.* **11**, 412–430.
Kuroda, Y., and Nagatani, T. (1965). *Dev. Biol.* **11**, 335–351.
Kutchai, H., and Steen, J. B. (1971). *Respir. Physiol.* **11**, 265–278.
Laird, A. K. (1966). *Growth* **30**, 263–275.
Lambson, R. O. (1970). *Am. J. Anat.* **129**, 1–20.
Langslow, D. R. (1972). *Comp. Biochem. Physiol. B* **43**, 689–701.
Langslow, D. R. (1975). *Br. Poult. Sci.* **16**, 329–333.
Langslow, D. R., and Lewis, R. J. (1972). *Comp. Biochem. Physiol. B* **43**, 681–688.
Lawrenz, N. K., and Johnson, L. G. (1970). *J. Embryol. Exp. Morphol.* **24**, 65–71.
Lebherz, H. G. (1972). *Dev. Biol.* **27**, 143–149.
Lee, P. C., and Fisher, J. R. (1971). *Dev. Biol.* **25**, 149–158.
Lee, W. H. (1951). *Anat. Rec.* **110**, 465–474.
Leeson, T. S., and Leeson, C. R. (1963). *J. Anat.* **97**, 585–595.
Leibenguth, F. (1969). *Z. Naturforsch., Teil B* **24**, 117–122.
Levere, R. D., and Granick, S. (1967). *J. Biol. Chem.* **242**, 1903–1911.
Levere, R. D., Kappas, A., and Granick, S. (1967). *Proc. Natl. Acad. Sci. U.S.A.* **58**, 985–990.
Levi, G. (1972). *Arch. Biochem. Biophys.* **151**, 8–21.
Levi, G., and Lattes, M. G. (1969). *Brain Res.* **13**, 579–594.
Levi-Montalcini, R., and Cohen, S. (1960). *Ann. N.Y. Acad. Sci.* **85**, 324–341.
Lindgren, J., Vaheri, A., and Ruoslahti, E. (1974). *Differentiation* **2**, 233–236.
Linsenmayer, T. F., Trelstad, R. L., Toole, B. P., and Gross, J. (1973). *Biochem. Biophys. Res. Commun.* **52**, 870–876.
Litke, L. L., and Low, F. N. (1975). *Am. J. Anat.* **142**, 527–530.
Liuzzi, A., and Angeletti, P. O. (1964). *Experientia* **20**, 512–513.
Llewellyn-Smith, I. J. (1974). *Proc. Aust. Biochem. Soc.* **7**, 67.
Low, R. B., and Rich, A. (1973). *Biochemistry* **12**, 4555–4559.
Lundy, H. (1969). *In* "The Fertility and Hatchability of the Hen's Egg" (T. C. Carter and B. M. Freeman, eds.), pp. 143–176. Oliver & Boyd, Edinburgh.
McGeer, E. G., and Maler, L. (1975). *Dev. Biol.* **47**, 464–465.
McGeer, E. G., Maler, L., and Fitzsimmons, R. C. (1974). *Dev. Biol.* **38**, 165–174.
McWhinnie, D. J. (1975). *Comp. Biochem. Physiol. A* **50**, 169–175.
McWhinnie, D. J., and Thommes, R. C. (1973). *J. Embryol. Exp. Morphol.* **29**, 515–527.
Malpoix, P. (1967). *Biochim. Biophys. Acta* **145**, 181–184.
Manasek, F. J., Reid, M., Vinson, W. C., Seyer, J. M., and Johnson, R. (1973). *Dev. Biol.* **35**, 332–348.
Maraud, R., Coulaud, H., and Stoll, R. (1969). *C. R. Seances Soc. Biol. Ses Fil.* **163**, 2125–2126.
Marchaim, U., and Kulka, R. G. (1967). *Biochim. Biophys. Acta* **146**, 553–559.
Marchisio, P. C., and Consolo, S. (1968). *J. Neurochem.* **15**, 759–764.
Martinez-Vargas, M. C., Gibson, D. B., Sar, M., and Stumpf, W. E. (1975). *Science* **190**, 1307–1308.
Masaki, T. (1974). *J. Biochem. (Tokyo)* **76**, 441–449.
Masaki, T. (1975). *J. Biochem. (Tokyo)* **77**, 901–904.
Masaki, T., and Yoshizaki, C. (1972). *J. Biochem. (Tokyo)* **71**, 755–757.
Masaki, T., and Yoshizaki, C. (1974). *J. Biochem. (Tokyo)* **76**, 123–131.
Mason, T. L., and Hooper, A. B. (1969). *Dev. Biol.* **20**, 472–478.
Medda, J. N., and Das, A. K. (1972). *Acta Histochem.* **43**, 115–118.
Medoff, J. (1967). *Dev. Biol.* **16**, 118–143.

Medoff, J., and Zwilling, E. (1972). *Dev. Biol.* **28**, 138–141.
Meyerhof, P. G., and Haley, L. E. (1975). *Biochem. Genet.* **13**, 457–470.
Mezei, C., and Hu, Y.-W. (1972). *J. Neurochem.* **19**, 2071–2081.
Mezei, C., Newburgh, R. W., and Hattori, T. (1971). *J. Neurochem.* **18**, 463–468.
Misson, B. H., and Freeman, B. M. (1972). *Respir. Physiol.* **14**, 343–352.
Mondal, H., Sutton, A., Chen, V.-J., and Sarkar, S. (1974). *Biochem. Biophys. Res. Commun.* **56**, 988–996.
Moog, F. (1965). *In* "The Biochemistry of Animal Development" (R. Weber, ed.), Vol. 1, pp. 307–365. Academic Press, New York.
Moog, F., and Grey, R. D. (1966). *Biol. Neonat.* **9**, 10–23.
Moscona, A. A. (1971). *In* "Hormones in Development" (M. Hamburgh and E. J. W. Barrington, eds.), pp. 169–189. Appleton, New York.
Mottet, N. K., and Hall, H. E. (1966). *J. Exp. Zool.* **162**, 295–300.
Mottet, N. K., and Jensen, H. M. (1968). *Exp. Cell Res.* **52**, 261–283.
Murota, S.-I., and Tamaoki, B.-I. (1967). *Biochim. Biophys. Acta* **137**, 347–355.
Murphy, L., Harsch, M., Mori, T., and Rosenbloom, J. (1972). *FEBS Lett.* **21**, 113–117.
Murray, H. A. (1925). *J. Gen. Physiol.* **9**, 1–37.
Myllylä, R., Risteli, L., and Kivirikko, K. I. (1975). *Eur. J. Biochem.* **52**, 401–410.
Myllylä, R., Risteli, L., and Kivirikko, K. I. (1976). *Eur. J. Biochem.* **61**, 59–67.
Nace, G. W. (1953). *J. Exp. Zool.* **122**, 423–448.
Narayanan, A. S., Page, R. C., and Martin, G. R. (1974). *Biochim. Biophys. Acta* **351**, 126–132.
Narbaitz, R. (1972). *Rev. Can. Biol.* **31**, 259–267.
Narbaitz, R. (1974). *Calcif. Tissue Res.* **16**, 339–342.
Narbaitz, R. (1975). *Gen. Comp. Endocrinol.* **27**, 122–124.
Needham, J. (1931). "Chemical Embryology," 3 vols. Cambridge Univ. Press, London and New York.
Needham, J. (1942). "Biochemistry and Morphogenesis." Cambridge Univ. Press, London and New York.
Nelson, P. A., Yarnell, G. R., and Wagle, S. R. (1966). *Arch. Biochem. Biophys.* **114**, 543–546.
Newman, T. L., and Low, F. N. (1973). *Am. J. Anat.* **136**, 407–426.
Noble, R. C., and Moore, J. H. (1964). *Can. J. Biochem.* **42**, 1729–1741.
Noble, R. C., and Moore, J. H. (1967a). *Can. J. Biochem.* **45**, 627–639.
Noble, R. C., and Moore, J. H. (1967b). *Can. J. Biochem.* **45**, 949–958.
Noble, R. C., and Moore, J. H. (1967c). *Can. J. Biochem.* **45**, 1125–1133.
Nwagwu, M., and Nana, M. (1974). *Dev. Biol.* **41**, 1–13.
Nwagwu, M., and Stevens, E. (1970). *J. Cell Biol.* **47**, 149a–150a.
Obinata, T. (1969). *Arch. Biochem. Biophys.* **132**, 184–197.
Obinata, T., Yamamoto, M., and Maruyama, K. (1966). *Dev. Biol.* **14**, 192–213.
O'Hare, M. J. (1973). *J. Embryol. Exp. Morphol.* **29**, 197–208.
Oppenheim, R. W. (1973). *In* "Studies on the Development of Behavior and the Nervous System" (G. Gottlieb, ed.), Vol. 1, pp. 163–244. Academic Press, New York.
Orkin, R. W. (1973). *J. Cell Biol.* **59**, 255a.
Oulton, M., and Mezei, C. (1973). *Lipids* **8**, 235–238.
Owczarzak, A. (1971). *Exp. Cell Res.* **68**, 113–129.
Ozawa, E. (1972). *Biol. Bull.* **143**, 431–439.
Pagram, J. R., and Parson, I. C. (1974). *Proc. Aust. Biochem. Soc.* **7**, 98.
Pane, G., Becchetti, E., and Carinci, P. (1974). *Int. J. Biochem.* **5**, 579–583.
Parsons, I. C. (1970). *Steroids* **16**, 59–65.
Paterson, B., and Strohman, R. C. (1972). *Dev. Biol.* **29**, 113–138.

Pearse, A. G. E., la Ham, Q. N., and Janigan, D. T. (1963). *Folia Histochem. Cytochem.* **1**, 409–421.
Pellegrino, E. D., and Biltz, R. M. (1972). *Calcif. Tissue Res.* **10**, 128–135.
Peter, H. W., Wiese, F., Graszynski, K., and Ahlers, J. (1975). *Dev. Biol.* **46**, 439–445.
Peterkofsky, B. (1968). *Arch. Biochem. Biophys.* **128**, 637–645.
Piddington, R. (1967). *Dev. Biol.* **16**, 168–188.
Piddington, R. (1970). *J. Embryol. Exp. Morphol.* **23**, 729–737.
Piddington, R. (1971). *J. Exp. Zool.* **177**, 219–228.
Piddington, R. (1973). *J. Exp. Zool.* **184**, 167–176.
Piras, M. M., Staneloni, R. J., Leiderman, B., and Piras, R. (1972a). *FEBS Lett.* **23**, 199–202.
Piras, M. M., Staneloni, R. J., Hernandez, M. N., Leiderman, B., and Piras, R. (1972b). *In:* "Biochemistry of the Glycosidic Linkage: An Integrated View" (R. Piras and H. G. Pontis, eds.), p. 459. Academic Press, New York.
Pollak, J. K., and Shorey, C. D. (1968). *Dev. Biol.* **17**, 536–543.
Pontremoli, S., and Grazi, E. (1968). *In* "Carbohydrate Metabolism and Its Disorders" (F. Dickens, P. J. Randle, and W. J. Whelan, eds.), Vol. 1, pp. 260–295. Academic Press, New York.
Popenoe, E. A., and Aronson, R. B. (1972). *Biochim. Biophys. Acta* **258**, 380–386.
Potter, J. D., and Herrmann, H. (1970). *Arch. Biochem. Biophys.* **141**, 271–277.
Pratt, R. M., and Terner, C. (1971). *Biochim. Biophys. Acta* **225**, 113–122.
Przybylski, R. J. (1967). *Gen. Comp. Endocrinol.* **8**, 115–128.
Pugh, E., and Sidbury, J. B. (1971). *Biochim. Biophys. Acta* **239**, 376–383.
Purves, R. D., and Vrbová, G. (1974). *J. Cell. Physiol.* **84**, 97–100.
Rabinovitch, A. L., and Gibson, M. A. (1972a). *Teratology* **5**, 199–218.
Rabinovitch, A. L., and Gibson, M. A. (1972b). *Teratology* **6**, 51–70.
Radha, E. (1975). *Br. Poult. Sci.* **16**, 263–267.
Radha, E., and Krishnamoorthy, R. V. (1973). *Comp. Biochem. Physiol. B* **45**, 847–865.
Rahn, H., and Ar, A. (1974). *Condor* **76**, 147–152.
Ramachandran, S., Klicka, J., and Ungar, F. (1969). *Comp. Biochem. Physiol.* **30**, 631–640.
Rifkind, A. B. (1974). *Biochim. Biophys. Acta* **338**, 164–169.
Rinaudo, M. T. (1966). *Enzymologia* **31**, 325–332.
Rinaudo, M. T., and Bruno, R. (1968). *Enzymologia* **34**, 45–50.
Rinaudo, M. T., and Giunta, C. (1967). *Enzymologia* **33**, 201–210.
Rinaudo, M. T., and Passano, C. (1972). *Boll. Soc. Ital. Biol. Sper.* **48**, 513–516.
Rinaudo, M. T., and Ponzetto, C. (1972). *Boll. Soc. Ital. Biol. Sper.* **48**, 518–521.
Rinaudo, M. T., Giunta, C., Bozzi, M. L., and Bruno, R. (1968). *Enzymologia* **36**, 321–331.
Risteli, L., Myllylä, R., and Kivirikko, K. I. (1976). *Biochem. J.* **155**, 145–153.
Romanoff, A. L. (1967). "Biochemistry of the Avian Embryo." Wiley, New York.
Romijn, C., and Lokhorst, W. (1960). *J. Physiol. (London)* **150**, 239–249.
Rose, F., and Shrago, E. (1975). *Int. J. Biochem.* **6**, 73–77.
Rosenbloom, J., Harsch, M., and Jimenez, S. A. (1973). *Arch. Biochem. Biophys.* **158**, 478–484.
Rudnick, D., and Waelsch, H. (1955). *J. Exp. Zool.* **129**, 309–326.
Rutter, W. J., Woodfin, B. M., and Weber, C. S. (1963). *Adv. Enzyme Regul.* **1**, 39–56.
Ryman, B. E., and Whelan, W. J. (1971). *Adv. Enzymol. Relat. Areas Mol. Biol.* **34**, 285–443.
Saleuddin, A. S. M., Kyriakides, C. P. M., Peacock, A., and Simkiss, K. (1976). *Comp. Biochem. Physiol. A* **54**, 7–12.

Salvatorelli, G. (1967). *J. Embryol. Exp. Morphol.* **17**, 359–265.
Schalekamp, M., Schalekamp, M., Van Goor, D., and Slingerland, R. (1972). *J. Embryol. Exp. Morphol.* **28**, 681–713.
Schjeide, O. A. (1963). *Prog. Chem. Fats Other Lipids* **6**, 251–289.
Searls, R. L. (1965a). *Dev. Biol.* **11**, 155–168.
Searls, R. L. (1965b). *Proc. Soc. Exp. Biol. Med.* **118**, 1172–1176.
Sedláček, J. (1972). *Physiol. Bohemoslov.* **21**, 597–602.
Sedláček, J. (1975a). *Physiol. Bohemoslov.* **24**, 229–237.
Sedláček, J. (1975b). *Physiol. Bohemoslov.* **24**, 305–310.
Sedláček, J. (1975c). *Experientia* **31**, 1170.
Seltzer, J. L., and McDougal, D. B. (1975). *Dev. Biol.* **42**, 95–105.
Semba, T., Tolnai, S., and Bélanger, L. F. (1966). *Electron. Microsc., Proc. Int. Congr., 6th, 1966* pp. 569–570.
Seyer, J. M., and Vinson, W. C. (1974). *Biochem. Biophys. Res. Commun.* **58**, 272–279.
Sheid, B., and Hirschberg, E. (1967). *Am. J. Physiol.* **213**, 1173–1176.
Shimada, Y., Piddington, R., and Moscona, A. A. (1967). *Exp. Cell Res.* **48**, 240–243.
Shimizu, K. (1972). *Dev. Growth Differ.* **14**, 281–295.
Shulman, H. J., and Meyer, K. (1968). *J. Exp. Med.* **128**, 1353–1362.
Simkiss, K. (1974). *Endeavour* **23**, 119–123.
Smith, E. E., Taylor, P. M., and Whelan, W. J. (1968). *In* "Carbohydrate Metabolism and Its Disorders" (F. Dickens, P. J. Randle, and W. J. Whelan, eds.), Vol. 1, pp. 89–138. Academic Press, New York.
Solomon, J. B. (1958). *Biochem. J.* **70**, 529–535.
Spooner, P. M., and Mainwaring, W. I. P. (1973). *Acta Endocrinol. (Copenhagen), Suppl.* **177**, 181.
Šťastný, F. (1971). *Brain Res.* **25**, 397–410.
Šťastný, F. (1972). *Physiol. Bohemoslov.* **21**, 479–484.
Stewart, M. E., and Terepka, A. R. (1969). *Exp. Cell Res.* **58**, 93–106.
Stoll, R., Faucounau, N., and Maraud, R. (1972). *C. R. Seances Soc. Biol. Ses Fil.* **166**, 858–861.
Stoolmiller, A. C., Horwitz, A. L., and Dorfman, A. (1972). *J. Biol. Chem.* **247**, 3525–3532.
Strandholm, J. J., Cardenas, J. M., and Dyson, R. D. (1975). *Biochemistry* **14**, 2242–2246.
Strittmatter, C. F. (1972). *Biochim. Biophys. Acta* **284**, 183–195.
Sugimoto, M., and Endo, H. (1971a). *Endocrinol. Jpn.* **18**, 457–461.
Sugimoto, M., and Endo, H. (1971b). *J. Embryol. Exp. Morphol.* **25**, 365–376.
Suzuki, O., Nagase, F., and Yagi, K. (1975). *Brain Res.* **93**, 455–462.
Sze, P. Y., Kuriyama, K., Haber, B., and Roberts, E. (1971). *Brain Res.* **26**, 121–130.
Taylor, T. G. (1963). *Biochem. J.* **87**, 7P.
Taylor, T. G., and Lewis, P. E. (1972). *J. Endocrinol.* **52**, xlvii.
Taylor, T. G., Shires, A., and Baimbridge, K. G. (1975a). *Comp. Biochem. Physiol. A* **52**, 515–517.
Taylor, T. G., Balderstone, O., and Lewis, P. E. (1975b). *J. Endocrinol.* **66**, 363–368.
Tazawa, H., Mikami, T., and Yoshimoto, C. (1971). *Respir. Physiol.* **13**, 160–170.
Terepka, A. R., Stewart, M. E., and Merkel, N. (1969). *Exp. Cell Res.* **58**, 107–117.
Tezuka, Y., Yokoe, Y., and Yamagami, K. (1974). *Annot. Zool. Jpn.* **47**, 74–83.
Thiebold, J. J. (1973). *C. R. Seances Soc. Biol. Ses Fil.* **167**, 338–339.
Thommes, R. C., and Aglinskas, A. S. (1966). *Gen. Comp. Endocrinol.* **7**, 179–185.
Thommes, R. C., and Firling, C. E. (1964). *Gen. Comp. Endocrinol.* **4**, 1–8.
Thommes, R. C., and Just, J. J. (1964). *Gen. Comp. Endocrinol.* **4**, 614–623.
Thommes, R. C., and Just, J. J. (1966). *Endocrinology* **79**, 1021–1022.

Thommes, R. C., and Mathew, G. (1969). *Physiol. Zool.* **42,** 311–319.
Thommes, R. C., and Pall, J. S. (1974). *Gen. Comp. Endocrinol.* **23,** 52–57.
Thommes, R. C., McCarter, C. F., and Nguyen, L. H. (1968). *Physiol. Zool.* **41,** 491–499.
Thompson, W. C., Buzash, E. A., and Heywood, S. M. (1973). *Biochemistry* **12,** 4559–4565.
Thorogood, P. V. (1973). *J. Embryol. Exp. Morphol.* **30,** 673–679.
Traut, T. W., and Petruska, J. A. (1976). *Biochim. Biophys. Acta* **418,** 73–80.
Trayer, I. P., and Perry, S. V. (1966). *Biochem Z.* **345,** 87–100.
Tullett, S. G., and Board, R. G. (1976). *Br. Poult. Sci.* **17** 441–450.
Tunnicliff, G., and Kim, S. U. (1973). *Brain Res.* **49,** 410–416.
Vandecasserie, C., Paul, C., Schnek, A. G., and Léonis, J. (1973). *Comp. Biochem. Physiol. A* **44,** 711–718.
Vinson, W. C., and Seyer, J. M. (1974). *Biochem. Biophys. Res. Commun.* **58,** 58–65.
Visschedijk, A. H. J. (1962). Ph.D. thesis, State University of Utrecht.
von der Mark, H., von der Mark, K., and Gay, S. (1976). *Dev. Biol.* **48,** 237–249.
Vos, J., Schadé, J. P., and Van Der Helm, H. J. (1967). *Prog. Brain Res.* **26,** 193–213.
Vuust, J. (1975). *Eur. J. Biochem.* **60,** 41–50.
Waddell, W. J. (1972). *Biochem. Biophys. Res. Commun.* **49,** 127–132.
Wagner, C., Payne, N. A., and Briggs, W. (1969). *Exp. Cell Res.* **55,** 330–338.
Wainwright, S. D. (1974). *Can. J. Biochem.* **52,** 149–154.
Wainwright, S. D., and Wainwright, L. K. (1966). *Can. J. Biochem.* **44,** 1543–1560.
Wainwright, S. D., and Wainwright, L. K. (1967a). *Can. J. Biochem.* **45,** 255–265.
Wainwright, S. D., and Wainwright, L. K. (1967b). *Can. J. Biochem.* **45,** 344–347.
Wainwright, S. D., and Wainwright, L. K. (1967c). *Can. J. Biochem.* **45,** 1483–1493.
Wainwright, S. D., and Wainwright, L. K. (1972). *Can. J. Biochem.* **50,** 1143–1145.
Wang, K.–M. (1968). *Comp. Biochem. Physiol.* **27,** 33–50.
Wang, L., Andrade, H. F., Silva, S. M. F., Simões, C. L., D'Abronzo, F. H., and Brentani, R. (1975). *Prep. Biochem.* **5,** 45–57.
Warshaw, J. B. (1972). *Dev. Biol.* **28,** 537–544.
Weber, R., ed. (1965). "The Biochemistry of Animal Development," Vol. 1. Academic Press, New York.
Weber, R., ed. (1967). "The Biochemistry of Animal Development," Vol. 2. Academic Press, New York.
Weller, E. M. (1966). *Proc. Soc. Exp. Biol. Med.* **122,** 264–268.
Wenger, E., Wenger, B. S., and Kitos, P. A. (1967). *J. Exp. Zool.* **166,** 263–270.
Weniger, J.-P., and Zeis, A. (1973). *Ann. Embryol. Morphog.* **6,** 219–228.
Weniger, J.–P., Mack, G., and Holder, F. (1975). *C. R. Hebd. Seances Acad. Sci., Ser. D* **280,** 1889–1891.
Werner, I., Peterson, G. R., and Shuster, L. (1971). *J. Neurochem.* **18,** 141–151.
Williams, J. (1967). *In:* "The Biochemistry of Animal Development" (R. Weber, ed.), Vol. 2, pp. 341–382. Academic Press, New York.
Willier, B. H. (1968). *Wilhelm Roux' Arch. Entwicklungmech. Org.* **161,** 89–117.
Wilson, J. R., and Allenspach, A. L. (1974). *Dev. Biol.* **41,** 288–300.
Wilt, F. H. (1968). *Biochem. Biophys. Res. Commun.* **33,** 113–118.
Wilt, F. H. (1974). *Ann. N.Y. Acad Sci.* **241,** 99–112.
Woessner, J. F., Bashey, R. I., and Boucek, R. J. (1967). *Biochim. Biophys. Acta* **140,** 329–338.
Wood, R. (1972). *Lipids* **7,** 596–603.
Yalovsky, U., Zelikson, R., and Kulka, R. G. (1969). *FEBS Lett.* **2,** 323–326.
Yarnell, G. R., Nelson, P. A., and Wagle, S. R. (1966). *Arch. Biochem. Biophys.* **114,** 539–542.
Zwilling, E. (1951). *Arch. Biochem. Biophys.* **33,** 228–242.

CHAPTER 4

Feather Keratins

Alan H. Brush

I. Introduction

Birds are often defined as "vertebrates with feathers." Feathers, in turn, consist of keratins, a special class of molecules characterized by their toughness, insolubility in most common solvents, and resistance to enzymatic digestion (Mercer, 1966). The molecules which form feathers are more than simple homopolymers and obtain their unique chemical properties from both their composition and macromolecular organization (Brush, 1974, 1975). Definitions of the term keratin have evolved along with our understanding of their chemical composition and physical organization (Fraser *et al.*, 1972). Keratins occur in a variety of avian tissues including the feathers, the skin and the beak. The frequently cited occurrence of Keratin in the gizzard is probably incorrect (Webb and Colvin, 1964). Although information continues to become available for a variety of tissues and structures, the remarks here will emphasize the feathers. This reflects their uniqueness in birds as well as the availability of the preponderance of information.

In this chapter I will discuss selected aspects of the chemistry and organization of feather keratins, the nonproteinaceous products, and some chemical aspects of development. The possible relationships of feathers to other epidermal structures, relationships between structure and function, and possible evolutionary relationships of the keratin structures of birds will be indicated where appropriate.

The terminology of the keratins has a long and often confusing history (Fraser *et al.*, 1972). Differences which provide the basis for the various classifications are basically structural but may reflect differences in synthetic mechanisms or the immediate chemical history of the particular preparation. The most comprehensive classification relies on X-ray diffraction patterns. Categorization according to morphological location has been attempted but is confusing. For purposes of this paper, I will follow the classification used by Fraser *et al.* (1972). The keratins are listed as α, β, or feather-types based on their high angle X-ray diffraction pattern. This approach emphasizes the uniqueness of the complex feather pattern. Thus, while the feather-keratin pattern has some characteristics in common with the β-keratin formed mechanically from α-keratin (Rogers and Filshie, 1963) the feather pattern is more complex. Keratins from reptilian scales have a similar X-ray diffraction pattern (Fraser *et al.*, 1972) and are designated feather-type (Baden *et al.*, 1974). Structural rather than chemical similarities are emphasized by this approach. The homologies at the molecular levels among these proteins have not yet been established but are currently an area of active interest (A. H. Brush, unpublished). All types of epidermal keratins are produced by keratinocytes.

In addition to an understanding of their chemical nature, other problems regarding the organization of feathers may be considered. These include the amount of genetic information necessary to produce the structural complexity at both the molecular and morphological levels, the nature and origins of the associated molecular heterogeneity, and the evolutionary history of the subunit structure. It is important to know how a single population of keratinocytes produces the array of proteins present and how the processes are controlled in both time and space. Finally there are the problems of the mechanisms by which the chemical structures are transformed into visible morphological differences, the relationship between structure and function, and the phylogenetic relationships of the various keratin-containing epidermal structures.

II. Feather Keratins

A. Chemistry

1. *Compositional Analysis*

Amino acid compositional analysis of contour feathers and feather parts, down, and scales of several gallinaceous birds is available (Crewether *et al.*, 1965; Kemp and Rogers, 1972; Busch, 1970). Feather keratins are relatively high in glycine, serine, proline, leucine, and glutamic acid. Acidic amino acids exceed the basic residues, and the overall pattern differs quantitatively from other fibrous proteins. When compared with mammalian hair keratins, feathers have relatively lower methionine and lysine and a higher proline content. The amino acid composition of feather parts (e.g., calamus, shaft, barbs, and medulla) are demonstratively different (Schroeder and Kay, 1955; Harrap and Woods, 1967; O'Donnell, 1973a; Kemp and Rogers, 1972; Busch, 1970). Preliminary statistical analysis (Brush, 1974) suggested that the composition of the various feather parts was significantly different as predicted from the mobility differences in the electrophoretic patterns of the solubilized proteins. Thus, the mobility differences are a result of structural differences rather than polymerization or other mechanical factors. Further, the difference between feather parts of a single feather (e.g., barbs and shafts) was greater than between the homologous parts in different species. The differences among species are limited to relatively few residues, mostly alanine, cystine, glycine, isoleucine, proline, and tyrosine. This implies that subunit structure is under genetic control and selection has determined the subunit distribution of the structural elements. The heterogeneity of the solubilized proteins (Section IIA,4) is a result of their production by a number of closely related genes. Feather keratins may constitute a multigene family (Hood *et al.*, 1975). These observations suggest selection for a protein structure suited to some particular function (Seifter and Gallop, 1966; Fraser *et al.*, 1972).

Information on the composition of fractionated feather protein subunits is available. The techniques of separation and amino acid analysis differ among the laboratories. Busch (1970) used ion-exchange chromatography on DEAE-cellulose eluting with a NaCl gradient, while Kemp and Rogers (1972) used preparative polyacrylamide gel electrophoresis (PAGE). Woods (1971) attempted fractionation of rachis SCM proteins of several species on DEAE-cellulose columns (pH = 7.6, KCl gradient elution) and O'Donnell (1973a) employed DEAE columns with Tris-EDTA buffers at pH 8.4 with 8.0 *M* urea. In no case was chromatographic separation complete and the overlap of peaks required additional steps, although O'Donnell's technique appeared the most satisfactory. The inability of a single step to separate all the keratin subunits was verified by electrophoretic comparison of the products (Brush, 1974). Data from different laboratories are also difficult to compare because of uncertainties regarding the identity of the actual subunits involved in amino acid analysis. Bands are usually identified by mobility differences, but without an internal standard or adequate comparisons it is impossible to establish the absolute identity of the various bands. Nevertheless, it is instructive to compare some of the compositional analysis of fractionated feather keratins. The Difference Index (DI) developed by Metzger *et al.* (1968) which appears to be linearly related to sequence differences (Fondy and Holohan, 1971) is useful (Table I).

The SCM components of feather keratins are probably individual products of a group of related genes (Kemp and Rogers, 1972; Brush, 1975). Determination of the homologies within this group is still imperfect. The elements in Table I were arranged according to their mobilities on PAGE in alkaline buffers. Normally, this is the reverse of the elution sequence of the same components from a DEAE column using salt gradient elution under neutral or alkaline pH buffer conditions. The analysis of amino acid composition by Metzger DI (Table IA) indicates that, within species, the differences are equivalent to those calculated for immunoglobin chains. Since the absolute differences are also small, generalities must be made cautiously. However, it is interesting to note that differences between subunits of the emu feather were smaller than in either of the gallinaceous species. This may reflect differences in feather morphology. All permutations of comparison among bands were not attempted and selection was arbi-

TABLE I
DIFFERENCE INDEX (METZGER *et al.*, 1968) OF FRACTIONATED FEATHER SCM PROTEINS

A. Intraspecific comparison[a]

Organism	Band/DI				References
Chicken	A_4 vs B_1	B_1 vs B_2	B_1 vs B_4	A_4 vs B_4	
	2.0	20.7	11.6	11.4	Kemp and Rogers, 1972
Emu	4–1	4–2	2–1		
	3.6	3.6	6.5		O'Donnell, 1973b
Turkey	VI–III	VI–II	III–II		
	9.3	9.4	5.8		Busch, 1970

B. Interspecific comparison[b]

	Chicken–turkey	Chicken–emu	Turkey–emu
High CMC	6.3	10.5	11.8
Low CMC	8.2	13.1	11.4

[a] Subunits arranged by decreasing mobility in PAGE. Nomenclature after original authors.

[b] Units defined by CMC content (see text).

trary, based simply on the magnitude of the mobility difference. This reflects compositional changes which favor charge differences. To compensate, two bands with adjacent mobility were compared. In some cases these gave larger differences implying that compositional differences can occur with only a minimal change in charge.

Interspecific comparisons were based on an operational definition of protein homology (Table IB). The calculation of the DI was based on the CM-cystine composition of fractions. The lowest and highest subunit was chosen for each species. The DI in both cases indicates the chicken and turkey proteins were more alike than either was to the emu. Further, the average difference between the two gallinaceous species and the emu were essentially identical. These data support the proposition that the feather keratins have undergone rather conservative evolution and are products of genes which shared a common ancestor. Compositional data also indicate similarity among feather and other epidermal structures (Kemp and Rogers, 1972; Brush, 1970).

Support of another type comes from immunological studies. Antisera made in rabbits to solubilized SCM proteins of chicken, duck, or rhea feathers react in double-diffusion experiments with SCM-keratins from many other orders and with each of the fractionated components (Brush, 1975). Preliminary quantitative microcomplement fixation data suggest close immunological distances among these molecules (A. H. Brush, unpublished).

2. *N-Terminal Sequences*

N-Terminal residues have been determined for feather keratins of relatively few species. No extensive comparisons of the N-terminal residues from subunits of the feather parts within a single species are available. Busch (1970) reported that the terminal residue from turkey calamus was heterogeneous, acetylated, and consisted of serine and CM-cystine. These residues were thought not to be repeated in the terminal sequence. In the goose *(Anser domesticus)* calamus O'Donnell (1971) reported only a single residue, Ac-serine. In the N-terminal sequence the next residues were CM-cystine and tyrosine. The same sequence occurred in goose rachis and emu calamus (O'Donnell, 1973b), but not in the Silver Gull *(Larus novae hollandiae)*. The differences in turkey and goose are presumably technical in nature. In the barbs of adult chicken feather and the embryonic chicken downy plumage the N-terminal residue is also Ac-serine. However, there is heterogeneity in the subsequent sequence. D. J. Kemp (in letter, 1972) has found either Phe or Tyr at position 3 and Asn at position 4. Further unpublished data from his laboratory indicate that as many as six N-terminal tryptic peptides occur in *Gallus* feathers. This would support the idea of different sequences among the SCM subunits. As data of this nature accumulate it will be possible to eliminate conclusively other possible sources of variability such as posttranslational modification, translational error, and conformational change.

3. *Peptide Mapping*

Maps of tryptic digests of the SCM derivatives of emu feather have been published (O'Donnell, 1973a). Separation of peptides on Sephadex G-50 columns followed by high voltage paper electrophoresis produced a large number of spots. In general, maps of calamus and rachi resembled each other more than did the maps from the barb extract. The differences may be greater in the end peptides of individual bands than in the center one (e.g., T_3 of O'Donnell's system). Differences were noticeable between the corresponding parts from different species. The bands of the calamus SCM extract of the gull and emu differed in their map patterns. Taken together these data support the hypothesis that each of the SCM units was a unique gene product and that the distribution of subunits differs among feather parts and between species.

4. *Amino Acid Sequence*

Two feather keratin monomers have been sequenced completely (O'Donnell, 1973b; O'Donnell and Inglis, 1974). These were band 3, from the emu and Silver Gull (Fig. 1). The SCM fraction, isolated by ion exchange on a DEAE column yielded five tryptic peptides. The sequences are reported here in full to encourage comparison with subsequent sequences.

No sequence data are available for other bands in the emu, although sequence differences are to be expected based on compositional analysis (Table I; Brush, 1974, 1975). Available sequence data provide a basis for only preliminary analyses which are listed below:

1. When compared to the peptides of turkey calamus produced by acid hydrolysis (Schroeder *et al.*, 1957) as many as 65/102 residues or 63.7% of the sequence may be in common. The DI based on the compositional analysis was approximately 15.
2. The chicken and emu may have a common sequence for at least the first 8 residues, but there were 4 differences between the emu and gull sequences.
3. Sequence comparison. Bands 3 from emu rachis and Silver Gull calamus were sequenced completely by O'Donnell and Inglis (1974).

```
                                                            T1
                         5                                  10                    15
Silver gull  NAc-Ala-Cys-Asn-Asp*-Leu-Cys-Thr/Gly-Pro-  -  -  -Cys-Gly-Pro-Thr-Pro-Leu-Ala-Asn-Ser-Cys-

Emu          NAc-Ser-Cys-Tyr-Asn-Pro-Cys-Leu-Pro-Arg-Ser-Ser-Cys-Gly-Pro-Thr-Pro-Leu-Ala-Asn-Ser-Cys-

                                               T2
              20                  ↓25      *       30↓                  35
Silver gull  Asn-Glu-Pro-Cys-Val-   -Arg-Gln-Cys-Glu-Ala-Ser-Arg-Val-Val-Ile-Gln-Pro-Ser-Thr-Val-

Emu          Asn-Glu-Pro-Cys-Leu-Phe-Arg-Gln-Cys-Gln-Asp-Ser-Thr-Val-Val-Ile-Glu*-Pro-Ser-Pro-Val-

                                               T3
              40                  45                  50                  55
Silver gull  Val-Val-Thr-Leu-Pro-Gly-Pro-Ile-Leu-Ser/Thr-Ser-Phe-Pro-Gln-Ser-Thr-Ala-Val-Gly-Gly-Ser-

Emu          Val-Val-Thr-Leu-Pro-Gly-Pro-Ile-Leu-Ser-Ser-Phe-Pro-Gln-Asn-Thr-Val-Val-Gly-Gly-Ser-

              60                  65                  70                  75
Silver gull  Ala-  -Ser-Ser/Ala-Ser/Ala-Val-Gly-Asn-Glu-Leu-Leu-Ala-Ser-Gln-Gly-Val-Pro-Tyr/Ile-Phe-Gly/Ser-Gly-

Emu          Ser-Thr-Ser-Ala-Ala-Val-Gly-Ser-   -Ile-Leu-Ser-Ser-Gln-Gly-Val-Pro-Ile-Ser-Ser-Gly-

                                                                    T4
              80                  85                  90    ↓              95
Silver gull  Gly/Tyr-Phe-Gly-Leu-Gly-Gly-Leu-   -Gly-Cys-Phe/Tyr-Ser-Gly-Arg-Arg-Gly-Cys-Tyr-Pro-Cys and 0

Emu          Gly-Phe-Asn-Leu-Ser-Gly-Leu-Ser-Gly-Arg-Tyr-Ser-Gly-Ala-Arg-   -Cys-Leu-Pro-Cys
```

FIG. 1. Comparison of the complete amino acid sequences of band 3 from Silver Gull feather calamus with that of the emu rachis (O'Donnell and Inglis, 1974).

Assuming that the electrophoretically similar bands were homologous, the sequences differed at about one-sixth of their residues (Fig. 1). Almost all the differences occurred in the end peptides, the same peptides where the cystine residues occurred. The crystalline, hydrophobic central peptide appeared to be structurally conservative. Several conclusions and speculations follow from these data: (a) The minor individual bands associated with the SCM-FKM units probably differ by only small amounts, perhaps simply in the number of amide groups. No sequence data are available to document the differences in the major bands within a species. (b) Sequence differences and distribution of residues between major, presumably homologous, SCM-FKM elements of various species is large enough to suggest that it is the result of selection. (c) The β-crystalline configuration may be limited to the internal one-third of the SCM-FKM molecule. This portion may be limited to the site of intermolecular recognition (Fraser, 1976).

4. The distribution of proline residues was not regular. This con-

firms the analysis of Busch (1970). The sequence data do not support the prediction of Schor and Krimm (1961) of a regular spacing or proline residues as predicted from X-ray structural analysis. However, the proline distribution is associated with aspects of the monomeric structure (Fraser, 1976).

5. The distribution of the cystine residues is not random. SCM-cystines were absent from peptide T_3 (64 residues) and consistently found in the end peptides T_1, T_2, T_4, and T_5 (38 residues). Because the subunits aggregate under conditions where the cystines are blocked, subunit recognition must occur by an alternate mechanism, perhaps located on the interior (T_3) peptide. However, the interactions of the disulfide linkages are probably responsible for stabilization of adjacent subunits during keratinogenesis.

6. Distribution of other amino acid residues. Because of potential differences in solubility, the polarity of individual amino acids may determine their distribution within a molecule (Fisher, 1964; Vogel and Zuckerkandl, 1972). The relative number of nonpolar groups, residues with limited solubility in the outer shell, will thus affect the overall size of a particular molecule (Gates and Fisher, 1971). From considerations such as polarity, volume, presence or absence of side groups, it was possible to establish correlations between composition, structure, and function in selected molecules (Sneath, 1966; Klapper, 1971). The relationship of molecular structure and the occurrence of amino acids in the helical and randomly coiled portions of a variety of proteins has been analyzed (Chou and Fasman, 1974a) and its usefulness in predicting the secondary structure of globular proteins evaluated (Chou and Fasman, 1974b). The role of specific residues in the determination of structure may be modified by a variety of other factors, and Robinson (1974) suggested that secondary modifications, such as deamination of specific residues may serve as a regulatory process. None of these techniques have been applied to feather keratins.

7. Repeating sequences. Although several double residues exist there were no Cys–Cys sequences in either species. This occurs in the α-keratins, ribonuclease, and insulin, but in few other molecules (Dayhoff, 1972). No apparent regular repetition of sequence has been detected in the keratins of contour feather, down or scale.

The influence of composition on structure and function in proteins is obviously a complex interaction of sequence and environment. Much of the available data for keratins were reviewed and discussed by Fraser and MacRae (1973). The amino acid composition and sequence must interact to establish the mechanical and thermal properties of the molecules. The composition obviously affects conformation (Anfinsen, 1973). However, no adequate mechanism to predict one from the other currently exists (Nagano, 1973, 1974; Chou and Fasman, 1974a,b). Amino acid composition must reflect the frequencies with which the twenty or so common residues arise by mutation. The incorporation of each potential change is a function of the chemical nature of the residue produced by the mutated codon and the selective forces acting on the particular molecule. The result is the evolution of a molecule uniquely suited to a particular function.

5. *Solubilized SCM Subunits*

One productive approach to the study of feather keratins has been the preparation of *S*-carboxymethyl (SCM-FKM) derivatives (Harrap and Woods, 1964). This technique (Gurd, 1972) is effective in solubilization and has proven reproducible in several laboratories. Solubilized proteins are studied subsequently by electrophoresis, ion-exchange chromatography, and other physical methods. The solubilized SCM extracts produce a large number of subunit components (Woods, 1971). The SCM subunits or components are charge isomers and differ in their primary chemical structure. The structural heterogeneity of the subunits, and their presumed genetic difference, was established by biochemical and immunological means (Brush, 1974). A theoretical approach to their evolution, function, and relations to structure has been presented (Brush, 1975, 1976a).

Alternative chemical methods for solubilization (Nagai and Nishikawa, 1970) are not as successful as carboxymethylation. The insolubilized feather is completely refractory to digestion by trypsin and chymotrypsin. There are no reports of the effectiveness of keratinases on feathers. Trypsin will digest the SCM derivatives. Cyanogen bromide cleavage (Witkop, 1968) is not useful due to the absence of methionine. A chemical approach using sodium hydrazide to cleave at proline residues has been developed (Busch, 1970).

The individual SCM-FKM components have a molecular weight of approximately 11,000 (Table II). On PAGE a large number of bands appear which presumably reflect differences in amino acid structure (Fig. 2). The patterns are useful in studies of feather structure (Brush, 1972), development (Kemp and Rogers, 1972), taxonomy (Brush, 1976b), and studies of synthetic mechanisms (Parkington *et al.*, 1973). Comparison of PAGE patterns involves scoring the presence and absence of bands. The number and the sharpness of individual bands is often increased by incorporation of urea into the gel. The sizes of the

TABLE II
SIZE ESTIMATES OF FEATHER KERATIN SCM COMPONENTS

Technique	Size	Reference
1 Sedimentation coefficient	11,000	Harrap and Woods, 1964
2 Osmotic pressure	10,400	Jeffery, 1969
3 Urea gel	10,500	Jeffery, 1970
4 SDS gels	11,000	Brush, 1974

subunits are reported to vary from 9000 to 12,000. A small amount of polymer of molecular weight of 37,000 to 40,000 has been reported (Busch, 1970). The origin and nature of this complex which occurs spontaneously in solution (Harrap and Woods, 1964) and its relation to the formation of the microfibrillar structure of the feather are not clear.

The SCM subunits prepared by ion-exchange chromatography or large-scale electrophoresis have proved to be a basic tool for a variety of approaches to studies on feather keratins and other epidermal structures. In their now classic work, Harrap and Woods (1964, 1967) provided the basic technical approach, and established some of the elemental observations regarding the distribution of patterns in feath-

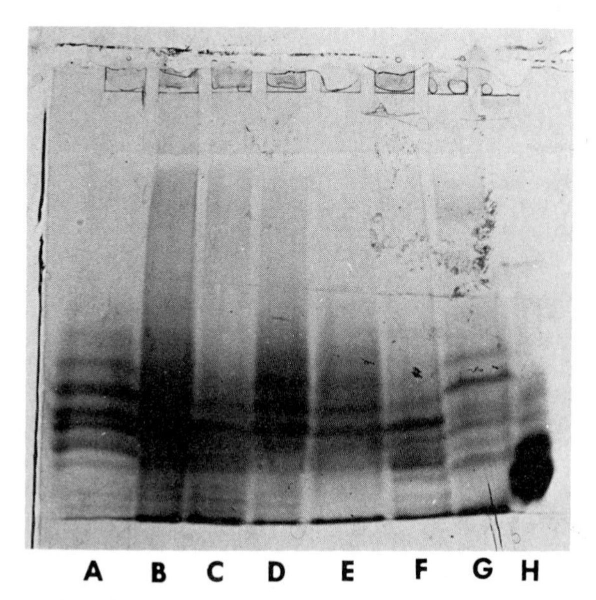

FIG. 2. Polyacrylamide gel electrophoretic patterns of SCM-keratins from feathers of birds representing several orders. (A) *Gallus;* (B) *Anas;* (C) *Anas;* (D) *Zenaidura;* (E) *Larus;* (F) *Aythya;* (G) *Turdus;* (H) blood serum.

ers, feather parts, and among taxonomic groups. This was extended in an analysis of the PAGE patterns of feather parts in the mutant and normal fowl (Brush, 1972). The differences in the mutants appeared to be one of organizational control rather than in specific synthetic mechanisms. The relationship of the SCM bands in feather structure has been studied immunologically (Brush, 1974) and sequence studies are currently underway (O'Donnell, 1973b; O'Donnell and Inglis, 1974). The mechanisms by which the components interact to form larger structures or react with pigment in growth are largely unexplored (Section II,C,2). The differences in PAGE pattern and immunological relationships of SCM proteins of down, scales, and skin are presently an area of interest (Brush, 1976a). Developmental aspects of the SCM components are discussed in Section II,C.

SCM subunits are considered the elemental units of feather structure at the molecular level. Their organization into larger, functional components is controlled genetically and the capacity exists for the production of a large variety of structural and functional forms (Section II,C,2). The degree of sequence difference among the monomer units within a species and between species are essentially unknown. The SCM-keratin associated with various epidermal structures in mammals, e.g., hair, quills, hoof, horn, baleen, fingernails, etc., also have complex and characteristic PAGE patterns (Gillespie, 1972). The components have a molecular weight comparable to that found for feathers (Table II) and are rich in glycine and tyrosine residues. The diversity of molecular structure and the resultant morphological characteristics are related to the proportion of the low-sulfur, high-sulfur, and high-tyrosine fractions present (Gillespie and Frenkel, 1974).

B. Organization

The influence of amino acid composition on molecular conformation was mentioned in Section II,A,4. At present it is not possible to classify the individual amino acid residues in regard to their contribution to the structure of a given molecule (Fraser and MacRae, 1973), although significant attempts are being made (Chou and Fasman, 1974a,b; Nagano, 1974). The structural organization of keratins has been explored by the techniques of X-ray crystallography but with little reference to compositional data or electrophoretic heterogeneity. The crystallographic pattern of feather keratin is one of the most complex of all biological molecules (Marwick, 1931; Fraser *et al.*, 1972). The complexity reflects both the structure of individual molecules and their three-dimensional interactions (Astbury and Marwick, 1932).

The solubilized feather keratins form spontaneously into a fibrillar structure from concentrated solution or when dried in thin films (Filshie and Rogers, 1962; Harrap and Woods, 1964). These units, microfibrils about 35 Å in diameter, have X-ray diffraction properties which resemble the original feather material. Several types of fibrous units have been obtained with feather SCM keratins which can be distinguished by X-ray diffraction patterns and fibrillar size (reviewed in Fraser *et al.*, 1972). Since the products are the result of interactions of roughly similar subunits, the observed differences in some cases may be due only to the conditions of polymerization.

The X-ray diffraction patterns of the intact feather calamus indicate a high degree of crystallinity which can be deformed by stretching (Astbury and Marwick, 1932) and is affected by heating in water or aqueous butanol (Bear and Rugo, 1951). The native conformation is almost certainly the β-pleated sheet configuration (Pauling and Corey, 1951; Dickerson and Geis, 1969) as suggested by IR absorption (Ambrose and Elliott, 1951; O'Donnell, 1973a). A number of complex models have been proposed for the molecular organization of feather keratins (Fraser *et al.*, 1972). The most thorough analysis takes into account X-ray and IR spectral data, and microscopic and chemical analysis to derive the final model (Rogers and Filshie, 1963; Fraser *et al.*, 1971; Fraser, 1976). The model consists of microfibrils which have two strands of antiparallel β-sheets twisted about a fourfold screw axis (Fig. 3). The absolute composition and orientation of individual polypeptide units is still not resolved. One attractive suggestion is that each pleated sheet unit is a single molecule and equivalent to one SCM subunit. The remaining non-β, two-thirds of the protein also has a native conformation capable of producing a rich low angle X-ray pattern. The chains loop in a serpentine fashion to form the pleated sheet while the non-β portions lie external to the microfibril framework and produce the amorphous matrix. The nonpleated portions may be the site of the intermolecular disulfide bonding.

X-ray diffraction patterns may vary among the various morphological portions of the feather (Rudall, 1947), although the majority of modeling has been based on studies of the calamus. The most orderly structure is found in the shaft, while other areas produce patterns of less clarity (Razumov *et al.*, 1959). This corresponds with the area of greatest structural strength. The relations between molecular structure and function have not been systematically explored. The implication is that structural changes involve a reorganization and possible redistribution of the polypeptide subunits. Often this reorganization is a reduction in the number of units which lie parallel to the long axis of

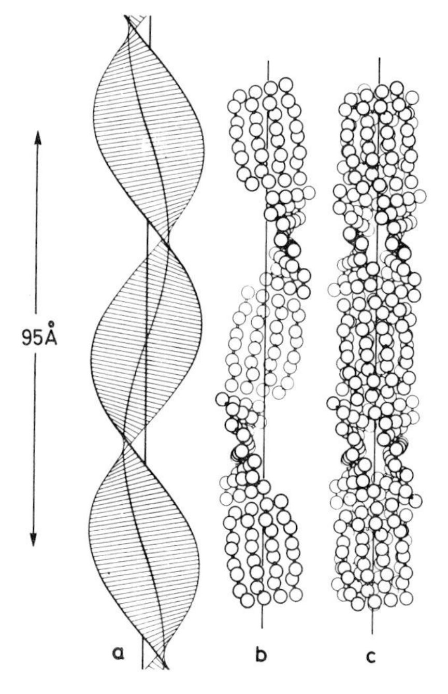

FIG. 3. Suggested molecular model for feather keratin (Fraser *et al.*, 1971). (a) Left-handed helical ruled surface of pitch 95 Å; (b) pleated sheet constructed of SCM units organized as dyad into microfibrils, strands of microfibrils conform to helical surface; (c) completed model consisting of two sets of complementary pleated sheets.

the particular feather part, but may also involve changes in cross-linkage of subunits. Combinations of such differences are reported for barbs and in homologous portions of contour and flight feathers (Razumov *et al.*, 1959). Differences in the same part between species were less marked than among parts in the same feather with different structure from the same species. This correlates with, and supports, the conclusions reached from amino acid composition analysis (Brush, 1974). No direct functional correlations have been made from these structural differences. Further, the presence of such differences in SCM subunit distribution should be visible in PAGE, but this hypothesis has not been thoroughly tested. Thus, although the complexity of the crystalline structure of feather keratin has been known for some time and probably modeled accurately, there is little interpretative material available regarding structural and functional relationships at this level.

The production and distribution of feather protein subunits is under genetic control and their organization influences the gross morphology of the feather (Brush, 1972). The relationship between genetic and molecular phenotypes is still not clear and represents a potentially productive area for investigation. The nature of the epidermal structures and their development are characterized by specific PAGE patterns (Kemp and Rogers, 1972). Presumably the pattern differences represent structural differences at all levels, e.g., amino acid sequences, SCM subunit, and microfibrillar organization. The sequence homologies of proteins in various structures of individual feathers, in the feathers of different birds, and among other epidermal structures have not been established. The problem might be approached through immunological techniques which supplement those of direct biochemical analysis. Studies of this sort could provide information regarding the amounts and nature of genetic information necessary to form the complex structural units from the simple, but heterogeneous, SCM subunits. Demonstration of the evolutionary relationships among various epidermal structures in birds and potentially between birds and reptiles is wanting. Similar data may be informative in systematic studies and in various aspects of developmental biology.

C. Production

The production of feather proteins occurs in several phases. The distinctions are not always completely clear and overlap in the timing and occurrence of the different processes certainly exists. The important processes include the synthesis of the keratin molecules, the chemical processes of keratinization, the attendant cytochemical changes, and the morphological changes associated with the development of individual structural components.

1. Biogenesis

The feather keratins are synthesized by special cells in the follicle termed keratinocytes. There are some differences in detail in the developmental sequence of embryonic down and skin (Matulionis, 1970) and some reports on the microstructural aspects may not consider keratin formation as a separate entity (Kischer, 1963). Nevertheless, despite the relative paucity of information a general pattern has emerged. The process of the development of epidermis in birds differs from both reptiles and mammals in some details (Fraser *et al.*, 1972). Based mainly on studies of the chicken it appears that little protein synthesis occurs in the presumptive epidermal structure before the thirteenth–fourteenth day of incubation. Polyribosomes, which consist of four ribosomal units, are present but remain nonfunctional (Bell *et al.*, 1965). The chemical analysis and X-ray diffraction of proteins present just prior to this point are unlike those in the adult (Bell and Thatachari, 1963). The production of keratin lags behind the appearance of morphological evidence of feather formation and does not begin until morphological development is essentially complete. The possible role of a lipoprotein framework in keratin development (Bell and Thatachari, 1963), based on changes in the X-ray patterns of the proteins induced by treatment with lipid solvents, has not been resolved. Subsequent work by Rogers and Dyer (cited in Fraser *et al.*, 1972) suggested that the effects were due to other sources such as degeneration of mitochondria, but evidence has not been presented.

The increase in production of keratin at 13–14 days is dramatic and the populations of SCM subunits change following a well-ordered sequence (Kemp and Rogers, 1972; Kemp *et al.*, 1974a). All major SCM-keratin subunits are present at this point. In embryonic feather the process is characterized by increased ribosomal activity, a decrease in water content, an increase in the S/N ratio of the proteins, the appearance of immunologically detectable feather proteins, increased incorporation of cystine, and the appearance of protofibrillar structure (Malt and Bell, 1965). Autoradiographic evidence indicates that before keratin synthesis begins, the cells lose the ability to synthesize DNA, but RNA synthesis continues. In the early stages of synthesis mitochondrial dehydrogenases are active (Ellis, 1964). Synthesis is presumably initiated by *N*-acetylserine which is the only N-terminal residue found in these proteins (Section II,A,2). The effects of other metabolites such as vitamin A (Fitton-Jackson and Fell, 1963) and various hormones on the production of keratins are reviewed by Fraser *et al.* (1972).

The microfibrils aggregate rapidly, especially under conditions of increased ionic strength, and assume the dimensions (Filshie and Rogers, 1962) and structural characteristics (Bell and Thatachari, 1963) of adult feather proteins. Specific PAGE patterns of the SCM subunits are retained (Kemp and Rogers, 1972).

An mRNA, which produced keratin products in a cell-free system indistinguishable from the native product, has been isolated (Parkington *et al.*, 1973). This 12 S RNA appeared to be translated with a high degree of fidelity in a cell-free system. The mRNA has a molecular weight of 250,000 and presumably contains long untranslated segments (Kemp *et al.*, 1974b). However, the available data do not account for the observation of electrophoretic heterogeneity of the SCM proteins as extracted from either the feather or produced artificially (Brush, 1974). Subsequent studies demonstrate a single phase of keratin synthesis, where the major proteins are present in similar relative amounts in all stages of development (D. J. Kemp *et al.*, 1974a). This is unlike the condition in hair or wool and as proposed previously for feathers (Malt and Bell, 1965).

The process at the biochemical level associated with the development of other structures in birds is less well known. In juvenile wing, feather and down, protein production occurs concurrently with increases in RNA activity in the sheath cells, alkaline phosphatase activity, and metachromic acid mucopolysaccharide levels (Koning and Hamilton, 1952; Koning, 1957). The functional roles of all the histochemical changes are not clear, but suggest a glycogen energy source for the synthetic activity (Koning, 1957; Matulionis, 1970). It is not possible to distinguish whether the glycogen is deposited intra- or extracellularly. The metabolism of mammalian, and presumably avian, skin is relatively low and may rely more on anaerobic pathways than on the TCA cycle. Energy production in mammalian skin may be from the oxidation of fatty acids, but comparative data are not available for birds.

2. *Folding and Assembly*

The production of SCM subunits of the keratins follows a well-defined sequence. Their organization into the characteristic filaments and subsequent organization into fibrillar units is less well understood. Folding of the filament is primarily a function of energetic considerations modified by specific sequence characteristics and the ionic environment. The formation of fibers must also be the product of complex physical, chemical and thermal interactions. However, little direct experimental data are available on these processes in birds.

After the initiation of synthesis the keratinocytes undergo a series of histochemical and morphological changes associated with the process of keratinization (Spearman, 1966; Fraser *et al.*, 1972). In the epidermis a matrix of filaments and amorphous ground substance appears in the cell periphery while lipids are concentrated in the center. The plasma membrane thickens and multigranular bodies of unknown function appear. Keratohyaline is present, but few keratin filaments are produced in these cells (Matoltsy, 1969). In feather keratinocytes the organization into microfibers probably coincides closely with the onset of synthesis. As growth continues the filaments enlarge and become associated with the plasma membrane. Eventually the fibers coalesce and the remaining cellular organelles are resorbed (Matulionis, 1970; Filshie and Rogers, 1962).

The problem of the identity of a keratin precursor is not yet resolved. Various investigations have suggested that the keratin fibers arise from either the tonofibrils (desmosomal tonofilaments) or keratohyaline bodies, both of which are preexisting entities in the cell. A third alternative is that the nature, origin, and composition of the precursor is still unknown and perhaps unique to the feather keratinocytes. More comparative data would be invaluable in attempts to resolve this problem as it has not been demonstrated convincingly that the ultrastructural events are the same in all keratinizing tissues.

Another area of interest are the mechanisms which determine the postsynthetic organization of keratin in the different feather parts. Not only will this determine many of the mechanical properties of the resultant feather, but the spacing of the fibers and the relations of keratinized and nonkeratinized spaces are important in certain types of color production (Dyck, 1971). Among mammalian keratin structures the functional characteristics may be determined by the arrangement and degree of cross-linkage among the various types of SCM subunits (Gillespie and Frenkel, 1974).

3. *Keratinization*

The process of keratinization has interested biologists for a long period. Aside from its intrinsic interest to biochemists and developmental biologists the avian integument provides an almost ideal experimental system for investigations of complex interactions of tissues in time and space. The general process of keratinization in vertebrates includes both biochemical and cytological correlates (Spearman, 1966; Flaxman, 1972). Histochemical studies indicate accumulation of appreciable amounts of RNA and alkaline phosphatase in the developing follicle. This is accompanied by a general dehydration of the tissue, and a shift from acidophilic to basophilic conditions. The latter is presumably the result of the increased RNA levels. In an extensive series of studies, Hamilton and his students (reviewed in Hamilton, 1966) concluded that alkaline phosphatase was of primary importance in feather organization, growth and differentiation. Most chemicals which block development were shown to inhibit this enzyme. The interactions of alkaline phosphatase with the mitochondria has broad implications and can affect both the developmental and synthetic processes. They also argue that enzyme levels explain, in part, the effects of some hormones on the tissues.

In general outline the process of keratinization in birds resembles that of other amniotic vertebrates, most closely the scaled skin of reptiles and mammalian tails (Cane and Spearman, 1967; Flaxman,

1972). The differences in the biochemistry and phenotypic expression in the system are, to a great degree, presumably the result of reorganization of control or regulatory pathways rather than the evolution of new systems or capacities (Brush, 1972; Fraser *et al.*, 1972).

Ultrastructural aspects of keratinization in chicks has recently received attention (Matoltsy, 1969; Matulionis, 1970). The process involves the movement of cells towards the outside of the epidermis accompanied by the appearance of lipid droplets and "multigranular" bodies. Subsequently the appearance of presumptive keratohyaline granules is associated with the degeneration of the cellular organelles which leaves only intracellular filaments, lipids, keratinohyaline, and a thickened plasma membrane. There is still not complete agreement on all the phases and processes involved (Matulionis, 1970). Fully keratinized cells in the lower stratum corneum simply contain a centralized lipid mass in a fibrous cortex. The cells of the outermost layer are thin and lose the lipid component. Details of the process are found in Fraser *et al.* (1972) and Lucus and Stettenhiem (1972). In addition to the changes within the cell at the tissue level the spacial organization of the epidermal structures are critical and probably function independently of other aspects of feather formation (Linsenmayer, 1972; Dhouailly, 1973; Novel, 1973). Keratinization in penguins resembles histologically and histochemically the processes in *Gallus*. Yet mechanisms must exist to account for the differences in feather structure, molting pattern, and for some of the adaptations for the highly aquatic existence characteristic of the group (Spearman, 1969).

Structural changes associated with keratin production are accompanied by biochemical changes. At least two processes occur, keratin synthesis and enzymatic cytolysis of cellular components (Spearman, 1966). The process of keratinization in developing feathers may (Matulionis, 1970) or may not (Fraser *et al.*, 1972) resemble those of mammalian cells. Biochemical changes are associated with structural changes in cellular organization. The early appearance of fibers and microtubules are correlated and Fraser *et al.* (1972) reviewed the general features of keratinocyte metabolic activity.

The formation of disulfide bonds by the cystine residues in feather keratins is of paramount importance in keratinization. The factors which control the initiation and sequence of linkage formation are not known. The possibility of the role of a "disulfide interchange" enzyme (Fuchs *et al.*, 1967) has not been examined. Disulfide bridges presumably link the SCM subunits forming the amorphous portion of the keratin macromolecular complex. The recognition site for monomer aggregation appears to be the internal pleated sheet portion.

4. Morphogenesis

The study of the morphological development of feathers has a long and distinguished history (Lillie, 1942; Hamilton, 1952; Rawles, 1960; Lucus and Stettenhiem, 1972). Classical studies which employed the light microscope and cytochemical approaches have been supplemented by ultrastructural work with the electron microscope (Kischer, 1963, 1968; Matulionis, 1970). The problems of the next higher level of organization, spatial distribution of the feathers on the skin surface, has also received attention (Sengel, 1971). The orientation of the early feather germs, which appear significantly in advance of the synthesis of feather specific proteins, is determined through interactions of the early feather germs with extracellular collagen and anchor filaments (Wessells and Evans, 1968). However, the forces which specify the orientation are not known. Experiments with tissue explants of both feather and scale forming tissue indicate that both the temporal sequence and spatial arrangements are controlled by the skin. The pattern does not require a specific "initiator row" indicating that isolated skin is capable of autonomous pattern formation. Nor was there evidence for a latent pattern. These data imply a high degree of independence of spatial arrangement from other processes (Linsenmayer, 1972). However, all the evidence regarding specific initiators is not in agreement (Fraser *et al.*, 1972). Rawles (1972) recently demonstrated tract specific difference in structure of down feathers. Further, tract morphology is often specific for various taxonomic groups (Clench, 1970). It is apparent that the developmental processes of the epidermis is under rather strong influence of the dermis, modified by its intrinsic capacity for specific responses (Fraser *et al.*, 1972; Linsenmayer, 1972).

Differentiation and morphogenesis of the epidermal structures is influenced by hormones (Voitkevich, 1966). Hormones capable of affecting general processes such as metabolic rate may affect both the morphological and biochemical events in development. For example, thyroxine levels are critical to normal morphological development (Wessells, 1961) and the chemical processes of keratinization (Lawrenz and Johnson, 1970). The relationship of morphology and pigment chemistry has been explored thoroughly (Strong, 1902; Watterson, 1942), and the observations have been extended to the ultrastructural level (Matulionis, 1970). The morphological observations reflect the biochemical changes typical of the epidermal cell as it becomes increasingly specialized and metabolically restricted. As mentioned previously much of the morphological change and arrangement into patterns precedes biochemical differentiation (Section II,B). In addition to the production of specific cellular products, the cell may also undergo modifications relevant to supportive metabolic process, i.e., energy production. However, no direct experimental evidence is available (Kischer, 1963; Matulionis, 1970) and the significance of all the changes in the cell membrane has not been established.

D. Nonproteinaceous Materials

Feathers contain a variety of nonproteinaceous molecules in addition to the keratins. These include the various pigments and adventitious molecules (Chapter 5). A small number of polypeptides, thought to resemble the insoluble keratins, and at least 16 amino acids have been identified in aqueous washes of feathers. In addition, polysaccharides, nucleotide breakdown products (adenine, guanine, xanthine, and uric acid), glucose, and ribose may be present (Gross, 1956). Much of this material is thought to be the by-product of the denuclearization of the epidermal cells associated with the process of keratinization. A lipid component which includes considerable quantities of cholesterol has been identified (Bolliger and Varga, 1960). These molecules, mainly stearates and free steroid, are also found in the blood and appear to be associated with keratinization. Since the relation to the molt cycle has not been carefully studied, one presumes that lipoids are associated with the general process of the replacement of the skin rather than specificially with the production of feathers in the follicular tissue. These products appear in several of the avian orders and there seems to be no direct correlation with the ecological relationships of the species (see Chapter 6). Thus, most authors eliminate waterproofing as a possible function of this class of molecules.

The waxes, alcohols, and fatty acids of the preen gland can be distinguished from the products of keratinization and have been isolated and characterized (see Jacob, Chapter 6 for review). The products differ in composition and relative abundance among related species and may have taxonomic significance. The gland is capable of synthesizing a variety of aliphatic waxes, cholesterol esters, triglycerides, and phospholipids. Composition of the product, in some species, is affected by diet. The functional roles of the preen gland secretion probably include waterproofing, maintenance of flexibility of the feathers and adsorption of particulate matter.

A series of heavy metals, especially mercury, has been reported in bird feathers. The incidence of mercury has undergone an apparent increase in parallel with the increased utilization of alkylated mercurial compounds in agriculture and various manufacturing processes

(Berg *et al.*, 1966). Mercury levels vary among species and are dependent on the position in the food chain. The distribution of other minerals has been used to indicate the recent ecological history of certain bird species (Hanson and Jones, 1968). This approach, using both wet chemical and spectrometric techniques for analysis, has been extended in attempts to trace the geographic origins of waterfowl (Kelsall and Calaprice, 1972). A variety of other techniques including neutron activation, emission spectrophotometry, and X-ray spectrometry have been applied to similar problems. Significant quantitative differences in a number of elements (Zn, Fe, Ca, K, P, and Cu) occur among species. Although still in a preliminary stage this approach may prove valuable in studies on mineral metabolism, aid in the localization of populations, and in species identification.

ACKNOWLEDGMENTS

Work on the structure and organization of feather keratins has been supported by grants from the National Science Foundation and the University of Connecticut Research Foundation.

REFERENCES

Ambrose, E. J., and Elliott, A. (1951). *Proc. R. Soc. London, Ser. A* **206**, 206.
Anfinsen, C. B. (1973). *Science* **181**, 223.
Astbury, W. T., and Marwick, T. C. (1932). *Nature (London)* **130**, 309.
Baden, H., Sviokla, S., and Roth, I. (1974). *J. Exp. Zool.* **187**, 287.
Bear, R. S., and Rugo, H. J. (1951). *Ann. N.Y. Acad. Sci.* **53**, 627.
Bell, E., and Thatachari, Y. T. (1963). *J. Cell Biol.* **16**, 215.
Bell, E., Humphreys, T., Slayter, H. S., and Hall, C. E. (1965). *Science* **148**, 1739.
Berg, W., Johnels, A., Sjöstrand, B., and Westermark, T. (1966). *Oikos* **17**, 71.
Bolliger, A., and Varga, D. (1960). *Aust. J. Exp. Biol.* **38**, 265.
Brush, A. H. (1972). *Biochem. Genet.* **7**, 87.
Brush, A. H. (1974). *Comp. Biochem. Physiol.* **48**, 661.
Brush, A. H. (1975). *Isozymes* **4**, 912.
Brush, A. H. (1976a). *Proc. Int. Ornithol. Congr., 16th, 1975*, 402.
Brush, A. H. (1976b). *J. Zool.* **179**, 467.
Busch, N. E. (1970). Ph.D. Dissertation, Iowa State University, Ames (University Microfilms, Ann Arbor, Michigan).
Cane, A. K., and Spearman, R. I. C. (1967). *J. Zool.* **153**, 337.
Chou, P. Y., and Fasman, G. D. (1974a). *Biochemistry* **13**, 211.
Chou, P. Y., and Fasman, G. D. (1974b). *Biochemistry* **13**, 222.
Clench, M. H. (1970). *Auk* **87**, 650.
Crewether, W. G., Fraser, R. D. B., Lennox, F. G., and Lindley, H. (1965). *Adv. Protein Chem.* **20**, 191.
Dayhoff, M. O., ed. (1972). "Atlas of Protein Sequence and Structure," Vol. 5. National Biomed. Res. Found., Washington, D.C.
Dhouailly, D. (1973). *J. Embryol. Exp. Morphol.* **30**, 587.
Dickerson, R. E., and Geis, L. (1969). "Structure and Action of Proteins." Harper, New York.

Dyck, J. (1971). *Z. Zellforsch. Mikrosk. Anat.* **115,** 17.
Ellis, R. A. (1964). *In* "The Epidermis" (W. Montagna and W. C. Lobitz, eds.), Chapter 8. Academic Press, New York.
Filshie, B. K., and Rogers, G. E. (1962). *J. Cell Biol.* **13,** 1.
Fisher, H. F. (1964). *Proc. Natl. Acad. Sci. U.S.A.* **51,** 1285.
Fitton-Jackson, S., and Fell, H. B. (1963). *Dev. Biol.* **7,** 394.
Flaxman, B. A. (1972). *Am. Zool.* **12,** 13.
Fondy, T. P., and Holohan, P. D. (1971). *J. Theor. Biol.* **31,** 229.
Fraser, R. D. B. (1976). *Proc. Int. Ornithol. Congr., 16th, 1975,* 443.
Fraser, R. D. B., and MacRae, T. P. (1973). "Conformation in Fibrous Proteins." Academic Press, New York.
Fraser, R. D. B., MacRae, T. P., Perry, D. A. D., and Suzuki, E. (1971). *Polymer* **12,** 35.
Fraser, R. D. B., MacRae, T. P., and Rogers, G. E. (1972). "Keratins. Their Composition, Structure and Biosynthesis." Thomas, Springfield, Illinois.
Fuchs, S., De Lorenzo, F., and Anfinsen, C. B. (1967). *J. Biol. Chem.* **242,** 398.
Gates, R. E., and Fisher, H. F. (1971). *Proc. Natl. Acad. Sci. U.S.A.* **68,** 2928.
Gillespie, J. M., (1972). *Comp. Biochem. Physiol.* **41B,** 723.
Gillespie, J. M., and Frenkel, M. J. (1974). *Comp. Biochem. Physiol.* **47B,** 339.
Gross, R. (1956). *Aust. J. Exp. Biol.* **34,** 65.
Gurd, F. R. N. (1972). *In* "Methods in Enzymology" (C. H. W. Hirs and S. N. Timasheff, eds.), Vol. 25, Part B, Chapter 34a. Academic Press, New York.
Hamilton, H. L. (1952). "Lillie's Development of the Chick," 3rd ed. Holt, New York.
Hamilton, H. L. (1966). *Adv. Biol. Skin* **8,** 313.
Hanson, H. C., and Jones, R. L. (1968). *Ill. Nat. Hist. Surv., Biol. Notes* **60.**
Harrap, B. S., and Woods, E. F. (1964). *Biochem. J.* **92,** 8.
Harrap, B. S., and Woods, E. F. (1967). *Comp. Biochem. Physiol.* **20,** 449.
Hood, L., Campbell, J. H., and Elgin, S. C. R. (1975). *Annu. Rev. Genet.* **9,** 305.
Jeffery, P. D. (1969). *Biochemistry* **8,** 5217.
Jeffery, P. D. (1970). *Aust. J. Biol. Sci.* **23,** 809.
Kelsall, J. P., and Calaprice, J. R. (1972). *J. Wildl. Manage.* **36,** 1088.
Kemp, D. J., and Rogers, G. E. (1972). *Biochemistry* **11,** 969.
Kemp, D. J., Dyer, P. Y., and Rogers, G. E. (1974a). *J. Cell Biol.* **62,** 114.
Kemp, D. J., Partington, G. A., and Rogers, G. E. (1974b). *Biochem. Biophys. Res. Comm.* **60,** 1006.
Kischer, C. W. (1963). *J. Ultrastruct. Res.* **8,** 305.
Kischer, C. W. (1968). *J. Morphol.* **125,** 185.
Klapper, M. H. (1971). *Biochim. Biophys. Acta* **229,** 557.
Koning, A. L. (1957). *Am. J. Anat.* **100,** 17.
Koning, A. L., and Hamilton, H. C. (1952). *Am. J. Anat.* **95,** 75.
Lawrenz, N. K., and Johnson, L. G. (1970). *J. Embryol. Exp. Morphol.* **24,** 65.
Lillie, F. R. (1942). *Biol. Rev. Cambridge Philos. Soc.* **17,** 247.
Linsenmayer, T. F. (1972). *Dev. Biol.* **27,** 244.
Lucus, A. M., and Stettenhiem, P. (1972). *U.S., Dep. Agric., Handb.* No. 362.
Malt, R. A., and Bell, E. (1965). *Nature (London)* **205,** 1081.
Marwick, T. C. (1931). *J. Text. Sci.* **4,** 31.
Matoltsy, A. G. (1969). *J. Ultrastruct. Res.* **29,** 438.
Matulionis, D. H. (1970). *Z. Anat. Entwicklung Gesch.* **132,** 107.
Mercer, E. H. (1966). *Adv. Biol. Skin* **9,** 556.
Metzger, H., Shapiro, M. B., Mosimann, J. E., and Vinton, J. E. (1968). *Nature (London)* **219,** 1166.
Nagai, Y., and Nishikawa, T. (1970). *Agric. Biol. Chem.* **34,** 1676.
Nagano, K. (1973). *J. Mol. Biol.* **75,** 420.

Nagano, K., (1974). *J. Mol. Biol.* **84**, 357.
Novel, G. (1973). *J. Embryol. Exp. Morphol.* **30**, 605.
O'Donnell, I. J. (1971). *Aust. J. Biol. Sci.* **24**, 179.
O'Donnell, I. J. (1973a). *Aust. J. Biol. Sci.* **26**, 401.
O'Donnell, I. J. (1973b). *Aust. J. Biol. Sci.* **26**, 414.
O'Donnell, I. J., and Inglis, A. S. (1974). *Aust. J. Biol. Sci.* **27**, 369.
Parkington, G. A., Kemp, D. J., and Rogers, G. E. (1973). *Nature (London), New Biol.* **246**, 33.
Pauling, L., and Corey, R. B. (1951). *Proc. Natl. Acad. Sci. U.S.A.* **37**, 251.
Rawles, M. E. (1960). *In* "Biology and Comparative Physiology of Birds" (A. J. Marshall, ed.), Vol. 1, Chapter 6. Academic Press, New York.
Rawles, M. E. (1972). *Proc. Natl. Acad. Sci. U.S.A.* **69**, 1136.
Razumov, L. L., Lemazhikhin, B. K., Lebedev, L. A., and Pen'kina, V. S. (1959). *Dokl. Akad. Nauk SSSR* **128**, 186.
Robinson, A. B. (1974). *Proc. Natl. Acad. Sci. U.S.A.* **71**, 885.
Rogers, G. E., and Filshie, B. K. (1963). *In* "Ultrastructure of Protein Fibers" (R. Borasky, ed.), p. 123. Academic Press, New York.
Rudall, K. M. (1947). *Biochim. Biophys. Acta* **1**, 549.
Schor, J., and Krimm, S. (1961). *Biophys. J.* **1**, 489.
Schroeder, W. A., and Kay, L. M. (1955). *J. Am. Chem. Soc.* **77**, 3901.
Schroeder, W. A., Kay, L. M., Munger, N., Martin, N., and Balog, J. (1957). *J. Am. Chem. Soc.* **79**, 2769.
Seifter, S. and Gallop, P. M., (1966). *In* "The Proteins." (H. Neurath, ed.), 2nd ed., Vol. 4, p. 155. Academic Press, New York.
Sengel, P. (1971). *Adv. Morphog.* **9**, 181.
Sneath, P. H. A. (1966). *J. Theor. Biol.* **12**, 157.
Spearman, R. I. C. (1966). *Biol. Rev. Cambridge Philos. Soc.* **41**, 59.
Spearman, R. I. C. (1969). *Z. Morphol. Tiere* **64**, 361.
Strong, R. M. (1902). *Bull. Mus. Comp. Zool.* **40**, 147.
Vogel, H., and Zuckerkandl, E. (1972). *Proc. 6th Berkeley Symp. Math. Stat. Prob., 1971*, Vol. 5, 155.
Voitkevich, A. A. (1966). "Feathers and Plumage of Birds." Sidgwick & Jackson, London.
Watterson, R. L. (1942). *Physiol. Zool.* **15**, 234.
Webb, T. E., and Colvin, J. R. (1964). *Can. J. Biochem.* **42**, 59.
Wessells, N. K. (1961). *Exp. Cell Res.* **24**, 131.
Wessells, N. K., and Evans, J. (1968). *Dev. Biol.* **18**, 42.
Witkop, B. (1968). *Science* **162**, 318.
Woods, E. F. (1971). *Comp. Biochem. Physiol.* **39A**, 325.

Note added in proof. The following papers have appeared subsequent to the preparation of this manuscript.

A. Structure of avian keratins.

Akahane, K., Murozono, S., Murayama, K. (1977). *J. Biochem.* **81**, 11.
Murayama, K., Akahane, K., and Murozono, S. (1977). *J. Biochem.* **81**, 19.
Walker, D., and Bridgen, J. (1976). *Eur. J. Biochem.* **67**, 283.
Walker, D., and Rogers, G. E. (1976). *Eur. J. Biochem.* **69**, 329.

B. Bird beak proteins.

Frenkel, M. J., and Gillespie, J. M. (1976). *Aust. J. Biol. Sci.* **29**, 467.

CHAPTER 5

Avian Pigmentation

Alan H. Brush

I. Introduction

Birds combine a number of physical and biological features which, when taken together, are unique. They exist in three-dimensional space and are often characterized by an exceptional degree of mobility both as individuals and as species. Birds are diurnal and rely heavily on the visual senses for orientation and communication. These characteristics are shared with the higher primates and enhance their attractiveness to humans. The ability of birds to fly and their gregariousness reinforces this relationship. The colors and plumage patterns of birds are often bright and striking. As a result, birds have attracted the attention of man since time immemorial.

In addition to the characteristics of flight, diurnal activity, and conspicuousness, birds maintain an intense level of metabolic activity, are highly seasonal in their reproductive periods and, in some species, engage in mass migratory movements. Each function interacts during the seasonal cycle and represents changes in specific activity, faunal composition, and behavior which can be identified by human observers. It is one thesis of this chapter that many of the diverse seasonal activities of birds are related to the plumage color and patterns of individuals and species. Such relationships may seem obvious and superficial, but upon reflection one realizes that colors play a surprisingly diverse and large number of roles in the biology of birds. What is axiomatic to the behaviorist may be a challenge to the biochemist and simple genetic information may have important taxonomic or ecological consequences.

The thrust of this chapter is to categorize the common biochromes of birds with special regard to their function. This is now feasible in light of the extensive information available on their general biochemistry. The major chemical pigments are limited to a few categories (Section II), and a considerable portion of avian color is derived from nonpigment sources (Section V). The physical basis of chemical colors is reviewed by Needham (1974). The adaptive processes are diverse, reflect various levels of control and often involve complex interrelations of structure and function. Nevertheless pigments may illustrate important evolutionary processes and provide an almost ideal system for examination of the biological roles of relatively simple molecular systems.

To the human observer, the most conspicuous functional roles of avian colors are in identification, communication, and protection. Each species of bird, and often each sex, is recognizable by unique combinations of color and pattern. This provides taxonomists and field workers with a standard means for identification and categorization. Even "primitive" societies have sophisticated visual identification systems for the animal species with whom they share the environment (Berlin, 1973). This is not surprising when one considers the interdependence of the two. The number of recognized avian species corresponds closely with the number recognized by professional ornithologists and this holds even in areas with extremely complex avifauna (Diamond, 1966). Identification from the birds' viewpoint is critical in intraspecific and mate recognition. This represents but one of the general functions of animal communication.

Interspecific communication involves many of the same elements. Combinations of colors and patterns are used to initiate a number of functions, define space, and establish individuality. Each plays an important role in an individual's function and in the maintenance of species integrity.

In addition to the color in feathers, various fleshy body parts are used in display. The behavioral role may be supplemented by a physiological one. In some closely related congeneric species (e.g., Boobies, warblers) leg and foot color varies and may therefore have a social function. The legs may also have a thermoregulatory function. The selection of the particular color used in display is not completely understood. The spectra characteristics of the avian eye are similar to those of mammals (Sillman, 1973) and presumably colors are perceived in the same way. But it is necessary to distinguish between the conspicuous nature of a particular display pattern and the color involved. Of course, the possibility of multiple functions exists also. In a conspicuous display of a small body area (e.g., legs, crown, nape, or wing bars), one might ask if the specificity of color is relevant to the display itself. The nature of the selective forces that determine whether red, yellow, blue, or white will appear in any given species is perhaps irrelevant ethologically, but may have metabolic constraints (or vice versa). The factors that determine visual effectiveness of display colors are not known.

The conspicuous visual effects of colors and their attendant patterns may be supplemented by other functions. For the melanins, these include increased radiation protection, thermoregulation, and others (Hadley, 1972). Specialized functions are reflected in their distribution. There are other, less conspicuous but perhaps no less important, functions of the carotenoids. Vitamin A, a carotenoid derivative, participates in the visual process. Lipid droplets which contain carotenoid pigments are important in color vision (Sillman, 1973). Carotenoids may be involved in the reproductive process, but their function in yolk remains unknown (Goodwin, 1954; Needham, 1974). Imbalances in vitamin A levels can produce diverse pathological conditions (Dingle and Lucy, 1965).

The obvious necessity to balance selective forces of the function and appearance of the integument and its related structures is not always appreciated. The complex phenotypic structures may be the result of a surprisingly few, closely related gene products whose distribution in time and space are controlled by a number of related regulatory genes (Chapter 4, this volume; Brush, 1975). It must be necessary for each species to evolve a strategy which balances the protective (either mechanical or metabolic) functions with other, visual requirements such as crypticity or communication. By extension, consideration of such problems, even for a single species, leads ultimately to broader biological considerations. The role of pigmentation in thermoregulation and energy metabolism may be incompatible with the requirements of signal functions. Brightly colored or highly patterned species, with the obvious implications of inter- and intraspecific communication, require an ecological situation with a certain degree of visibility. By the same token, the most conspicuous state of the individual (e.g., males in alternate plumage) must be relatively free of predation. Similarly, there is a high selective pressure on cryptic colors in eggs and fledgling young to render them less visible. This may preclude other possible functions of a pigmented integument. One such function, certainly in selected environments, would be protection from incident solar radiation. One solution to this dilemma has been different functions of pigments as compared to the skin or internal organs (for recent review, see Hamilton, 1973).

The colors of individual birds are not distributed uniformly throughout a given population. Seasonal changes, sexual dichromatism and marked polymorphism exist in many populations (see Section III). The adaptative nature of colors has been studied in various vertebrates (see Nevo, 1973; Webster and Burns, 1973). Seasonal plumage changes are well known and often closely associated with reproductive, migratory, or other major features of the annual cycle. At least ten specific functions can be attributed to species-specific plumage characters (Hamilton and Barth, 1962). Functions as diverse as correlations with the degree of environmental stability and reduction of interspecific and intraspecific hostility have been proposed.

II. Pigments

Avian pigments fall into three general chemical categories: melanins, carotenoids, and porphryins. The discussion here is limited to those aspects of their biology which have recently attracted the attention of avian biologists.

A. Melanins

The melanins are distributed widely among plants, animals, fungi, and bacteria. Their role in the biology of the skin, function in integumental derivatives, association with the nervous system (Cowell and Weston, 1970) and various neoplasms have given them wide interest in biology and medicine. The synthesis of melanins from the amino acid tyrosine is well known and is interconnected with a number of other important biochemical pathways (reviewed by Jimbow *et al.*, 1976). The most important intermediate in this respect is dopaquinone, the product of the action of tyrosinase on dopa. Other physiological active molecules such as epinephrine and norepinephrine share this precursor.

The broad distribution of melanin among organisms has been given special significance by comparative biochemists. Unlike the other biopolymers typical of living forms (e.g., proteins, polysaccharides, and polynucleotides) which are characteristically synthesized by a dehydration condensation mechanism, melanins are cross-linked by a nonenzymatic process involving free radicals (Kenyon and Steinman, 1969). This may represent a primitive type of synthetic mechanism. The degree of melanization in organisms is extremely variable and depends on the life cycle stage, racial characteristics, and selective pressures (Kittlewell, 1973).

Melanins are found in every order of birds and chemically resemble those present in reptiles (Lubow, 1963). Avian melanocytes, like those of other vertebrates, have their embryonic origin in the neural crest cells (Rawles, 1948). However, some workers differentiate among dermal, mesenteric, and epidermal types on the basis of locality (Hamilton, 1940). The general biology of the melanocyte was reviewed by Bagnara and Hadley (1973) and specific aspects of melanophore structure and function in birds are found in Lucus and Stettenheim (1972). One scheme proposed for the interrelations of melanin-containing cells is that of Mottaz and Zelickson (1967) shown below.

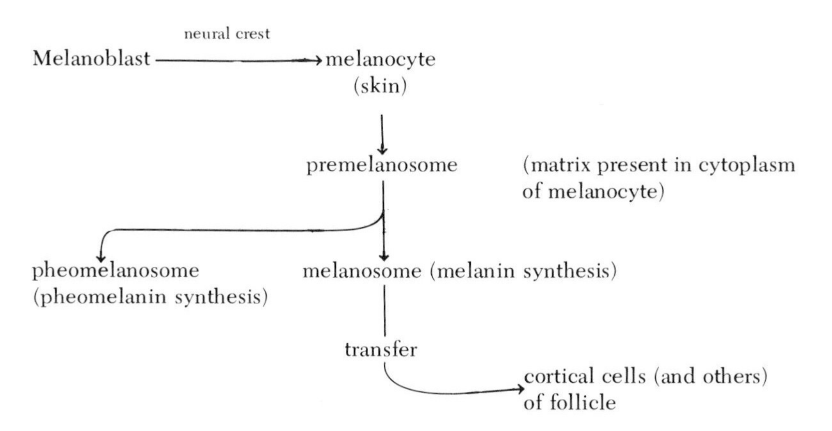

The general chemistry (Fitzpatrick *et al.*, 1967) and histochemistry (Lillie, 1969) of the melanins indicate that at least two types, eumelanin and pheomelanin, occur in birds. Both types occur as distinct granules and, in some birds, the structure often has a marked correlation with color differences. The feather melanins of most species are poorly characterized chemically. Rod-shaped melanosomes are associated with black and indicate the presence of eumelanin. Spherical or ovoid melanosomes contain pheomelanins and are associated with red and yellow colors, especially in the plumage. This distinction is not precise, however, the colors must be analyzed in order to establish their nature. The melanins can be distinguished chemically, by the granule shape, solubility, and fluorescence (reviewed in Fox and Ververs, 1960), differences which suggest more than simple differences in the degree of polymerization or binding with protein molecules. Operationally, differences are found in their solubility in acids (Nickerson, 1946), absorption spectrum, and chemical content (Bohren *et al.*,

1943). Eumelanins may have a variety of colors depending on the degree and mechanism of polymerization.

The metabolic pathway to melanin is known accurately from tyrosine to indole-5,6-quinone (Fig. 1). Subsequent steps to melanins are obviously under genetic control as determined by a number of studies with various albinistic mutants and species-specific differences may exist. The major pathway which leads to melanins involves the enzymatic production of dopa-quinone from tyrosine by tyrosinase. Eumelanins are subsequently produced by polymeriza-

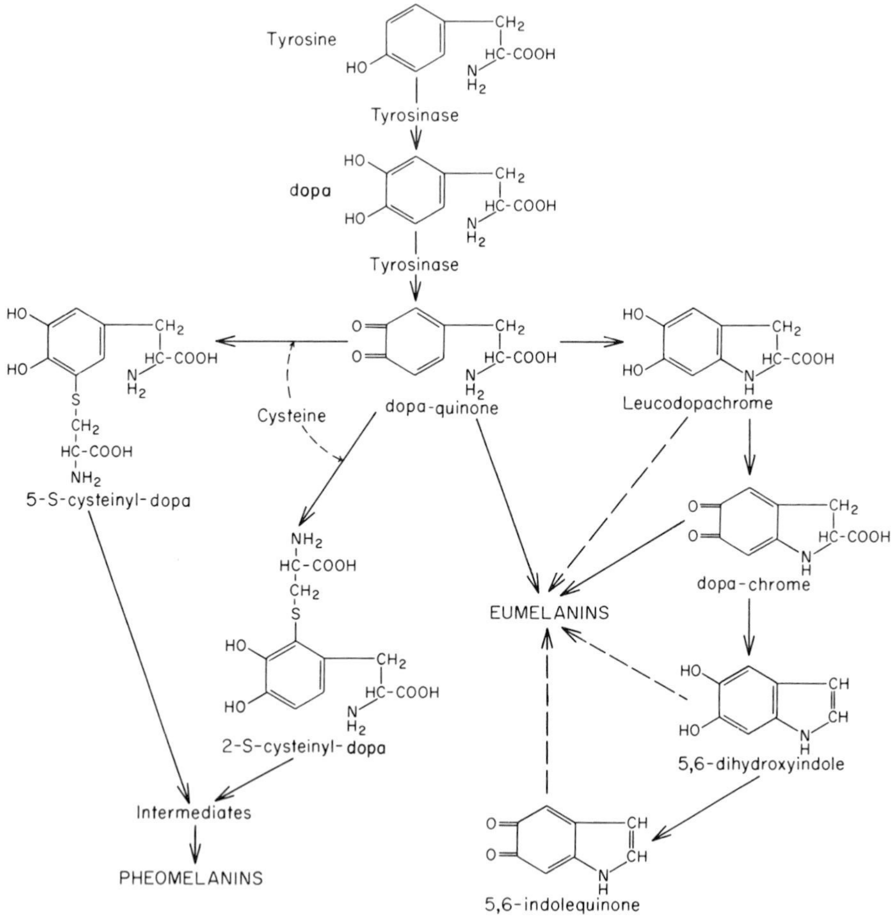

FIG. 1. Suggested metabolic pathways of melanin metabolism.

tion of indole-5,6-quinone. However, eumelanin may not be a pure homopolymer. Some studies suggest the presence of indole units with different oxidative states as well as other tyrosine metabolites. The chemical stability of the eumelanin polymer makes it extremely difficult to study with anything less than radical chemical procedures. The available evidence indicates that eumelanins are probably heteropolymers which consist of the cross-linking of a variety of different monomers. These include quinones, quinols, 2′-carboxyl-quinol derivatives, and a number of unspecified pyrrole carboxylic acids. The monomers are linked by a variety of covalent linkages (C—C—; —C—O—C—; —C—O—O—C—) and hydrogen bonds. Eumelanins are not converted to pheomelanins.

Avian pheomelanins have been fractionated chromatographically (Somes and Smyth, 1965) and direct identification, *in vivo,* may be possible (Juhn, 1964). In addition to solubility and spectral differences, pheomelanins differ chemically from eumelanin. Specifically, differences appear in the presence of sulfur in pheomelanins. This could be due to alterations in the fate of dopa-quinone (Porta and Nicolaus, 1967; Bagnara and Hadley, 1973). One possibility involves a reaction of dopa-quinone with cysteine to form 5-*S*-cysteinyl-dopa and 2-*S*-cysteinyl-dopa (Fig. 2A). These products are subsequently oxidized to form the pheomelanins. The pheomelanins may be identical chemically to a group of widely distributed, iron containing pigments called trichosiderins (Fig. 2B) (Flesch *et al.*, 1966). Pheomelanins are responsible for a range of yellow, orange, and red shades in plumage of birds. Both eumelanin and pheomelanin pathways may occur in the same individual and are thus not mutually exclusive. The possibility of the interconversion of the two pathways has not been studied in birds.

Tyrosinase plays a critical role in melanin synthesis. Although no primary sequence data are available, the chemical biology of this enzyme appears to be similar in organisms as diverse as mushrooms and mammals. Tyrosinase has been characterized as a tetramere (=homomere?) with a molecular weight of 125,000. It occurs in multiple molecular forms in several species (Jolley, 1966; Jolley *et al.*, 1969) and is inhibited by compounds that complex copper (Duckworth and Coleman, 1970). The tyrosinase involved in melanin synthesis is located in the melanosomes of the melanopore and in birds is associated exclusively with subcellular particulate matter. Synthetic activity of the enzyme may be controlled by the glutathione reductase system which regulates glutathione levels in pigment cells (Chian and Wil-

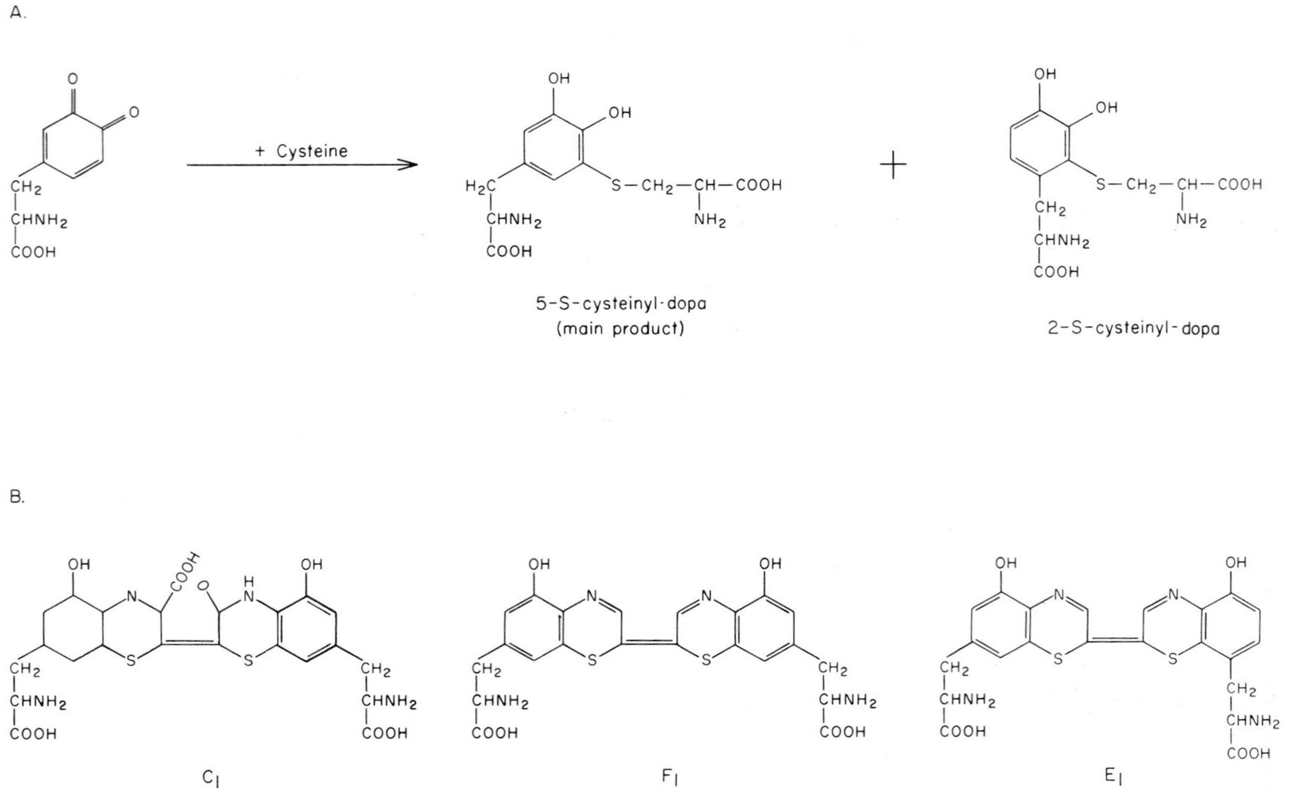

A.

+ Cysteine

5-S-cysteinyl-dopa (main product)

2-S-cysteinyl-dopa

B.

C_I F_I E_I

FIG. 2. (A) Possible mechanism for the formation of pheomelanin precursor. (B) Suggested formulas of the trichsiderins (Flesch *et al.*, 1966).

gram, 1966). Changes in tyrosinase activity are associated with aging, irradiation, and other insults to the skin of mammals (Szabó, 1965). In addition to tyrosinase, significant activity of the enzymes of glycolytic pathways, the TCA cycle, lipid and amino acid metabolism may be detected in melanocytes (Adachi, 1967). Hormonal control of the enzyme is still a puzzle in birds (Section IV). The avian pituitary lacks a pars intermedia, although there are reports of melanocyte-stimulating hormone (MSH) activity in the adenohypophysis (Tixier-Vidal and Follett, 1973; Chapter 7, this volume) and levels may be higher in strains which produce more pigmentation. The tyrosinase in the feather follicle differs from that of the adrenal cortex but seems to be similar to the molecule found in other epidermal structures (Hall and Okazaki, 1966). Activity was increased by stimulation with interstatial cell stimulating hormone (ICSH), at least in some species, and was refractory to treatment with mammalian α-MSH. Details of the hormonal control of tyrosinase and its relationship to the process of melanin production in birds is still unclear (Hall, 1969a).

Melanocytes all differentiate from a common pool of melanoblasts. Subsequently each produces only a single type of melanin. Since the cells possess the same genotype, environmental factors are presumably responsible for the nature of the final metabolic product. Extrinsic factors such as nutrition, hormonal level, and growth rates may affect melanin synthesis. Brumbaugh (1967) demonstrated the interaction of genotypic and environmental influences on the *E* locus in chickens and melanoblast differentiation. Ultrastructural events in the avian melanocyte are similar to those in the epidermis and retinal tissue of other vertebrates (Hamilton, 1940; Ruprecht, 1971). Characteristic differences exist in the details of structural events associated with the production of the two classes of melanin molecules (Brumbaugh, 1968). These may reflect biochemical differences in the synthetic pathways. Cliffman (1964) suggested that sulfhydryl compounds were associated with specific synthetic processes. Specifically, high levels were correlated with both tyrosinase and dopa uptake in pheomelanin when compared with eumelanin synthesis. Possibly the increased sulfhydryl concentration inhibited the process in the premelaninosome which oxidizes and polymerizes the melanin precursor. The mechanisms involving the sulfhydryls are not known but may involve a reaction with the tyrosinase copper. Another option, as mentioned earlier, is the stoichiometry of the dopa-quionone reaction. One source of sulfhydryls may be cysteine (Fig. 2A). A possible mode of action is to affect hormone action directly. Brumbaugh (1968) suggested an alternative explanation based on the interaction of sulfhydryls with the structural elements of the matrix of the melanosomes to reduce cross-linking and subsequent strand formation. The resulting structural disorganization produces pheomelanin rather than eumelanin.

A variety of functions have been proposed for the melanin pigments in birds (Fox, 1976). One complex includes the direct pigmentation of the epidermal structures, alone or in combination with other pigments, and participation in production of structural colors (Section V). These are basically visual functions and play significant roles in communication, identification, or concealment. Another possible role relative to the photic environment is the absorption of incident radiation in the infrared range of wavelengths and thus protection from solar radiation. This provides protection from excessive heat gain or may supplement metabolic heat and play a role in thermal regulation (Heppner, 1970; Ohmart and Lasiewski, 1970). Recently another, but related role, as a reversible switch was suggested (McGinness *et al.*, 1974; McGinness and Proctor, 1973). This intracellular mechanism proposes conversion of the vibrational or rotational energy of molecules into heat which would then be reradiated to the environment. This provides a reversal to the mechanism of energy absorption from the environment. Melanin is unusual in this respect in that it would develop an absorbing or radiating capacity depending on the energetic levels of the cell relative to the environment. Melanins may also act as a sink (scavenger) for free radicals metabolically produced within the cell or from external energy sources.

B. Carotenoids

The various carotenoids used by birds are not synthesized *de novo* but are derived from dietary sources (Goodwin, 1950). The pigments ultimately deposited, however, are not limited to dietary pigments as birds are capable of significant metabolic modification of the dietary precursors. Most frequently the modification involves addition of oxygen to the cyclic end groups of the basic carotene molecule forming hydroxy and keto derivatives. Various aspects of carotenoid metabolism in birds have been reviewed recently by Thommen (1971) and Brush (in press). Included in these reviews are discussions of the biochemical evidence of the metabolic pathways involved, distribution of carotenoids among birds, transportation and fractionation, sources of individual and species variability, and some systematic and evolutionary considerations. A variety of carotenoids are found in the feathers, plasma, and soft tissues of birds. No recognizable pattern regarding the taxonomic distribution of pathways or the occurrence of carotenoid pigments in feathers exists. Thus the inclusion of carotenoids in integumentary structures presumably evolved a number of times in birds.

The influence of diet on pigmentation is now widely established, but only a few carefully controlled feeding experiments have been attempted (Brush and Power, 1976; Brush, in press). Detailed studies of the precursors in feed and the resultant colors produced in the flesh and yolk of commercially bred chickens are available (Baurenfeind, 1971). In addition to diet, the carotenoids displayed in the plumage are under hormonal control (Sections III and IV; Witschi, 1961). There are still a number of unresolved problems in regard to such processes as the production of sexual dichromatism, seasonal coloration, and color polymorphisms. Feeding and plucking experiments on the male House Finch *(Carpodacus mexizanus)* showed the plumage color in this polymorphic, sexually dimorphic species to be a complex interplay of dietary pigment level and composition and hormonal level (Brush and Power, 1976). It appears, by inference, that the biochemistry is often simple. That is, only a relatively few enzymatic steps are involved and simple hormonal control is suggested. However, only a few cases have been described adequately. Considerably less is known about the mechanisms of pigment transportation, intermediary metabolism, and follicular specificity. There is essentially no information available on the carotenoproteins or lipoproteins of birds. Both molecules are important in transportation and specificity. Only the crudest biochemical investigations are available on pigments in mutant and hybrid situations (Brush, 1970; Fox, 1974). No data are available on the postdeposition modification of carotenoids.

Carotenoids are responsible for many of the bright red, orange, and yellow hues of feathers. They are deposited as lipid-containing granules in the follicular cells of the developing feather (Lucus and Stettenheim, 1972). In combination with structural elements, especially blue, they may produce various shades of green. No green carotenoids are definitely known from birds despite the undocumented assertion that they exist in several orders (Auber, 1957). The green facial feathers of the common eider *(Somatteria)* yield a yellow pigment on extraction in alkaline ethanol or acetone. The spectra in hexane was like other xanthophylls. However, the pigment becomes intensely green on transfer to pyridine (A. H. Brush, unpublished). The nature of the pigment is unknown.

A fluorescent pigment from the feathers of some Psittacidacines has been reported (Volker, 1937). The crude chemical evidence available

(Fox and Ververs, 1960) eliminates several possibilities, but is not sufficient for identification of the pigment. The possibility of an extrinsic origin of these pigments has not been disproved.

C. Porphyrins (Tetrapyrroles)

Porphyrins have been reported from the feathers of at least one quarter of all the orders of birds. They occur in large quantities in the downy plumage and adult feathers of owls and bustards (Volker, 1939). The pigment of the Strigi and *Lophotis* has been identified as coproporphyrin III (Fig. 3) (With, 1967a). The mechanism of deposition and relation to feather structure was elucidated by Thiel (1968). Other avian porphyrins include protoporphyrin IX (1,3,5,8-tetramethyl-2,4-divinylporphin-6,7-diproionic acid) and uroporphyrin I (porphin-1,3,5,7-tetraacetic-2,4,6,8-tetrapropionic acid). Generally these molecules are labile when exposed to light. Their distribution in feathers tends to be limited to those areas protected from direct sunlight. Eggshells of many birds contain protoporphyrins (Fischer and Linder, 1925). The eggshell pigments are relatively photostable, widely distributed, and often identified by their fluorescence. The stability of the pigment in some species may be due to protein binding.

Turacin, a copper-containing derivative of uroporphyrin III, occurs in members of the Musophagidae (With, 1967b). The copper atom increases the solubility, reduces the fluorescence, and prevents interactions with proteins to form stable hemelike chromogens. Unlike the iron-containing analogue heme, turacin is inactive catalytically (Keilin, 1951). Turacin can be synthesized under essentially physiological conditions from uroporphyrin III and copper (Keilin and

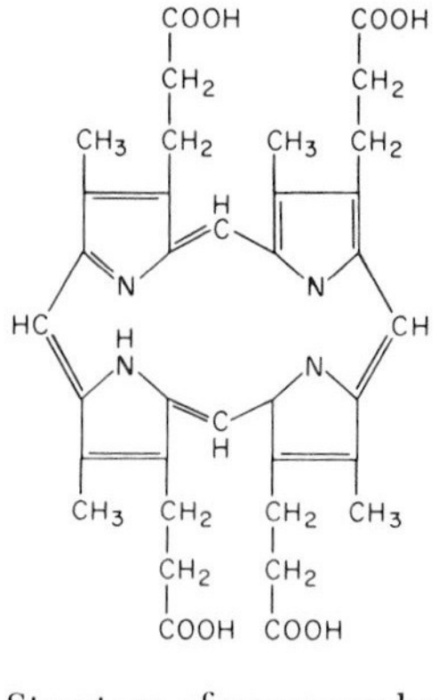

FIG. 3. Structure of coproporphyrin III.

McCosker, 1961). An additional pigment, turacoverdin, has been identified in the green feathers of *Turaco*. This pigment may be an oxidized form of turacin (Fox, 1976) although definitive chemical investigations have not been done. Turacoverdin, contains copper, not iron, and always occurs in association with turacin (Moreau, 1958). J. Dyck (personal communication) extracted a pigment from a Jacana with a visible spectrum similar to turacoverdin. Greenish pigments found in other species *(Rollolus, Corythaeola)* have not been characterized chemically.

The classically cited source of the tetrapyrroles is the catabolism of hemoglobin by the liver. The pyrrole nuclei are used subsequently in the production of bile salts. Porphyrins are also synthesized in large quantities from the amino acid glycine by the following pathway:

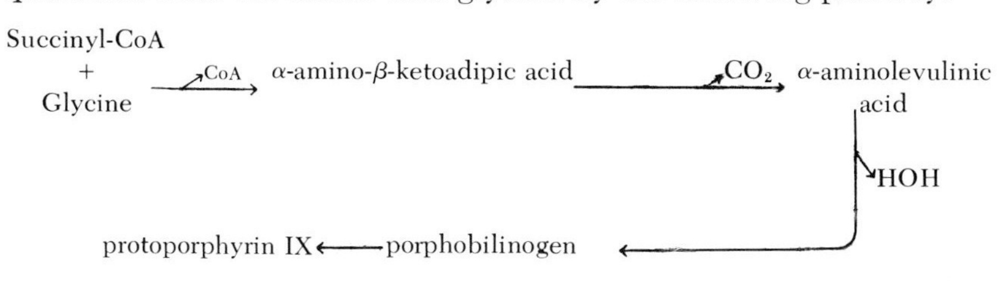

Incorporation of an iron atom in protoporphyrin IX produces heme, the functional chromatogen of hemoglobin, myoglobin, and cytochromes.

Studies on avian erythrocytes indicate that uroprophyrin is converted to protoporphyrin by oxidase enzymes associated with the mitochondrial fraction (Thiel, 1968). There are no studies which identify the relative contributions of the two potential synthetic pathways. The pigments in eggshells (protoporphyrin, coproporphyrin, pentacarboxylic porphyrins, uroporphyrins, and others) are synthesized in the oviduct (With, 1974). This interpretation differs from the older hypothesis that eggshell tetrapyrroles are simply metabolic breakdown products (Thiel, 1968). The suggested metabolic functions of porphyrin deposition are to detoxify copper or to excrete catabolites of tetrapyrroles (Keilin and McCosker, 1961). Evidence for function as a photoreceptor molecule has also been presented (reviewed by Fox, 1976).

III. Location of Pigments

The distribution of biochromes may be as important in the species biology as the nature of the pigments and their functional chemistry. It is important to consider distribution as, among other points, it illustrates the ways in which quite unrelated molecules function in integrated biochemical systems. By the same token, it illustrates the diverse ways in which similar problems are solved by distantly related organisms. Viewed from this perspective, a number of basic biological problems present themselves. While I do not propose solutions or explanations for all of them, they may be seminal to further thinking in this area.

Pigmentation is widespread in avian tissues. The presence of the heme proteins, myoglobin, cytochrome, and hemoglobin imparts color to many internal organs and provides substantial functional support of various metabolic functions. Hemoglobin is the major molecular species involved in the transport of blood gases (Chapter 10, this volume). Hemoglobins commonly occur in multiple molecular forms in birds (Ghosh, 1965) which have both structural and perhaps important functional differences (Vandecasserie *et al.*, 1973). The presence of myoglobin in skeletal muscles provides a molecular basis for many physiological correlations (George and Berger, 1966). The gizzard (posterior muscular stomach) contains a thick muscular wall with a high myoglobin concentration. This is marked in graniverious species and presumably correlates with long bouts of anaerobic function during digestion. Hemoglobins often impart a red or rust color to epidermal structures.

Carotenoids are found in a variety of internal tissues including the visual retina and corpora leutia. Carotenoids also occur in both the skin and yolk. Pigments of the yolk vary from pale yellow to bright vermilion (Goldsmith, 1965) depending on the nature and concentration of the pigments present. Yolk and skin color is of special interest to commercial producers of fowl (Bauernfeind *et al.*, 1974). Consumer tastes vary and have been a subject of concern for breeders. The importance of feeding regimens for commercial birds is now well established and a growing body of information of the importance of dietary carotenoids exists for zoo and cage birds (Brush, in press).

The epidermal derivatives of birds also contain conspicuous deposits of pigment. The scaly portion of the legs and feet (podotheca) frequently contain melanin. Melanized skin over the general body surface occurs less frequently. It is usually stated that the reason for the lack of pigementation is that the protection offered by the feathers eliminates the necessity of pigments. There are two interesting exceptions. In the silky fowl, a mutant with unusual feather structure (Hutt, 1949; Brush, 1972), large areas of the skin contain extensive melanin deposits. These are not present in other chicken breeds. The functional significance is not known. In the Roadrunner *(Geococcyx californianus*), the dorsal skin surface has a complex melanic pattern. Physiological evidence indicates that these areas are used as heat absorbing sites as a supplement to metabolic thermal regulation (Ohmart and Lasiewski, 1970). This interpretation may also apply to the frequent reports of increased melanization in nestling birds. The problem was discussed in detail by Hamilton (1973).

The eggshells of birds represent a complex array of pigments and patterns. The pigments appear mainly as tetrapyrroles (see Section II,3). The older names oocyan (=bilirubin) and ooporphyrin (=protoporphyrin) are no longer used. Pigments occur in various combinations or singly and may be distributed through the entire shell structure. The pigments are secreted by the oviduct wall during shell formation. The resultant colors and patterns vary widely (Kennedy and Ververs, 1976). The eggshell pattern is characteristic of the species, but there is no apparent taxonomic value associated with eggshell color. However, ecological correlations do occur (Lack, 1958). Most are associated with increased visibility in hole-nesting species and crypticity in others. Despite the many correlations of color and pattern with specific ecological features significant exceptions may exist (Yapp, 1970). The pigments often are considered waste products (Welty, 1975; Thiel, 1968) presumably deposited originally by accident. This argument assumes that the primitive condition in birds was a white eggshell. Obviously, since cryptically colored eggs are common, subsequent selection has occurred. In this sense the pigments have assumed a new function and cannot be considered waste products.

Both carotenoids and melanins are found in the beak. This conspicuous organ is used for a number of manipulative and signal functions. In the feet, both signal function and thermal regulatory roles have been recognized. The pigments are deposited in these structures by mechanisms identical to those of other epidermal structures. Color variation in the various soft parts of birds may be extensive (Buckley and Buckley, 1970). Often the variation appears to be continuous within populations. This implies complex selective pressures and possibly different roles than in the case of marked color polymorphisms. A variety of soft structures, such as skin flaps, knobs, etc. are colored.

The most thoroughly studied pigment systems in birds are associated with the feathers. Melanins are present in all orders and carotenoids occur in fully one-half the orders. Adult plumages are better known than those of juvenile birds, a situation that may simply reflect the interest of collectors, although it is widely accepted that juvenile plumages tend to lack carotenoids in favor of melanins. Pigments are deposited during feather growth and development and become an integral part of the structure (Desselberger, 1930; Lucus and Stettenheim, 1972). Pigments and pigment pattern have been considered an important taxonomic character as they determine the "Gestalt" appearance of the bird. Arguments still continue over the role of color variation within and between populations, its systematic and evolutionary significance (Short, 1976). Subspecies have been established on the basis of plumage color, the genetic validity of which is often questionable. Few definitive studies exist on the relation of plumage color characteristics and diet, hormonal influence, and individual variability.

Sexual dimorphism and color polymorphism are described for many avian species. In the sexually and seasonally dichromatic Scarlet Tanager *(Piranga olivaca)* the pigment color change is due to a simple reaction:

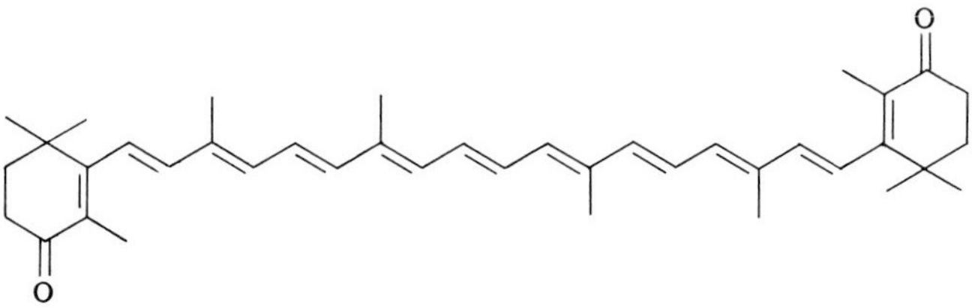

O

O

Alternate (4, 4′-diketo-β-carotene)

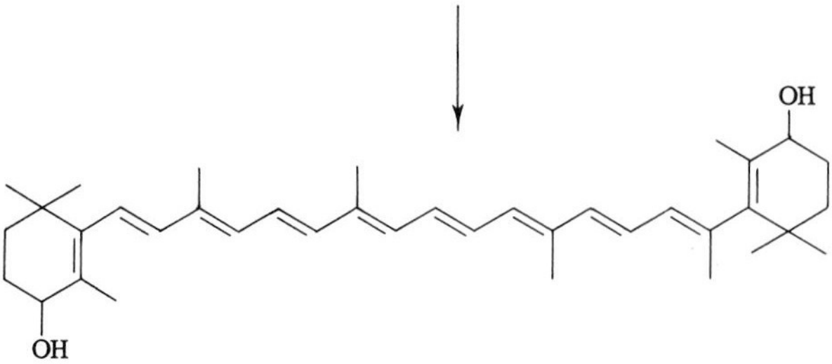

OH

OH

Basic (4, 4′-dihydroxy-β-carotene)

which is presumably mediated by a single enzyme (Brush, 1967b). Plumage polymorphisms occur in many species and are only partly analyzed. The origin, maintenance, and function of color polymorphisms is an extensive field in need of critical review.

Color polymorphisms in Ross' Goose *(Anser rossii)* and the Mute Swan *(Cygnus olor)* are the result of a single gene with alternate alleles (Cooke and Ryder, 1971; Muro *et al.*, 1968). Color phases in a Screech Owl *(Otus asio)* are under similar genetic control (Hrubant,

1955). The phenotype variation in feral populations of the Rock Dove *(Columba livia)* is well known and color phases in the Arctic Skua have been analyzed (O'Donald, 1972). Population differences in the Red-eyed Vireo plumage have been demonstrated by reflectance spectra, but no apparent clines exist (Barlow and Williams, 1971). The relation among reflectance spectra, pigment content, and feather structure is discussed below (Section V). The biochemical basis of a complex plumage pattern and polymorphism in the Gouldian Finch *(Poephila gouldiae)* has been elucidated (Brush and Seifried, 1968). The Canary *(Serinus serinus)* has a variety of plumage patterns presumably under genetic controls while the colors may be influenced by diet. Polymorphisms as a result of a combination of structural colors and pigments also occur (Section V). Recent work on the Sooty-capped Tanager *(Chlorospingus pileatus)* demonstrated a chemical basis for the two color morphs and correlated their distribution with environmental factors (Johnson and Brush, 1972). In congeneric warblers *(Vermivora)* color differences were attributed to carotenoid concentration differences. The plumage coloration corresponded broadly to the color of the preferred nesting habitat and was presumably part of the reproductive isolating mechanism. Similar examples are found in numerous other passerine species (Brush and Johnson, 1976). In an especially penetrating study of the distribution and causes of color variation in *Zosterops borbonica,* Gill (1973) demonstrated the existence of genetic polymorphism and the association of morphs with environmental gradients. However, the complex relationships of physical, chemical, and genetic interactions responsible for the production of plumage colors and patterns and the associated selective forces still remain relatively unexplored.

IV. Control

It is obvious from the above that the most conspicuous display of avian pigments are in the feathers. The catalogue of the pigments present in birds is essentially complete. However, the systems which control plumage color, determine the relationship to other physiological parameters, and integrate colors with the life cycle are not as thoroughly understood. Indeed, we have adequate evidence from only relatively few species. Nevertheless, these studies have provided models for subsequent experimental work.

The cells responsible for the production of feathers and other epidermal structures are exquisitely sensitive to their biotic and chemical environment. Thus molt cycles, color, and structure respond to various hormonal stimuli (Ralph, 1969). Patterns of individual feathers and the general plumage pattern interact in a complex manner (Cohen, 1966), with much of the control provided by hormones. A variety of mechanisms are involved and are best understood for melanocytes. The amounts of melanin present is determined by the activity of the enzyme tyrosinase (Section II). Tyrosinase activity is subject to regulation by ICSH, estrogens, MSH, and adrenocorticotropin (Hall, 1969a,b). In the case of MSH and ACTH, the similarity in action on melanocytes is presumably due to similarity in the amino acid sequence (Hoffman, 1962). Other hormones such as thyroxine, androgens, and the catecholamines may affect melanocytes directly or modify the action of other hormones. Hormones are responsible for most cases of sexual dimorphism in plumage coloration (Hohn and Cheng, 1967; Greij, 1973) and affect most phases of plumage growth (Voitkevich, 1966).

The complex developmental, biochemical, and structural processes involved in feather production are subject to genetic modification and mutation. With some pigments, e.g., carotenoids, diet becomes an influential factor. A variety of plumage aberrations have been described in birds. These include melanism (Gross, 1965a), albinism (Ross, 1963; Gross, 1965b), erythrism, xanthochroism (Harrison, 1963), and structural changes. All types are probably widely distributed and appear to be under simple genetic control although definitive breeding data are available for only a few isolated cases. The pathways for deposition of carotenoids and melanins and the related structural aspects are under separate genetic control (Brush and Seifried, 1968). Thus, species albinistic or partly albinistic relative to melanin may show normal carotenoid pigmentation. Cases of excessive pigment deposition, e.g., "melanism," "erythrism," and "xanthochroism" are reported but their biochemical basis is uninvestigated.

The effects of hormones have been studied traditionally by combinations of plucking, gland extirpation, and hormone injection (Witschi, 1961). This approach has been productive, especially in our understanding of melanin biology, but many gaps exist (Bagnara and Hadley, 1973). Less is known of the hormonal controls of carotenophores. The brighter colors are, almost without exception, associated with male plumages. The influence of pigment on keratin structure, the biochemical nature of pigment mutants, hormones and pigment cell interactions, and the influence of hormones on nonmelanic pigments are all areas of possible fertile research.

V. Structural (Nonpigment) Colors

The combination of pigments with color-producing structures often results in an extension of the color effect and textures not obtainable by either acting independently. Colors in plumage can be produced by physical means, which act independently of chemical pigmentation. When combined with pigment the spectrum of color production is enhanced and many of the more vividly colored species are a result of such combinations. The primary sources of nonpigment color effects are the familiar physical phenomena of iridescence produced by thin-layer interference and noniridescent scattering, the Tyndall or Rayleigh effect. Structural elements may be combined with pigments to provide modified colors or increase the effectiveness of either element.

The simplest nonpigment colors are black and white. White is the result of the reflection of incident light. No pigment is involved. Microscopic air spaces within the keratin structure and the numerous surfaces of the rami and barbules provide the reflective surface. Black is produced by the absorption of incident light in the feathers. Black feathers range from shiny, glossy to deep, soft, velvety intensities.

In the feathers and skin of many species colors are produced by the dispersal of fine colloidal aggregate in an unpigmented matrix which overlays darker, often pigmented, layers. Color production by such dispersion is often called Tyndall or Rayleigh effect. Recent analysis indicate the process involved is due to back scattering from a highly organized macromolecular structure and is more akin to interference phenomenon (Raman, 1936; Dyck, 1971a,b). The reflective mechanism is often modified to produce blue colors. The fine structure of keratin is augmented by a layer of melanin placed as absorbing layer behind the reflecting keratin-air structure. The long wave portion of the impinging light passes through the structure and is absorbed in the melanin, only the shorter blue wavelengths are reflected. When viewed by transmitted light, such feathers appear reddish, or brown. Blue structures of this type are combined with red pigments to produce magenta (Auber, 1958), purple (Brush, 1969), or with yellow pigment to produce green plumages (Brush and Seifried, 1968; Dyck, 1969). Structural nonpigment colors are modified by mechanical destruction or filling the spaces with fluids of the same refractive index as the keratin structure.

Iridescent colors in plumages are the product of surface interference phenomena which occur on specialized surfaces within the feather.

The phenomenon is highly developed in the hummingbirds (Trochilidae) and the sunbirds (Nectariniidae), but occurs in many other groups. The structure of iridescent feathers has been analyzed thoroughly (Greenewalt *et al.*, 1960a,b; Durrer, 1962; Durrer and Villiger, 1962, 1966, 1967). Iridescence, reflection, scattering, and chemical pigments can all occur on a single bird and even in combinations in individual feathers. The microscopic and physical basis for physically produced color was reviewed by Dyck (1976).

The presence or absence of pigment, its chemical nature and the structural modifications necessary to produce nonpigmented color may each be under separate genetic control. One such example is the Budgerigar, *Melopsittacus undulatus* (Fox and Ververs, 1960). Interactions between pigments and their keratin substrate can effect both the color and the structure of the feathers (Desselberger, 1933; Olson, 1970). Recent interest in the study of colors by reflectance spectrophotometry (Dyck, 1969) has proven valuable to taxonomists (Bowers, 1956, 1959) and the approach extends to ecological factors such as background matching or other environmental parameters (Bowers, 1960; Brush and Johnson, 1976). Precise identification of individual pigments is not possible by reflectance spectra alone because the interactions of pigments with the substrate cause spectral shifts. Differences *in vivo* and *in vitro* of spectral maxima in the visible range have been demonstrated for feather carotenoids in several species (Dyck, 1969; Johnson and Brush, 1972). Reflectance spectra in the IR (e.g., 1200–2000 nm) are similar for both melanin and carotenoid pigmented feathers, but differences appear in the very near IR (e.g., 800–1100 nm) (Dyck, 1969, Fig. 12). The functional significance of these differences is not resolved and care must be taken not only to include appropriate controls but also to consider plumage location and structure in such studies.

Feathers which contain carotenoids frequently show structural modifications (Desselberger, 1930; Brush and Seifreid, 1968; Brush, 1967a; Olson, 1970). The changes act to intensify the brilliance of the pigment in display (Johnson and Brush, 1972). These relationships are important functionally. Binding pigments to feather keratin presumably increases their stability, as is known for visual chromatophores. It is especially important to provide photostability. Many physical and chemical properties of pigments are modified by binding (Bellin, 1965). In addition, carotenoids form aggregations which affect their chemistry in solution (Buchwald and Jencks, 1968). Astaxanthin aggregates in aqueous solvents significantly changes the λ_{max}. The relationships of carotenoid binding to tissue specificity and the structure of feathers is totally unstudied. Serum carotenoproteins are found in birds (Cheesman *et al.*, 1967) and serum lipoproteins play an active role in pigment transport (Trams, 1969). No information is available regarding the mechanisms or controls of these interactions. Binding of carotenoids to feather keratin produces spectral shifts, changes in feather structure, and affects the brightness of the display (Dyck, 1969; Johnson and Brush, 1972).

The occurrence of adventitious coloration on feathers has been a matter of concern to ornithologists for some time (Kennard, 1918). The most common source of advantitious colors is stain from water or dust bathing. The deposition of iron oxides, soot, and a variety of other substances on the feathers has been documented in over 100 species (Berthold, 1967; Berthold and Rau, 1968). Iron oxide staining on the plumage of the Great White Pelican, *Pelecanus onocrotalus,* is confined to the distal portion of the feathers. The mechanism of the binding reaction between the insoluble feather protein and insoluble metal oxide is unknown (Baxter and Urban, 1970).

In addition to stains with inorganic materials, nonmolting sources of color change include abrasion, stain from preen oil, and photochemical changes of pigments. These are all limited to the feather surface. Claims for color pattern change without molt but of internal origin, e.g. aposochromatic change, remain unsubstantiated.

ACKNOWLEDGMENTS

Investigations in avian pigment biology in my laboratory have received support from the University of Connecticut Research Foundation and the National Science Foundation. It is a pleasure also to thank all the many people who have supplied materials, ideas, and pointed out some of the more stimulating problems.

REFERENCES

Adachi, K. (1967). *Adv. Biol. Skin* **8,** 223.
Auber, L. (1957). *Ibis* **99,** 463.
Auber, L. (1958). *Ibis* **100,** 57.
Bagnara, J. T., and Hadley, M. E. (1973). "Chromatophores and Color Change." Prentice-Hall, Englewood Cliffs, New Jersey.
Barlow, J. C., and Williams, N. (1971). *Can. J. Zool.* **49,** 417.
Bauernfeind, J. C., Brubacher, G. B., Klaui, H. M., and Marusich, W. L. (1971). *In* "Carotenoids" (O. Isler, ed.), Chapter XI. Birkhaeuser, Basel.
Baxter, R. M., and Urban, E. K. (1970). *Ibis* **112,** 336.
Bellin, J. S. (1965). *Photochem. Photobiol.* **4,** 33.
Berlin, B. (1973). *Annu. Rev. Ecol. Syst.* **4,** 259.
Berthold, P. (1967). *Zool. Jahrb., Abt. Syst. Oekol. Geogr. Tiere* **93,** 505.
Berthold, P., and Rau, R. (1968). *Z. Zellforsch. Mikrosk. Anat.* **85,** 492.
Bohren, B. B., Conrad, R. M., and Warren, D. C. (1943). *Am. Nat.* **77,** 481.
Bowers, D. E. (1956). *Syst. Zool.* **5,** 147.

Bowers, D. E. (1959). *Condor* **61,** 38.
Bowers, D. E. (1960). *Condor* **62,** 91.
Brumbaugh, J. A. (1967). *J. Exp. Zool.* **166,** 11.
Brumbaugh, J. A. (1968). *Dev. Biol.* **18,** 375.
Brush, A. H. (1967a). *Wilson Bull.* **79,** 322.
Brush, A. H. (1967b). *Condor* **69,** 549.
Brush, A. H. (1969). *Condor* **71,** 431.
Brush, A. H. (1970). *Comp. Biochem. Physiol.* **36,** 785.
Brush, A. H. (1972). *Biochem. Genet.* **7,** 87.
Brush, A. H. (1975). *Isozymes* **4,** 901.
Brush, A. H. (1977). *In* "Carotenoid Technology" (J. C. Baurenfiend, ed.). Academic Press, New York (in press).
Brush, A. H., and Johnson, N. K. (1976). *Condor* **78,** 412.
Brush, A. H., and Power, D. M. (1976). *Auk* **93,** 725.
Brush, A. H., and Seifried, H. (1968). *Auk* **85,** 416.
Buchwald, M., and Jencks, W. P. (1968). *Biochemistry* **7,** 834.
Buckley, P. A., and Buckley, F. G. (1970). *Auk* **87,** 1.
Chian, L. T. Y., and Wilgram, G. F. (1966). *Science* **155,** 198.
Cheesman, D. F., Lee, W. L., and Zagalsky, P. F. (1967). *Biol. Rev. Cambridge Philos. Soc.* **42,** 131.
Cliffman, G. (1964). *Exp. Cell Res.* **35,** 590.
Cohen, J. (1966). *Adv. Morphog.* **5,** 1.
Cooke, F., and Ryder, J. P. (1971). *Evolution* **25,** 483.
Cowell, L. A., and Weston, J. A. (1970). *Dev. Biol.* **22,** 670.
Desselberger, H. (1930). *J. Ornithol.* **78,** 328.
Diamond, J. (1966). *Science* **151,** 1102.
Dingle, J. T., and Lucy, J. A. (1965). *Biol. Rev. Cambridge Philos. Soc.* **40,** 422.
Duckworth, H. W., and Coleman, J. E. (1970). *J. Biol. Chem.* **245,** 1613.
Durrer, H. (1962). *Verh. Naturforsch. Ges. Basel* **73,** 304.
Durrer, H., and Villiger, W. (1962). *Rev. Suisse Zool.* **69,** 801.
Durrer, H., and Villiger, W. (1966). *J. Ornithol.* **107,** 1.
Durrer, H., and Villiger, W. (1967). *Z. Zellforsch. Mikrosk. Anat.* **81,** 445.
Dyck, J. (1969). *Dan. Ornith. Foren. Tidsskr.* **60,** 49.
Dyck, J. (1971a). *Z. Zellforsch. Mikrosk. Anat.* **115,** 17.
Dyck, J. (1971b). *K. Dan. Vidensk. Selsk., Biol. Skr.* **18,** 1.
Dyck, J. (1976). *Proc. Int. Ornithol. Congr., 16th, 1975,* p. 426.
Fischer, H., and Linder, F. (1925). *Hoppe-Seyler's Z. Physiol. Chem.* **142,** 141.
Fitzpatrick, T. B., Miyamoto, M., and Ishikawa, K. (1967). *Adv. Biol. Skin* **8,** 1.
Flesch, P., Esoda, E. J., and Katz, S. (1966). *J. Invest. Dermatol.* **47,** 595.
Fox, D. L. (1962). *Comp. Biochem. Physiol.* **6,** 1.
Fox, D. L. (1974). *Comp. Biochem. Physiol.* **48B,** 295.
Fox, D. L. (1976). "Animal Biochromes and Structural Colours," 2nd ed. Univ. of California Press, Berkeley.
Fox, H. M., and Ververs, G. (1960). "The Nature of Animal Colors." Univ. of Washington Press, Seattle.
George, J. C., and Berger, A. J. (1966). "Avian Myology." Academic Press, New York.
Ghosh, J. (1965). *Comp. Biochem. Physiol.* **16,** 341.
Gill, F. B. (1973). *Am. Ornithol. Union, Monogr.* No. 12, 1.
Goldsmith, T. H. (1965). *Postilla* **91,** 1.
Goodwin, T. W. (1950). *Biol. Rev. Cambridge Philos. Soc.* **25,** 391.

Goodwin, T. W. (1954). "Carotenoids: Their Comparative Biochemistry." Chem. Publ. Co., New York.

Greenewalt, C. H., Brandt, W., and Friel, D. D. (1960a). *J. Opt. Soc. Am.* **50,** 1005.

Greenewalt, C. H., Brandt, W., and Friel, D. D. (1960b). *Am. J. Physiol.* **104,** 249.

Greij, E. D. (1973). *Auk* **90,** 533.

Gross, A. O. (1965a). *Bird-Banding* **36,** 67.

Gross, A. O. (1965b). *Bird-Banding* **36,** 240.

Hadley, M. E. (1972). *Am. Zool.* **12,** 63.

Hall, P. F. (1969a). *Gen. Comp. Endocrinol., Suppl.* **2,** 456.

Hall, P. F. (1969b). *Aust. J. Dermatol.* **10,** 125.

Hall, P. F., and Okazaki, K. (1966). *Biochemistry* **5,** 1202.

Hamilton, H. L. (1940). *Anat. Rec.* **78,** 525.

Hamilton, T. H., and Barth, R. H. (1962). *Am. Nat.* **96,** 129.

Hamilton, W. J., III. (1973). "Life's Color Code." McGraw-Hill, New York.

Harrison, C. J. O. (1963). *Bull. Br. Ornithol. Club* **83,** 90.

Heppner, F. (1970). *Condor* **72,** 50.

Hoffman, K. (1962). *Annu. Rev. Biochem.* **31,** 213.

Hohn, E. O., and Cheng, S. C. (1967). *Gen. Comp. Endocrinol.* **8,** 1.

Hrubant, H. E. (1955). *Am. Nat.* **89,** 223.

Hutt, F. B. (1949). "Genetics of the Fowl." McGraw-Hill, New York.

Jimbow, K., Quevedo, W. C., Fitzpatrick, T. B., and Szabó, G. (1976). *J. Invest. Dermatol.* **67,** 72.

Johnson, N. K., and Brush, A. H. (1972). *Syst. Zool.* **21,** 245.

Jolley, R. L. (1966). *Adv. Biol. Skin* **8,** 269.

Jolley, R. L., Nelson, R. M., and Robb, D. A. (1969). *J. Biol. Chem.* **244,** 1593 and 3251.

Juhn, M. (1964). *Nature (London)* **202,** 507.

Keilin, J. (1951). *Biochem. J.* **49,** 544.

Keilin, J., and McCosker, P. J. (1961). *Biochim. Biophys. Acta* **52,** 424.

Kennard, F. H. (1918). *Auk* **35,** 123.

Kennedy, G. Y., and Ververs, H. G. (1976). *Comp. Biochem. Physiol.* **55,** 117.

Kenyon, D. H., and Steinman, G. (1969). "Biochemical Predestination." McGraw-Hill, New York.

Kittlewell, H. B. D. (1973). "The Evolution of Melanism: The Study of a Recurring Necessity." Oxford Univ. Press, London and New York.

Lack, D. (1958). *Ibis* **100,** 145.

Lillie, R. D. (1969). *In* "Pigments in Pathology" (M. Wolman, ed.), p. 327. Academic Press, New York.

Lubow, E. (1963). *J. Ornithol.* **104,** 68.

Lucus, A. M., and Stettenheim, P. (1972). U.S., *Dep. Agric., Handb.* **362.**

McGinness, J., and Proctor, P. (1973). *J. Theor. Biol.* **39,** 677.

McGinness, J., Corry, P., and Proctor, P. (1974). *Science* **183,** 853.

Moreau, R. E. (1958). *Ibis* **100,** 238.

Mottaz, J. H., and Zelickson, A. S. (1967). *Adv. Biol. Skin* **9,** 471.

Muro, R. E., Smith, L. T., and Kupa, J. J. (1968). *Auk* **85,** 504.

Needham, A. E. (1974). "The Significance of Zoochromes." Springer-Verlag, Berlin and New York.

Nevo, E. (1973). *Evolution* **27,** 353.

Nickerson, M. (1946). *Physiol. Zool.* **19,** 66.

O'Donald, P. (1972). *Nature (London)* **238,** 403.

Ohmart, R. D., and Lasiewski, R. C. (1970). *Science* **172,** 67.

Olson, S. (1970). *Condor* **72,** 424.
Porta, G., and Nicolaus, R. A. (1967). *Adv. Biol. Skin* **8,** 323.
Ralph, C. L. (1969). *Am. Zool.* **9,** 521.
Raman, C. V. (1936). *Proc. Indian Acad. Sci., Sect. A* **1,** 1.
Rawles, M. E. (1948). *Physiol. Rev.* **28,** 382.
Ross, C. C. (1963). *Cassinia* **47,** 2.
Ruprecht, K. W. (1971). *Z. Zellforsch. Mikrosk. Anat.* **112,** 396.
Short, L. L. (1976). *Proc. Ornithol. Congr., 16th,* 1975, p. 185.
Sillman, A. J. (1973). *Avian Biol,* **3,** 349.
Somes, R. G., and Smyth, J. R. (1965). *Poult. Sci.* **44,** 276.
Szabó, G. (1965). *In* "Biology of the Skin and Hair Growth" (A. G. Lyne and B. F. Short, eds.), p. 705. Angus & Robertson, Sidney, Australia.
Thiel, H. (1968). *Zool. Jahrb., Abt. Syst. (Oekol.), Geogr. Biol.* **95,** 147.
Thommen, H. (1971). *In* "Carotenoids" (O. Isler, ed.), p. 637. Birkhaeuser, Basel.
Tixier-Vidal, A., and Follett, B. K. (1973). *Avian Biol.* **3,** 110.
Trams, E. G. (1969). *Comp. Biochem. Physiol.* **28,** 1177.
Vandecasserie, C., Paul, C., Schnek, A. G., and Leonis, J. (1973). *Comp. Biochem. Physiol.* **44A,** 711.
Voitkevich, A. A. (1966). "Feathers and Plumage of Birds." Sidgwick & Jackson, London.
Volker, O. (1939). *Hoppe-Seyler's Z. Physiol. Chem.* **258,** 1.
Volker, O. (1937). *J. Ornithol.* **85,** 136.
Webster, T. P. and Burns, J. M. (1973). *Evolution* **27,** 368.
Welty, J. (1975). "The Life of Birds," 2nd ed. Saunders, Philadelphia, Pennsylvania.
With, T. K. (1967a). *J. Ornithol.* **108,** 480.
With, T. K. (1967b). *Nature (London)* **179,** 824.
With, T. K. (1974). *Biochem. J.* **137,** 597.
Witschi, E. (1961). *In* "Biology and Comparative Physiology of Birds" (A. J. Marshall, ed.), Vol. 2, p. 115. Academic Press, New York.
Yapp, W. B. (1970). "The Life and Organization of Birds." Arnold, London.

CHAPTER 6

Uropygial Gland Secretions and Feather Waxes

Jürgen Jacob

I. Uropygial Gland Secretions and Other Feather Lipids

A. Introduction

Integumental glands are common in all classes of animals and are related to different types of secretory systems. Pheromones or olfactory active substances are produced predominantly in apocrine or eccrine secreting glands, whereas protective materials are secreted by the holocrine gland type. The activity of production and the types of lipids produced in the integument differ remarkably between taxa. Reptilian skin is poor in lipids whereas mammals generally possess considerable amounts of skin surface lipids produced mainly by the sebaceous glands. Another minor source of lipid material is the keratinizing epidermal cells. These lipoidal products can be distinguished by their compositions in man (Grimmer *et al.*, 1971; Jacob and Grimmer, 1973b; Nicolaides, 1974). Among mammals the composition varies significantly on the ordinal level, and this seems to be true for other animal classes also.

The avian skin differs from that of other vertebrates by three facts: the substitution of hairs by feathers, the absence of sweat glands, and the absence of the large number of sebaceous glands. Instead of the latter, birds possess a holocrine uropygial (preen) gland on the dorsal surface of the tail (see Section I,B), which produce predominantly ester waxes (see Section I,C,2), the compositions of which are characteristic for single orders or even families (see Section I,C,3). In some

TABLE I
INTEGUMENTAL GLAND PRODUCTS OF SOME ANIMAL CLASSES

Animal class	Gland	Predominant secretion product
Insects	Wax gland	Hydrocarbons (ester waxes)
Birds	Uropygial gland	Ester waxes (triglycerides)
Mammals (man)	Sebaceous gland	Triglycerides (squalene, cholesterol esters, ester waxes, cholesterol, alcohols)

bird species other lipid-producing areas may occur in the integument (see Section I,D). Gland types and secretion products in some well-investigated classes are summarized in Table I.

The uropygial waxes and their constituents are complex mixtures of homologous and isomeric compounds, the separation of which is difficult. Improved analytical techniques of separation (gas–liquid-chromatography = GLC) and structure elucidation (mass spectrometry = MS), have solved most of the problems leading to a large number of investigations which will be summarized in this review. Earlier reviews from a more chemical viewpoint have been published (Jacob, 1976a,b).

B. Anatomy and Histology of the Uropygial Gland

The uropygial gland (synonym: glans uropygii, glans uropygialis, oil gland, preen gland) is a bilobed organ located at the rump in the dorsal caudal tract at the bottom of the tail feathers. Both lobes are surrounded by a connective tissue capsule (capsula glandis uropygialis) and separated from each other by the interlobular septum (septum interlobulare) which ends in the muscular isthmus in the papilla (Fig. 1). Each lobe possesses a large number of tubules in which the secretion or the fat-containing cells are produced and from which the waxes are transported into the cavity (cavitas glandis uropygialis). From there the secretion passes through the uropygial duct (ductus glandis uropygialis) and leaves the gland through the pores (porus ductus uropygialis). It then can be distributed over the plumage by the bird's bill. Often a circlet of fine down feathers can be observed around the gland's tip (circulus uropygialis). Details of the anatomy of the preen gland have been given by Lucas and Stettenheim (1972) and by Paris (1913) who described glands of different species.

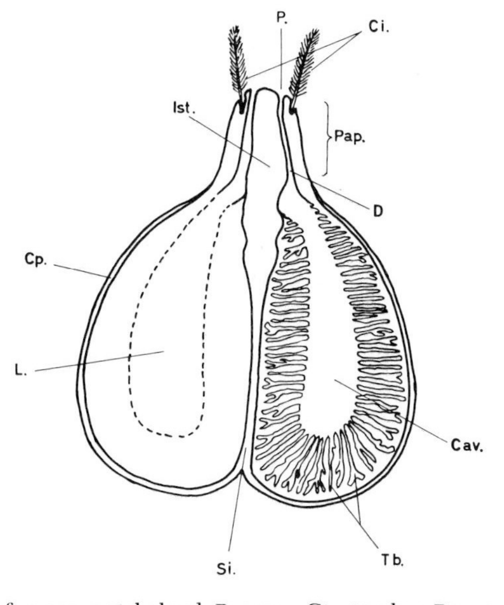

FIG. 1. Section of an uropygial gland. P, porus; Ci, circulus; Pap, papilla; Ist, isthmus; D, ductus; Cav, cavitas; Tb, tubuli; Si, septum interlobulare; L, lobus; Cp, capsula.

Histological and histochemical investigations on geese, ducks, and chicken have distinguished three zones in the tubuli (Lennert and Weitzel, 1951, 1952; Lucas and Stettenheim, 1972; Stern, 1905), although some authors prefer to recognize only two zones (Cater and Lawrie, 1950; Lunghetti, 1907). The three zones (Fig. 2) were termed peripheral, intermediate, and inner by Lennert and Weitzel (1952). Zone I (outer or peripheral zone) contains fat particles but shows only

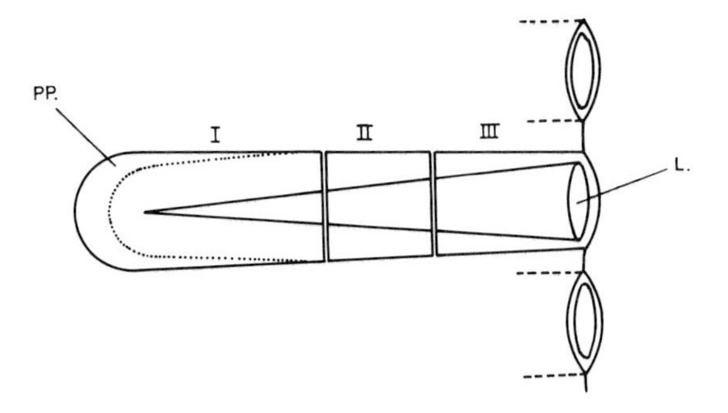

FIG. 2. Diagram of a section through a preen gland tubule (Lennert and Weitzel, 1952). Zone I, peripheral; Zone II, intermediate; Zone III, inner zone; L, lumen; PP, peripheral portion.

little osmophilic reaction, the reaction is strongest in the peripheral part. This zone often is called "sebaceous zone" and the wax constituents are synthesized here. Zone II (intermedial zone) shows high osmophilic activity. The triglycerides synthesized here are normally not a part of the uropygial gland secretion but fulfill other functions inside the gland. This zone often is considered to be indistinguishable from Zone III. In Zone III (inner zone) extensive cell decomposition is observed. It is still an open question whether this zone participates in wax synthesis. Both acid phosphatase and unspecific esterase activity is high in Zones II and III. The tubular lumen diameter increases from Zone I to Zone III (Fig. 2). Some work has been done on the enzyme topography of the preen gland (Lennert and Weitzel, 1952; Cater and Lawrie, 1950, 1951; Ishida *et al.*, 1971, 1973). Dehydrogenases, acid phosphatase, and unspecific esterase predominate in Zone III, whereas no esterase and acid phosphatase were detected in Zone I.

Some species lack a uropygial gland, e.g., *Struthio,* Rheiformes, Dromaiidae, Casuaridae, Otididae, and some Psittaciformes species *(Amazona, Ara)*. Other groups possess only rudimentary uropygial glands, e.g., Columbiformes, Ardeidae, and Rhynchochitidae. All these birds, however, possess powder downs which also are observed in the Tinamiformes, Ciconiiformes, Gruiformes, and in the family Artamidae (wood swallows) of the order Passeriformes. Perhaps these powder downs fulfill the functions of the preen wax in these species.

C. Uropygial Gland Secretions of Birds

1. Methods

The free flowing uropygial gland secretion can be obtained by scraping the preen tip of the living animal when simultaneously squeezing the gland area; this technique is recommended for large animals living in captivity (Odham, 1963; Bertelsen, 1973). It is more risky in the case of small animals. Most investigations have been performed with extracts of homogenized glands of freshly killed animals. The extirpation of the gland must be done carefully because the fat tissue which surrounds the preen gland is rich in triglycerides which might contaminate the preen secretion. Extraction is performed with chloroform/methanol (2 : 1, v : v). After addition of 1 volume of water the lipid remains in the hypophasic layer.

Thin-layer chromatography (TLC) is used in the initial examination of the crude lipids. In uropygial gland secretions monoester waxes predominate, but diester and triester waxes as well as triglycerides (as genuine secretions) and hydrocarbons have been observed. Different solvent systems for TLC of preen lipids have been reported (Jacob, 1975a). Lipids must be separated into single lipid classes before investigation by GLC and MS techniques. This is done by preparative TLC, or by column chromatography on silica gel (9–14% water content) if more than 1 mg is available. Details are described in various papers (Jacob and Zeman, 1972a; Jacob and Grimmer, 1973b; Poltz and Jacob, 1974a; Jacob, 1975a).

The purified lipids usually are complex mixtures which cannot be separated satisfactorily by GLC because of their low volatility. Lipids therefore must be cleaved to their constituent fatty acids and alcohols in the case of waxes. That can be achieved through acid reesterification by boiling with 5% methanolic HCl (Jacob and Grimmer, 1967) or BF_3-methanol (Morrison and Smith, 1964). Diester and triester waxes, containing hydroxy acids are best treated with 1 *N* methanolic NaOH (alkaline saponification) in order to avoid the production of artifacts. As the resulting free alcohols yield mass spectra which are difficult to interpret, they are converted to their acetates, trimethylsilyl ethers (Langer *et al.*, 1958), or oxidized with CrO_3 in tertiary butanol, acetic acid, and cyclohexane (Jacob and Zeman, 1970a; Jacob, 1975a) to the corresponding fatty acids which are subsequently esterified with methanolic HCl. This procedure produces two fractions of methyl esters from the original wax which can be identified by GLC/MS technique.

Diester waxes which are common in Galliformes species yield methyl esters and diols after reesterification. These diols can be analyzed by GLC/MS directly (Haahti and Fales, 1967; Jacob and Grimmer, 1970b,c) as their acetonides (Saito and Gamo, 1970, 1972; Jacob and Grimmer, 1970c) or TMS-ethers derivative (Sawaya and Kolattukudy, 1972) or, after oxidative cleavage with IO_4'/MnO_4' and esterification as methyl esters (Jacob and Grimmer, 1970b,c). Discrimination between *threo*- and *erythro*-diols can be achieved by TLC on silica gel plates impregnated with 5% boric acid (Morris, 1963; Thomas *et al.*, 1965; Hansen *et al.*, 1969).

Uropygial wax constituents are predominantly complex homologous and isomeric series of branched fatty acids and alcohols, although simple waxes containing only one acid and alcohol have been observed. Compounds with up to 5 methyl branches in the molecule have been detected and components with four methyl branches are common in some orders. The direct identification of the waxes by means of GLC had been successful only in relatively few species. The complexity of the waxes and their constituents requires high efficient GLC columns with at least 15,000 theoretical plates. For this purpose capillary columns (Jacob and Zeman, 1971b) or packed 10-m glass columns are recommended (Jacob and Zeman, 1972a). Chromatograms from both column types are given in Figs. 3 and 4.

The merits of the ECL values (equivalent chain length, relative retention time) for the identification of fatty acid methyl esters cannot be overestimated and a large body of data is now available for comparison of methyl esters (Nicolaides, 1971; Ackman, 1972; Jacob, 1975a; Jacob and Poltz, 1975a). A methyl substitution at a certain position of the carbon chain yields a definite shift of the retention time compared with the unsubstituted compound. Additional branches result in almost equal additional increments which must be added to the first as illustrated in Fig. 5.

The identification of lipid constituents is best achieved by mass spectrometry and a large body of material on mass spectrometry of lipids has been published. Although occasionally mass spectrometry of

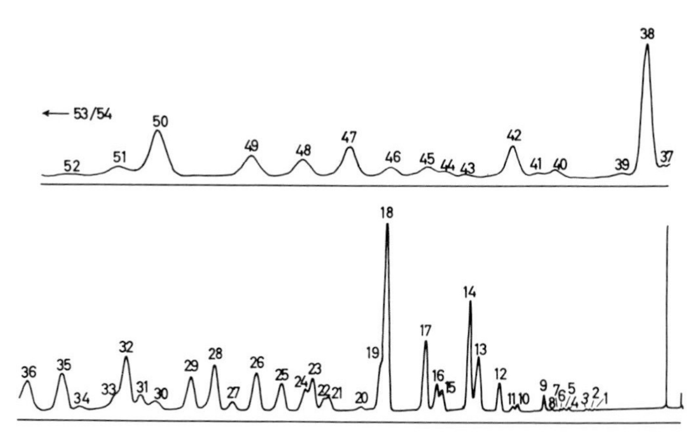

FIG. 3. GLC of wax fatty acid methyl esters of the uropygial gland secretion of the Coot *(Fulica atra)* (Jacob and Zeman 1972a). 50-m Capillary column with OV 101, split 1 : 20, column temperature 180°C, carrier gas 1.4 ml N_2/minute.

Nos. 3, 9, 13, 19, 27 — 2-methyl-C_{11-15}
Nos. 4, 10, 15, 21, 30, 40, 50 — 2,6-dimethyl-C_{11-17}
Nos. 5, 11, 16, 22, 30, 40 — 2,8-dimethyl-C_{11-16}
Nos. 8, 12, 17, 23, 41, 50 — 2,10-dimethyl-C_{11-17}
Nos. 25, 34, 44 — 2,12-dimethyl-C_{14-16}
Nos. 14, 18, 26, 35, 45, 53 — 2,6,10-trimethyl-C_{12-17}
Nos. 28, 37, 46 — 2,6,12- and 2,8,12-trimethyl-C_{14-16}
Nos. 24, 32, 42, 51 — 4,8- and 4,10-dimethyl-C_{14-17}
Nos. 33, 42, 51 — 4,12-dimethyl-C_{15-17}
Nos. 29, 38, 47, 54 — 4,8,12-trimethyl-C_{14-17}
No. 49 — 4,8,14-trimethyl-C_{16}
Nos. 1, 2, 6, 7, 20, 36, 39, 43, 48, 52 — unidentified

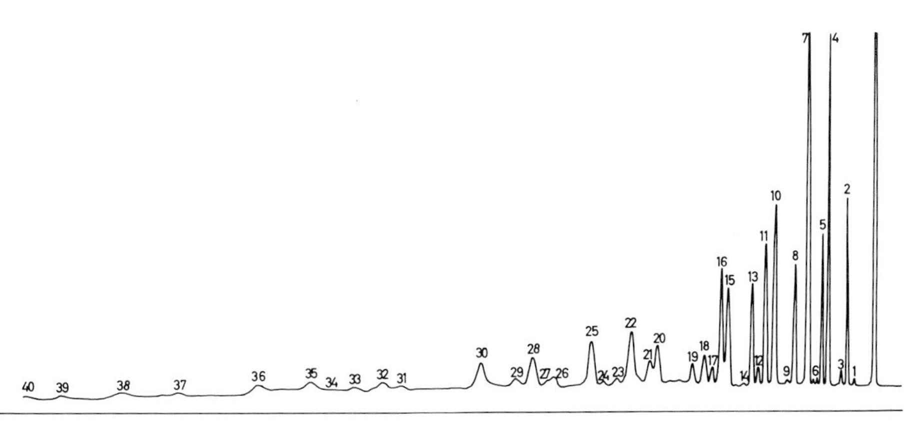

FIG. 4. GLC of the wax fatty acid methyl esters from the uropygial gland secretion of the Herring Gull (*Larus argentatus*). (Zeman and Jacob, 1972). 10-m Glass column packed with 5% OV 101 impregnated GasChrom Q, column temperature 180°C.

Nos. 2, 6, 9, 14	n-$C_{8,10,11,12}$
Nos. 1, 3, 5, 8, 13, 19	2-methyl-C_{7-12}
Nos. 4, 7, 10, 15, 20, 26, 31, 37	2,6-dimethyl-C_{8-15}
Nos. 11, 16, 21, 27, 32, 38	2,8-dimethyl-C_{10-14}
Nos. 22, 28, 33	2,10-dimethyl-$C_{12,13}$
Nos. 12, 17, 23, 29, 34	2,4,8-trimethyl-C_{10-14}
Nos. 18, 24	2,6,8-trimethyl-$C_{11,12}$
Nos. 25, 30, 35, 39	2,6,10-trimethyl-C_{12-15}
Nos. 36, 40	2,6,12-trimethyl-$C_{14,15}$

unhydrolized waxes had been successful, e.g., *Passer domesticus* (Jacob and Zeman, 1970a) and Ploceidae (Poltz and Jacob, 1973), generally the methyl esters resulting from methanolysis and subsequent oxidation of the wax alcohols are best used for MS identification. Mass spectra of methyl-substituted fatty acid esters have been discussed extensively and results are presented in several reviews (Ryhage and Stenhagen, 1960a,b, 1963; Zeman and Scharmann, 1972, 1973; Zeman and Jacob, 1973a; Jacob and Poltz, 1975a; Jacob, 1975a, 1976a,b). Mass spectra of ethyl-substituted fatty acid esters were reported as constituents of the preen waxes of Strigiformes (Jacob and Poltz, 1974a), Paridae (Poltz and Jacob, 1974a), and Podicipediformes (Jacob, 1977c). Propyl- and butyl-substituted fatty acid methyl esters occur in the secretion of some Strigiformes (Jacob and Poltz, 1974a; Jacob, 1974). Details on the methyl esters of 3-hydroxy acids are given by Ryhage and Stenhagen (1960c), Jacob and Zeman (1972b), Jacob and Grimmer (1975a), and data on alkyl hydroxy malonic acids, which are common in the uropygial gland lipids of birds in various orders are reported by Jacob and Grimmer (1973a). Mass spectra of 1,2-diols

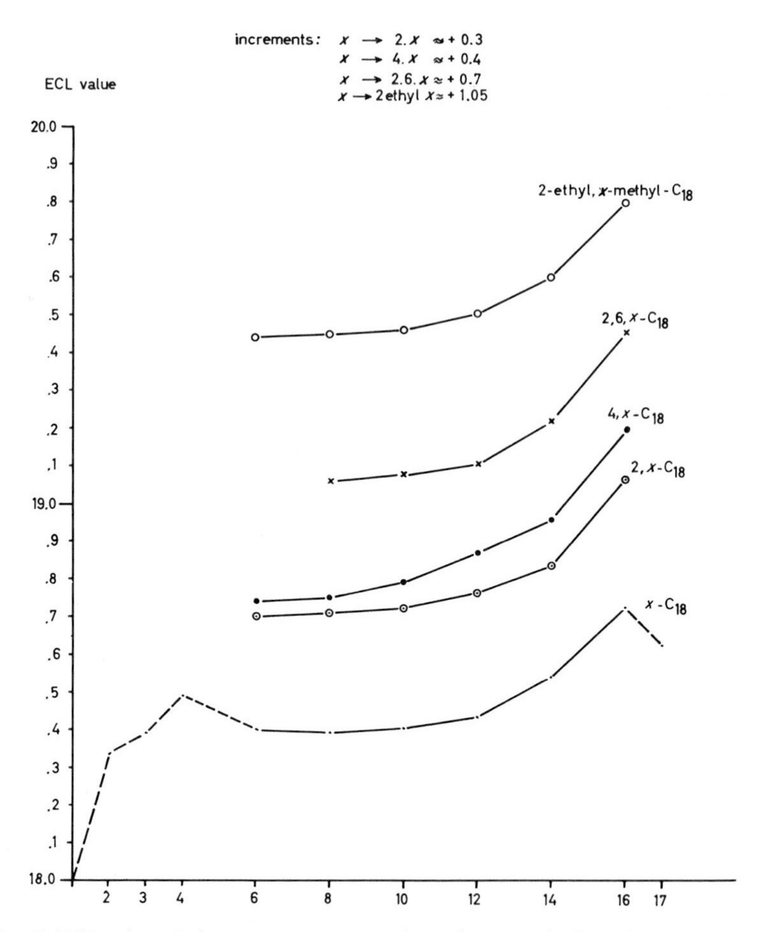

FIG. 5. ECL values (relative retention times) in relation to the branching position. The number of C atoms of the main chain is 18 in all cases.

detected in different species and of 2,3-diols from Galliformes species and their derivatives have been presented (Haahti and Fales, 1967; Jacob and Grimmer, 1970b,c; Saito and Gamo, 1970, 1972; Sawaya and Kolattukudy, 1972).

2. *Types of Waxes and their Distribution in Different Orders and Families*

Although monoester waxes predominate among the lipids of uropygial gland secretion, other lipids have been detected (Table II).

TABLE II
PREEN GLAND SECRETION LIPIDS AND THEIR DISTRIBUTION

Lipid	Distribution (species)
Hydrocarbons	
Squalene	Anseriformes (*Anseranas semipalmata, Cairina moschata*)
Paraffins (saturated, mono- and diunsaturated)	Podicipediformes (*Podiceps rolland, P. occipitalis*)
Monoester waxes	
Primary alcohols containing monoester waxes	Sphenisciformes
	Podicipediformes
	Procellariiformes
	Ciconiiformes
	Phoenicopteriformes
	Anseriformes
	Falconiformes
	Gruiformes
	Charadriiformes
	Lariformes
	Psittaciformes
	Cuculiformes
	Strigiformes
	Coraciiformes
	Piciformes
	Passeriformes
Secondary alcohols containing monoester waxes	*Nycticorax nycticorax*
Diester waxes	
Hydroxy acids containing diester waxes	Columbiformes
	Laniidae
1,2-Diols containing diester waxes	Gruiformes
	Piciformes
	Passeriformes
2,3-Diols containing diester waxes	Galliformes
Triester waxes	
Alkylhydroxymalonic acids containing lipids	Anseriformes
	Gruiformes
	Piciformes
	Passeriformes
Triglycerides	
As gland constituents	In all birds
As secretion	Ardeidae

Additional information on the quantitative composition of the uropygial wax constituents is given in Tables III and IV, and further details are available in the original papers or published reviews (Jacob, 1976a,b).

TABLE III

QUANTITATIVE COMPOSITION OF PREEN WAX ACIDS (MONOESTER WAXES) FROM THE HITHERTO INVESTIGATED SPECIES[a]

Species	n-	2-	3-	4-	6-	Iso-	Ante-iso-	other mono-	2,4-	2,x-	2,(ω − 1) 2,(ω − 2)	3,x	4,x-	Other di-	2,4,x-	2,x,y-	3,x,y- (3,x,y,z-)
Spheniscus magellanicus	—	19.1	—	37.6		—	—	2.0	—	25.4	—	—	9.0	—	—	6.9	—
Pygoscelis papua	—	1.7	67.3	6.7		—	—	—	—	9.9	—	9.0	3.5	—	—	—	—
Eudyptes crestatus	tr.	1.8	67.2	10.7		—	—	2.1	—	5.6	—	8.1	—	—	—	1.2	—
Procellaria aequinoctialis	—	tr.	22.2	4.1	2.0	—	—	—	5.3	—	—	30.9	—	—	18.0	8.0	7.7
Puffinus griseus	—	—	65.9	—	—	—	—	—	—	1.3	—	23.1	2.2	—	—	—	—
Pelecanoides urinatrix	—	—	50.0	6.0	—	—	—	—	—	—	—	29.4	8.4	—	—	—	—
Pachyptila belcheri	—	0.3	31.7	43.7	1.0	—	—	1.0	—	7.1	—	2.7	12.5	—	—	—	—
Fulmarus glacialis	—	0.4	53.3	22.6	4.0	—	—	—	0.5	6.0	—	8.8	0.4	—	—	—	—
Garrodia nereis	16.7	25.4	2.6	36.7	6.7	—	—	—	—	4.3	—	—	7.1	—	—	—	—
Diomedea melanophris	—	—	—	—	—	—	—	—	17.5	1.8	—	2.9	—	—	62.3	5.1	—
Podiceps occipitalis	—	—	22.0	—	—	—	—	—	—	4.2	—	38.3	—	—	4.5	11.4	6.9
Podiceps rolland	1.1	0.2	tr.	—	—	—	—	—	—	1.0	—	5.8	—	—	0.1	2.8	(35.7) 3.4
Ardea cinerea	100.0	—	—	—	—	—	—	—	—	—	—	—	—	—	—	—	—
Ciconia ciconia	100.0	—	—	—	—	—	—	—	—	—	—	—	—	—	—	—	—
Theristicus caudatus	2.1	—	—	—	—	—	—	—	—	—	—	—	—	—	—	—	—
Threskiornis aethiopicus	0.2	46.9	—	—	—	—	—	—	—	40.3	0.9	—	0.3	—	2.3	7.8	—
Scopus umbretta	11.5	8.5	—	53.7		—	—	—	—	4.1	—	—	15.9	5.1	—	—	—
Phoenicopterus ruber	—	—	—	—	—	—	—	—	—	35.7	—	—	3.8	—	53.4	7.1	—
Anseranas semipalmata	93.5	6.5	—	—	—	—	—	—	—	—	—	—	—	—	—	—	—
Dendrocygna viduata	—	—	—	—	—	—	—	—	—	—	—	—	—	—	56.8	—	—
Coscoroba coscoroba	—	—	—	—	—	—	—	—	—	—	—	—	—	—	100.0	—	—
Cygnus cygnus	—	—	—	—	—	—	—	—	—	—	—	—	—	—	100.0	—	—
Cygnus columbianus	—	—	—	—	—	—	—	—	—	—	—	—	—	—	100.0	—	—
Cygnus melanocoryphus	—	—	—	—	—	—	—	—	—	—	—	—	—	—	44.5	—	—
Cygnus olor	—	—	—	—	—	—	—	—	—	—	—	—	—	—	86.5	—	—
Cygnus atratus	—	—	—	—	—	—	—	—	78.3	—	—	—	—	—	21.7	—	—
Anser anser	—	—	—	—	—	—	—	—	—	—	—	—	—	—	—	—	—
Anser a. f. domesticus	—	—	—	—	—	—	—	—	—	—	—	—	—	—	—	—	—
Anser fabalis	—	—	—	—	—	—	—	—	—	—	—	—	—	—	89.3	—	—
Anser caerulescens	—	—	—	—	—	—	—	—	—	—	—	—	—	—	100.0	—	—
Anser indicus	—	—	—	—	—	—	—	—	2.0	—	—	—	—	—	87.3	—	—
Branta leucopsis	—	—	—	—	—	—	—	—	—	—	—	—	—	—	100.0	—	—
Tadorna ferruginea	—	—	—	—	—	—	—	—	—	—	—	—	—	—	95.4	—	—
Tadorna tadorna	—	—	—	—	—	—	—	—	—	—	—	—	—	—	96.3	—	—

Anas platyrhynchos	12.7	43.3	—	43.9	—	—	—	—	—	—	—	—	—	—	—	—	—
Anas pl. f. domesticus	2.5	44.2	—	50.2	—	—	—	—	—	—	—	—	—	—	—	—	—
Anas strepera	—	72.5	—	27.5	—	—	—	—	—	—	—	—	—	—	—	—	—
Anas clypeata	—	94.5	—	—	—	—	—	—	—	2.9	—	—	2.6	—	—	—	—
Tachyeres patachonicus	13.6	9.0	—	7.3	—	—	—	—	—	12.0	—	—	14.1	—	15.1	7.8	—
Tachyeres pteneres	34.4	3.9	—	12.5	—	—	—	10.6	2.4	—	—	—	15.5	—	—	—	—
Cairina moschata	—	—	—	—	—	—	—	—	—	—	—	—	—	—	100.0	—	—
Aythya ferina	—	—	—	—	—	—	—	—	1.3	—	—	—	—	—	84.8	—	—
Aythya fuligula	—	6.3	—	—	—	—	—	—	1.5	21.9	—	—	—	—	52.7	10.2	—
Somateria mollissima	4.3	—	—	—	—	—	—	—	1.3	31.8	—	—	4.9	—	39.6	8.6	—
Mergus serrator	—	—	—	—	—	—	—	—	—	—	—	—	—	—	100.0	—	—
Mergus albellus	—	—	—	—	—	—	—	—	1.4	1.2	—	—	—	—	97.4	—	—
Melanitta nigra	—	—	—	—	—	—	—	—	5.0	—	—	—	—	—	95.0	—	—
Alcedo atthis	—	—	—	—	—	—	—	—	—	—	—	—	—	—	—	—	—
Falco tinnunculus	0.9	37.1	—	—	—	—	—	—	—	36.1	8.9	—	—	—	—	15.7	—
Falco columbarius	5.3	29.1	—	1.2	—	—	—	0.5	—	35.8	11.6	—	0.3	—	—	16.2	—
Accipiter nisus	—	20.7	—	—	—	—	—	—	—	34.5	36.2	—	—	—	—	8.6	—
Accipiter gentilis	1.6	12.7	—	0.7	—	—	—	0.9	—	34.5	21.1	—	1.1	—	—	23.8	—
Buteo buteo	0.5	0.8	—	—	—	—	—	0.1	—	24.8	15.9	—	—	—	—	49.9	—
Rallus aquaticus	—	4.9	—	—	—	—	—	—	—	35.2	2.6	—	0.7	—	—	55.4	—
Gallinula chloropus	—	12.9	—	1.1	—	—	—	—	—	12.8	3.6	—	35.5	—	—	13.1	—
Porzana porzana	—	3.4	—	0.6	—	—	—	—	—	10.1	2.3	—	48.0	—	—	12.5	—
Fulica atra	—	4.9	—	—	—	—	—	—	—	20.4	2.3	—	13.3	—	—	25.1	—
Haematopus ostralegus	1.3	60.9	—	1.3	1.0	—	—	—	—	18.0	15.2	—	1.8	—	—	—	—
Tringa totanus	0.9	13.3	—	0.3	—	—	—	—	—	47.4	14.0	—	2.6	—	—	13.4	—
Calidris canutus	tr.	30.5	—	11.8	0.2	—	—	—	—	33.3	12.4	—	5.9	—	—	2.3	—
Calidris alpina	1.3	11.3	—	0.6	0.4	—	—	—	—	50.6	10.8	—	0.3	—	—	13.4	—
Fratercula arctica	100.0	—	—	—	—	—	—	—	—	—	—	—	—	—	—	—	—
Uria aalge	41.8	0.4	—	2.0	0.7	—	—	0.7	—	—	—	—	21.2	—	0.4	—	—
Cepphus grylle	49.2	2.5	—	1.5	0.2	—	—	0.2	—	17.8	—	—	8.6	—	1.1	16.7	—
Alca torda	15.0	9.7	—	8.1	12.9	—	—	2.3	—	—	—	—	23.0	—	—	—	—
Thinocorus rumicivorus	4.0	14.2	0.2	—	—	—	—	—	—	48.4	7.3	—	—	—	—	24.7	—
Gallinago gallinago	1.9	41.0	—	—	—	—	—	—	—	38.8	9.9	—	—	—	—	—	—
Rissa tridactyla	8.5	46.6	—	—	—	—	—	—	—	32.4	—	—	—	—	—	—	—
Larus ridibundus	37.6	5.2	—	—	—	—	—	—	—	24.9	—	—	—	—	5.6	26.7	—
Larus fuscus	25.4	10.2	—	—	—	—	—	—	—	41.5	—	—	—	—	9.6	13.3	—
Larus argentatus	15.7	14.4	—	—	—	—	—	—	—	52.4	—	—	—	—	3.0	14.5	—
Melopsittacus undulatus	20.9	15.2	—	—	—	32.7		14.1	—	5.0	—	—	12.1		—	—	—
Loriculus sp.	1.0	—	—	—	—	78.6	16.4	2.7	—	—	—	—	—	—	—	—	—
Psittacula eupatria	9.8	—	—	—	—	74.4	3.4	—	—	—	—	—	4.9	—	—	—	—

TABLE III (*Continued*)

Species	*n*-	2-	3-	4-	6-	Iso-	Ante-iso-	other mono-	2,4-	2,*x*-	2,(ω − 1) 2,(ω − 2)	3,*x*	4,*x*-	Other di-	2,4,*x*-	2,*x*,*y*-	3,*x*,*y*- (3,*x*,*y*,*z*-)
Psittaculirostris desmarestii	3.1	—	—	0.3	—	—	—	—	—	—	—	—	1.3	—	—	3.6	—
Chalcopsitta atra	—	1.7	—	7.7	—	—	—	—	—	4.1	6.5	—	49.4	16.8	—	3.0	—
Agapornis fisheri	—	—	—	59.0	—	—	—	—	—	—	—	—	36.7	—	—	—	—
Asio otus	—	—	—	—	—	—	—	—	—	—	—	—	—	—	—	—	—
Bubo bubo	2.6	1.7	—	—	—	—	—	—	—	26.5	—	—	—	—	—	2.5	—
Tyto alba	—	—	59.3	—	—	—	—	—	—	—	—	39.7	—	—	—	—	1.0
Cuculus canorus	5.6	—	76.8	8.9	—	—	—	—	—	—	—	4.1	—	—	—	—	—
Dryocopus martius	—	—	54.2	—	—	—	—	—	—	—	—	39.5	—	—	—	—	2.7
Picus viridis	—	—	64.0	—	—	—	—	—	—	—	—	34.5	—	—	—	—	0.8
Coloeus monedula	35.7	64.3	—	—	—	—	—	—	—	—	—	—	—	—	—	—	—
Corvus frugilegus	—	100.0	—	—	—	—	—	—	—	—	—	—	—	—	—	—	—
Corvus corax	—	85.3	—	—	—	—	—	—	—	14.4	—	—	—	—	—	—	—
Corvus corone corone	—	—	—	34.8	3.2	—	—	—	—	—	—	—	42.3	—	—	—	—
Corvus corone cornix	—	—	—	34.2	2.7	—	—	—	—	—	—	—	43.0	—	—	—	—
Pyrrhocorax graculus	—	37.4	—	—	—	—	—	—	—	62.3	—	—	—	—	—	—	—
Pica pica	1.7	34.2	—	—	—	—	—	—	—	47.0	—	—	—	—	—	17.1	—
Garrulus glandarius	0.5	9.8	—	—	—	—	—	—	—	59.1	—	—	—	—	—	30.2	—
Oriolus oriolus	1.3	9.1	—	3.7	—	—	—	—	—	64.2	—	—	—	—	—	10.4	—
Bombycilla garrulus	—	30.6	—	—	—	—	—	—	—	61.2	1.9	—	—	—	—	6.3	—
Sturnus vulgaris	15.2	30.1	—	30.1	—	—	—	—	—	23.5	—	—	—	—	—	—	—
Gracula religiosa	—	1.2	—	43.9	—	—	—	—	—	5.7	—	—	33.7	—	—	13.9	—
Acrocephalus melanopogon	0.4	5.9	—	—	—	—	—	—	—	64.6	—	—	—	—	—	28.9	—
Acrocephalus schoenebaenus	0.7	8.2	—	—	—	—	—	—	—	49.2	—	—	—	—	—	38.8	—
Acrocephalus palustris	1.1	1.6	—	—	—	—	—	—	—	24.0	—	—	—	—	—	50.8	—
Acrocephalus scirpaceus	0.9	3.7	—	—	—	—	—	—	—	19.6	—	—	—	—	—	43.2	—
Acrocephalus arundinaceus	0.8	—	—	—	—	—	—	—	—	23.8	—	—	—	—	—	44.5	—
Sylvia atricapilla	1.4	10.6	—	—	—	—	—	—	—	46.8	—	—	—	—	—	33.7	—
Sylvia curruca	2.7	45.2	—	—	—	—	—	—	—	50.6	—	—	—	—	—	1.5	—
Sylvia communis	1.2	1.5	—	—	—	—	—	—	—	42.0	—	—	—	—	1.0	48.2	—
Certhia familiaris	5.1	33.2	—	—	—	—	—	—	—	54.0	—	—	—	—	—	7.7	—
Certhia brachydactyla	3.8	75.2	—	—	—	—	—	—	—	20.1	—	—	—	—	—	0.9	—
Leiothris lutea	0.5	73.5	—	—	—	—	—	—	—	21.8	—	—	—	—	—	4.2	—

Mesia argentauris	47.1	51.5	—	—	—	—	—	—	—	1.4	—	—	—	—	—	—	—
Paradoxornis gularis	2.2	12.4	—	—	—	—	—	—	—	44.0	—	—	—	—	—	36.3	—
Paradoxornis webbianus	—	1.1	—	—	—	—	—	—	—	40.3	—	—	—	—	—	48.0	—
Regulus regulus	1.3	26.8	—	—	—	—	—	—	—	68.5	—	—	—	—	—	3.4	—
Siva cyanouroptera	tr.	16.0	—	—	—	—	—	—	—	54.2	—	—	—	—	—	29.8	—
Remiz pendulinus	0.9	13.2	—	—	—	—	—	—	—	54.3	—	—	—	—	—	31.6	—
Aegithalos caudatus	14.2	31.6	—	—	—	—	—	—	—	50.0	—	—	—	—	—	4.2	—
Panurus biarmicus	0.3	2.6	—	—	—	—	—	—	—	49.8	—	—	—	—	—	47.3	—
Parus ater	5.9	—	—	—	—	—	—	—	—	1.0	—	—	—	—	—	—	—
Parus melanolophus	6.5	—	—	—	—	—	—	—	—	2.3	—	—	—	—	—	—	—
P. ater x P. melanolophus	1.3	—	—	—	—	—	—	—	—	1.5	—	—	—	—	—	—	—
Parus maior	2.1	—	—	—	—	—	—	—	—	—	—	—	—	—	—	—	—
Parus caeruleus	1.7	—	—	—	—	—	—	—	—	0.5	—	—	—	—	—	—	—
Parus atricapillus montanus	0.8	—	—	—	—	—	—	—	—	0.9	—	—	—	—	—	—	—
Parus atricapillus rhenanus	0.2	—	—	—	—	—	—	—	—	0.7	—	—	—	—	—	—	—
Parus xanthogenys	1.0	—	—	—	—	—	—	—	—	0.2	—	—	—	—	—	—	—
Emberiza schoeniclus	—	—	89.4	—	—	—	—	—	—	—	—	10.3	—	—	—	—	—
Emberiza bruniceps	—	—	87.3	—	—	—	—	—	—	—	—	12.7	—	—	—	—	—
Fringilla coelebs	8.4	—	88.6	—	—	—	—	—	—	—	—	3.0	—	—	—	—	—
Fringilla montifringilla	tr.	—	94.2	—	—	—	—	—	—	—	—	5.4	—	—	—	—	—
Serinus serinus	—	—	96.6	—	—	—	—	—	—	—	—	1.4	—	—	—	—	—
Carduelis spinus	2.4	—	88.5	—	—	—	—	—	—	—	—	9.1	—	—	—	—	—
Carduelis chloris	—	—	100.0	—	—	—	—	—	—	—	—	—	—	—	—	—	—
Carduelis flammea	1.6	—	96.3	—	—	—	—	—	—	—	—	2.1	—	—	—	—	—
Carduelis cannabina	—	—	100.0	—	—	—	—	—	—	—	—	—	—	—	—	—	—
Loxia curvirostra	0.8	—	97.3	—	—	—	—	—	—	—	—	1.9	—	—	—	—	—
Pyrrhula pyrrhula	—	—	100.0	—	—	—	—	—	—	—	—	—	—	—	—	—	—
Passer domesticus	—	—	100.0	—	—	—	—	—	—	—	—	—	—	—	—	—	—
Passer montanus	—	—	76.4	—	—	—	—	—	—	—	—	17.7	—	—	—	—	—
Ploceus cucullatus	—	—	—	—	—	—	—	—	100.0	—	—	—	—	—	—	—	—
Ploceus subaureus	—	—	—	—	—	—	—	—	100.0	—	—	—	—	—	—	—	—
Ploceus galbula	—	—	—	—	—	—	—	—	100.0	—	—	—	—	—	—	—	—
Ploceus sp.	—	—	—	—	—	—	—	—	100.0	—	—	—	—	—	—	—	—
Quelea quelea	—	—	—	—	—	—	—	—	0.4	—	—	—	—	—	99.6	—	—
Cinclus cinclus	—	—	—	—	—	—	—	—	—	—	—	—	—	—	100.0	—	—

TABLE III (*Continued*)

Species	4,*x*,*y*-	2,4,*x*,*y*-	2,6,*x*,*y*-	4,*x*,*y*,*z*-	Penta-	2-Ethyl (diethyl)	4-Ethyl- (4e-*x*-methyl)	2-Propyl- (2-butyl-)	2-Ethyl-*x*-methyl	4-Ethyl-*x*-methyl	2-Ethyl-*x*,*y*-dimethyl	Unidentified
Spheniscus magellanicus	—	—	—	—	—	—	—	—	—	—	—	—
Pygoscelis papua	—	—	—	—	—	—	—	—	—	—	—	1.9
Eudypzes crestatus	—	—	—	—	—	—	—	—	—	—	—	3.3
Procellaria aequinoctialis	—	—	—	—	—	—	—	—	—	—	—	1.8
Puffinus griseus	—	—	—	—	—	—	—	—	—	—	—	7.5
Pelecanoides urinatrix	—	—	—	—	—	—	—	—	—	—	—	6.2
Pachyptila belcheri	—	—	—	—	—	—	—	—	—	—	—	—
Fulmarus glacialis	—	—	—	—	—	—	—	—	—	—	—	4.0
Garrodia nereis	—	—	—	—	—	—	—	—	—	—	—	0.5
Diomedea melanophris	—	—	—	—	—	—	—	—	—	—	—	10.4
Podiceps occipitalis	—	—	—	—	—	4.6	—	—	6.3	—	—	1.8
Podiceps rolland	4.4	4.7	—	—	—	3.0	—	—	12.0	—	17.4	8.4
Ardea cinerea	—	—	—	—	—	—	—	—	—	—	—	—
Ciconia ciconia	—	—	—	—	—	—	—	—	—	—	—	—
Theristicus caudatus	—	—	—	—	—	22.4 (3.0)	33.7	—	16.5	11.4	—	10.9
Threskiornis aethiopicus	—	—	—	—	—	—	—	—	—	—	—	1.3
Scopus umbretta	—	—	—	—	—	—	—	—	—	—	—	1.2
Phoenicopterus ruber	—	—	—	—	—	—	—	—	—	—	—	—
Anseranas semipalmata	—	—	—	—	—	—	—	—	—	—	—	—
Dendrocygna viduata	—	43.2	—	—	—	—	—	—	—	—	—	—
Coscoroba coscoroba	—	—	—	—	—	—	—	—	—	—	—	—
Cygnus cygnus	—	—	—	—	—	—	—	—	—	—	—	—
Cygnus columbianus	—	—	—	—	—	—	—	—	—	—	—	—
Cygnus melanocoryphus	—	54.2	—	—	—	—	—	—	—	—	—	1.3
Cygnus olor	—	13.5	—	—	—	—	—	—	—	—	—	—
Cygnus atratus	—	—	—	—	—	—	—	—	—	—	—	—
Anser anser	—	100.0	—	—	—	—	—	—	—	—	—	—
Anser a. f. domesticus	—	100.0	—	—	—	—	—	—	—	—	—	—
Anser fabalis	—	10.7	—	—	—	—	—	—	—	—	—	—
Anser caerulescens	—	—	—	—	—	—	—	—	—	—	—	—
Anser indicus	—	10.7	—	—	—	—	—	—	—	—	—	—
Branta leucopsis	—	—	—	—	—	—	—	—	—	—	—	—

Tadorna ferruginea	—	4.6	—	—	—	—	—	—	—	—	—	—
Tadorna tadorna	—	3.7	—	—	—	—	—	—	—	—	—	—
Anas platyrhynchos	—	—	—	—	—	—	—	—	—	—	—	0.1
Anas pl. f. domesticus	—	—	—	—	—	—	—	—	—	—	—	3.1
Anas strepera	—	—	—	—	—	—	—	—	—	—	—	—
Anas clypeata	—	—	—	—	—	—	—	—	—	—	—	—
Tachyeres patachonicus	4.7	—	—	—	—	—	—	—	—	—	—	16.4
Tachyeres pteneres	3.5	—	—	—	—	—	—	—	—	—	—	17.2
Cairina moschata	—	—	—	—	—	—	—	—	—	—	—	—
Aythya ferina	—	13.9	—	—	—	—	—	—	—	—	—	—
Aythya fuligula	—	3.6	—	—	—	—	—	—	—	—	—	3.8
Somateria mollissima	—	3.6	—	—	—	—	—	—	—	—	—	5.9
Mergus serrator	—	—	—	—	—	—	—	—	—	—	—	—
Mergus albellus	—	—	—	—	—	—	—	—	—	—	—	—
Melanitta nigra	—	—	—	—	—	—	—	—	—	—	—	—
Alcedo atthis	—	93.5	—	—	6.5	—	—	—	—	—	—	—
Falco tinnunculus	—	—	—	—	—	—	—	—	—	—	—	1.3
Falco columbarius	—	—	—	—	—	—	—	—	—	—	—	—
Accipiter nisus	—	—	—	—	—	—	—	—	—	—	—	—
Accipiter gentilis	—	—	—	—	—	—	—	—	—	—	—	3.6
Buteo buteo	—	—	3.9	—	—	—	—	—	—	—	—	4.1
Rallus aquaticus	—	—	1.2	—	—	—	—	—	—	—	—	—
Gallinula chloropus	21.0	—	—	—	—	—	—	—	—	—	—	—
Porzana porzana	23.1	—	—	—	—	—	—	—	—	—	—	—
Fulica atra	28.2	—	—	—	—	—	—	—	—	—	—	5.8
Haematopus ostralegus	—	—	—	—	—	—	—	—	—	—	—	0.5
Tringa totanus	—	—	—	—	—	—	—	—	—	—	—	8.1
Calidris canutus	—	—	—	—	—	—	—	—	—	—	—	3.6
Calidris alpina	—	—	—	—	—	—	—	—	—	—	—	11.3
Fratercula arctica	—	—	—	—	—	—	—	—	—	—	—	—
Uria aalge	16.7	—	—	—	—	—	—	—	—	—	—	16.1
Cepphus grylle	1.7	—	—	—	—	—	—	—	—	—	—	0.5
Alca torda	17.0	—	—	—	—	—	—	—	—	—	—	12.0
Thinocorus rumicivorus	—	—	—	—	—	—	—	—	—	—	—	1.2
Gallinago gallinago	—	—	—	—	—	—	—	—	—	—	—	2.8
Rissa tridactyla	—	—	—	—	—	—	—	—	—	—	—	12.5
Larus ridibundus	—	—	—	—	—	—	—	—	—	—	—	—
Larus fuscus	—	—	—	—	—	—	—	—	—	—	—	—
Larus argentatus	—	—	—	—	—	—	—	—	—	—	—	—

TABLE III (*Continued*)

Species	4,*x*,*y*-	2,4,*x*,*y*-	2,6,*x*,*y*-	4,*x*,*y*,*z*-	Penta-	2-Ethyl (diethyl)	4-Ethyl- (4e-*x*-methyl)	2-Propyl- (2-butyl-)	2-Ethyl- *x*-methyl	4-Ethyl- *x*-methyl	2-Ethyl- *x*,*y*-dimethyl	Uniden-tified
Melopsittacus undulatus	—	—	—	—	—	—	—	—	—	—	—	—
Loriculus sp.	—	—	—	—	—	—	—	—	—	—	—	1.3
Psittacula eupatria	—	—	—	—	—	—	—	—	—	—	—	7.5
Psittaculirostris desmarestii	16.1	—	46.3	23.9	—	—	—	—	—	—	—	5.4
Chalcopsitta atra	10.1	—	—	—	—	—	—	—	—	—	—	0.7
Agapornis fisheri	—	—	—	—	—	—	—	—	—	—	—	4.3
Asio otus	—	—	—	—	—	24.4	—	{10.0 (55.6)}	—	—	—	10.0
Bubo bubo	—	—	—	—	—	42.3	—	—	9.9	—	—	14.5
Tyto alba	—	—	—	—	—	—	—	—	—	—	—	—
Cuculus canorus	—	—	—	—	—	—	—	—	—	—	—	4.6
Dryocopus martius	—	—	—	—	—	—	—	—	—	—	—	3.6
Picus viridis	—	—	—	—	—	—	—	—	—	—	—	0.7
Coloeus monedula	—	—	—	—	—	—	—	—	—	—	—	—
Corvus frugilegus	—	—	—	—	—	—	—	—	—	—	—	—
Corvus corax	—	—	—	—	—	—	—	—	—	—	—	0.3
Corvus corone corone	11.2	—	—	—	—	—	—	—	—	—	—	8.5
Corvus corone cornix	11.1	—	—	—	—	—	—	—	—	—	—	9.0
Pyrrhocorax graculus	—	—	—	—	—	—	—	—	—	—	—	0.3
Pica pica	—	—	—	—	—	—	—	—	—	—	—	—
Garrulus glandarius	—	—	—	—	—	—	—	—	—	—	—	0.4
Oriolus oriolus	—	—	—	—	—	—	—	—	—	—	—	11.3
Bombycilla garrulus	—	—	—	—	—	—	—	—	—	—	—	—
Sturnus vulgaris	—	—	—	—	—	—	—	—	—	—	—	1.1
Gracula religiosa	1.6	—	—	—	—	—	—	—	—	—	—	—
Acrocephalus melanopogon	—	—	—	—	—	—	—	—	—	—	—	0.2
Acrocephalus schoenebaenus	—	—	—	—	—	—	—	—	—	—	—	3.1
Acrocephalus palustris	—	—	20.0	—	—	—	—	—	—	—	—	2.5
Acrocephalus scirpaceus	—	—	31.4	—	—	—	—	—	—	—	—	1.2
Acrocephalus arundinaceus	—	—	30.0	—	—	—	—	—	—	—	—	0.9
Sylvia atricapilla	—	—	7.2	—	—	—	—	—	—	—	—	0.3
Sylvia curruca	—	—	—	—	—	—	—	—	—	—	—	—
Sylvia communis	—	—	5.7	—	—	—	—	—	—	—	—	0.4

Certhia familiaris	—	—	—	—	—	—	—	—	—	—	—	—
Certhia brachydactyla	—	—	—	—	—	—	—	—	—	—	—	—
Leiothris lutea	—	—	—	—	—	—	—	—	—	—	—	—
Mesia argentauris	—	—	—	—	—	—	—	—	—	—	—	—
Paradoxornis gularis	—	—	5.1	—	—	—	—	—	—	—	—	—
Paradoxornis webbianus	—	—	10.6	—	—	—	—	—	—	—	—	—
Regulus regulus	—	—	—	—	—	—	—	—	—	—	—	—
Siva cyanouroptera	—	—	—	—	—	—	—	—	—	—	—	—
Remiz pendulinus	—	—	—	—	—	—	—	—	—	—	—	—
Aegithalos caudatus	—	—	—	—	—	—	—	—	—	—	—	—
Panurus biarmicus	—	—	—	—	—	—	—	—	—	—	—	—
Parus ater	—	—	—	—	—	43.9	—	—	49.1	—	—	0.1
Parus melanolophus	—	—	—	—	—	75.2	—	—	14.5	—	—	1.5
P. ater x P. melanolophus	—	—	—	—	—	60.3	—	—	35.4	—	0.6	0.9
Parus maior	—	—	—	—	—	68.7	—	—	28.6	—	0.2	0.4
Parus caeruleus	—	—	—	—	—	49.2	—	—	46.8	—	1.6	0.2
Parus atricapillus montanus	—	—	—	—	—	56.3	—	—	40.9	—	1.1	—
Parus atricapillus rhenanus	—	—	—	—	—	70.0	—	—	28.9	—	0.2	—
Parus xanthogenys	—	—	—	—	—	74.2	—	—	24.4	—	—	0.2
Emberiza schoeniclus	—	—	—	—	—	—	—	—	—	—	—	0.3
Emberiza bruniceps	—	—	—	—	—	—	—	—	—	—	—	—
Fringilla coelebs	—	—	—	—	—	—	—	—	—	—	—	—
Fringilla montifringilla	—	—	—	—	—	—	—	—	—	—	—	0.4
Serinus serinus	—	—	—	—	—	—	—	—	—	—	—	2.0
Carduelis spinus	—	—	—	—	—	—	—	—	—	—	—	—
Carduelis chloris	—	—	—	—	—	—	—	—	—	—	—	—
Carduelis flammea	—	—	—	—	—	—	—	—	—	—	—	—
Carduelis cannabina	—	—	—	—	—	—	—	—	—	—	—	—
Loxia curvirostra	—	—	—	—	—	—	—	—	—	—	—	—
Pyrrhula pyrrhula	—	—	—	—	—	—	—	—	—	—	—	—
Passer domesticus	—	—	—	—	—	—	—	—	—	—	—	—
Passer montanus	—	—	—	—	—	—	—	—	—	—	—	5.9
Ploceus cucullatus	—	—	—	—	—	—	—	—	—	—	—	—
Ploceus subaureus	—	—	—	—	—	—	—	—	—	—	—	—
Ploceus galbula	—	—	—	—	—	—	—	—	—	—	—	—
Ploceus sp.	—	—	—	—	—	—	—	—	—	—	—	—
Quelea quelea	—	—	—	—	—	—	—	—	—	—	—	—
Cinclus cinclus	—	—	—	—	—	—	—	—	—	—	—	—

[a] Numbers indicate position of substitution. References see Table IV.

TABLE IV

QUANTITATIVE COMPOSITION OF THE PREEN WAX ALCOHOLS FROM VARIOUS BIRDS (MONOESTER WAXES ONLY)

Species	Un-branched	Mono-methyl-subst.	Di-methyl-subst.	Tri-methyl-subst.	Tetra-methyl-subst.	Others	References
Spheniscus magellanicus	1.4	59.6	37.8	—	—	1.2	Jacob (1976e)
Pygoscelis papua	—	35.6	61.2	1.4	—	1.8	Jacob (1976e)
Eudyptes crestatus	1.0	43.5	54.5	0.8	—	0.2	Jacob (1976e)
Procellaria aequinoctialis	—	29.8	45.9	20.2	—	4.1	Jacob (1976d)
Puffinus griseus	2.9	44.0	34.8	4.1	—	14.2	Jacob (1976d)
Pelecanoides urinatrix	43.2	43.9	9.8	—	—	3.1	Jacob (1976d)
Pachyptila belcheri	43.2	52.5	3.4	—	—	0.9	Jacob (1976d)
Fulmarus glacialis	40.5	41.7	1.4	—	—	16.4	Jacob and Zeman (1971a)
Garrodia nereis	69.2	30.8	—	—	—	—	Jacob (1976d)
Diomedea melanophris	5.2	37.4	38.8	10.9	—	7.7	Jacob (1976d)
Podiceps occipitalis	4.9	31.4	56.7	6.1	—	0.9	Jacob (1977c)
Podiceps rolland	2.3	5.9	18.9	55.2	7.0	10.7	Jacob (1977c)
Ardea cinerea	100.0	—	—	—	—	—	Poltz and Jacob (1974b)
Ciconia ciconia	100.0	—	—	—	—	—	J. Jacob (unpublished results, 1975)
Theristicus caudatus	55.7	38.9	3.6	—	—	1.8	J. Jacob (unpublished results, 1975)
Threskiornis aethiopicus	—	54.7	44.3	1.0	—	—	J. Jacob (unpublished results, 1975)
Scopus umbretta	87.1	12.2	0.7	—	—	—	J. Jacob (unpublished results, 1975)
Phoenicopterus ruber	87.3	12.7	—	—	—	—	Bertelsen (1970)
Anseranas semipalmata	97.8	2.2	—	—	—	—	Edkins and Hansen (1972)
Dendrocygna viduata	100.0	—	—	—	—	—	Odham (1967b)
Coscoroba coscoroba	100.0	—	—	—	—	—	Bertelsen (1973)
Cygnus cygnus	79.6	20.4	—	—	—	—	Bertelsen (1973)
Cygnus columbianus	74.8	25.2	—	—	—	—	Bertelsen (1973)
Cygnus melanocoryphus	80.3	19.7	—	—	—	—	Bertelsen (1973)
Cygnus olor	73.0	27.0	—	—	—	—	Odham (1965)
Cygnus atratus	64.9	35.1	—	—	—	—	Edkins and Hansen (1972)

Anser anser	99.1	0.9	—	—	—	—	Jacob and Glaser (1975)
Anser anser f. domesticus	100.0	—	—	—	—	—	Odham (1963)
Anser fabalis	88.4	11.6	—	—	—	—	Jacob and Glaser (1975)
Anser caerulescens	87.1	12.9	—	—	—	—	Jacob and Glaser (1975)
Anser indicus	95.0	5.0	—	—	—	—	Jacob and Glaser (1975)
Branta leucopsis	98.4	1.6	—	—	—	—	Odham (1967b)
Tadorna ferruginea	98.3	1.7	—	—	—	—	Odham (1966)
Tadorna tadorna	98.6	1.4	—	—	—	—	Odham (1966)
Anas platyrhynchos	62.0	37.3	—	—	—	0.7	Odham (1964)
Anas platyrhynchos f. domesticus	40.9	52.1	—	—	—	7.0	Odham (1967c)
Anas strepera	65.6	34.4	—	—	—	—	Jacob and Glaser (1975)
Anas clypeata	60.2	39.8	—	—	—	—	Jacob and Glaser (1975)
Tachyeres patachonicus	52.3	42.0	5.3	—	—	0.4	Jacob (1977a)
Tachyeres pteneres	64.5	34.3	0.9	—	—	0.3	Jacob (1977a)
Cairina moschata	100.0	—	—	—	—	—	Odham (1967b)
Aythya ferina	88.5	11.5	—	—	—	—	Jacob and Glaser (1975)
Aythya fuligula	72.7	27.3	—	—	—	—	Jacob and Zeman (1970b)
Somateria mollissima	94.9	5.1	—	—	—	—	Odham (1967b)
Mergus serrator	100.0	—	—	—	—	—	Odham (1967b)
Mergus albellus	99.0	1.0	—	—	—	—	Jacob and Glaser (1975)
Melanitta nigra	77.6	22.4	—	—	—	—	Jacob and Zeman (1972c)
Alcedo atthis	4.0	—	—	—	94.7	1.3	Jacob (1976c)
Falco tinnunculus	32.1	56.6	10.3	—	—	1.0	Jacob and Poltz (1975a)
Falco columbarius	10.3	68.5	21.2	—	—	—	Jacob and Poltz (1975a)
Accipiter nisus	13.8	65.1	19.1	—	—	2.0	Jacob and Poltz (1975a)
Accipiter gentilis	3.4	41.3	52.0	—	—	3.3	Jacob and Poltz (1975a)
Buteo buteo	1.0	41.6	48.0	4.0	—	5.4	Jacob and Poltz (1975a)
Rallus aquaticus	54.0	41.0	4.3	—	—	0.7	Jacob and Poltz (1975b)
Gallinula chloropus	76.3	21.0	2.7	—	—	—	Jacob and Poltz (1975b)
Porzana porzana	80.8	17.7	1.5	—	—	—	Jacob and Poltz (1975b)
Fulica atra	60.6	25.6	—	—	—	13.8	Jacob and Zeman (1971b)
Haematopus ostralegus	49.6	48.1	—	—	—	2.3	Jacob and Poltz (1973a); Karlsson and Odham (1969)

TABLE IV (*Continued*)

Species	Un-branched	Mono-methyl-subst.	Di-methyl-subst.	Tri-methyl-subst.	Tetra-methyl-subst.	Others	References
Tringa totanus	46.6	32.4	16.6	—	—	4.4	Jacob and Poltz (1973a)
Calidris canutus	57.8	35.7	3.7	—	—	2.8	Jacob and Poltz (1973a)
Calidris alpina	47.1	38.1	8.8	—	—	6.0	Jacob and Poltz (1973a)
Fratercula arctica	100.0	—	—	—	—	—	Jacob and Grimmer (1970a)
Uria aalge	60.3	28.4	5.3	—	—	6.0	Jacob and Zeman (1973)
Cepphus grylle	49.0	15.8	22.5	—	—	12.7	Jacob and Zeman (1973)
Alca torda	43.4	26.2	13.2	—	—	17.2	Jacob and Zeman (1973)
Thinocorus rumicivorus	43.6	46.0	10.4	—	—	—	J. Jacob (unpublished results, 1975)
Gallinago gallinago	0.9	71.8	25.9	—	—	1.4	J. Jacob (unpublished results, 1975)
Rissa tridactyla	46.3	34.7	14.9	—	—	4.1	Jacob and Zeman (1972a)
Larus ridibundus	64.7	18.0	13.3	4.0	—	—	Zeman and Jacob (1972, 1973b)
Larus fuscus	54.1	18.1	19.6	8.2	—	—	Zeman and Jacob (1972, 1973b)
Larus argentatus	49.4	16.4	24.8	9.4	—	—	Zeman and Jacob (1972, 1973b)
Melopsittacus undulatus	93.0	7.0	—	—	—	—	Jacob and Poltz (1974c)
Loriculus sp.	3.7	94.3	—	—	—	2.0	J. Jacob (unpublished results, 1975)
Psittacula eupatria	8.1	85.1	—	—	—	6.8	J. Jacob (unpublished results, 1975)
Psittaculirostris desmarestii	3.8	16.8	20.8	58.6	—	—	J. Jacob (unpublished results, 1975)
Chalcopsitta atra	14.3	42.3	43.4	—	—	—	J. Jacob (unpublished results, 1975)
Agapornis fisheri	55.9	32.6	9.6	—	—	1.9	J. Jacob (unpublished results, 1975)
Asio otus	28.8	65.5	—	—	—	5.7	Jacob and Poltz (1974a)
Bubo bubo	23.0	47.5	29.5	—	—	—	Jacob and Poltz (1974a)
Tyto alba	5.0	59.4	30.1	0.6	—	4.9	Jacob and Poltz (1974a)
Cuculus canorus	2.8	89.0	4.4	—	—	3.8	Jacob and Poltz (1972)
Dryocopus martius	70.9	26.6	2.5	—	—	—	Jacob and Poltz (1974b)
Picus viridis	4.6	61.4	34.0	—	—	—	Jacob and Poltz (1974b)
Coloeus monedula	87.5	12.5	—	—	—	—	Jacob and Grimmer (1973c)

Corvus frugilegus	100.0	—	—	—	—	—	Jacob and Glaser (1970)
Corvus corax	6.9	88.6	3.6	—	—	0.9	Poltz and Jacob (1974c)
Corvus corone corone	59.7	39.3	—	—	—	1.0	Jacob and Grimmer (1973c)
Corvus corone cornix	58.2	40.7	—	—	—	1.1	Jacob and Grimmer (1973c)
Pyrrhocorax graculus	87.0	12.7	—	—	—	0.3	Poltz and Jacob (1974c)
Pica pica	67.1	29.8	2.9	—	—	0.2	Poltz and Jacob (1974c)
Garrulus glandarius	11.1	64.7	23.6	—	—	0.6	Poltz and Jacob (1974c)
Oriolus oriolus	34.6	52.9	8.0	—	—	4.5	J. Jacob (unpublished results, 1975)
Bombycilla garrulus	13.8	75.6	10.6	—	—	—	J. Jacob (unpublished results, 1975)
Sturnus vulgaris	81.5	17.2	—	—	—	1.3	J. Jacob (unpublished results, 1975)
Gracula religiosa	—	—	—	92.9	—	7.1	J. Jacob (unpublished results, 1975)
Acrocephalus melanopogon	70.8	21.7	7.0	—	—	0.5	Poltz and Jacob (1975)
Acrocephalus schoenebaenus	12.8	14.7	31.3	40.8	—	0.4	Poltz and Jacob (1975)
Acrocephalus palustris	16.5	18.5	39.6	23.2	—	2.2	Poltz and Jacob (1975)
Acrocephalus scirpaceus	25.6	19.0	33.6	21.3	—	0.5	Poltz and Jacob (1975)
Acrocephalus arundinaceus	22.8	25.7	20.7	24.6	4.9	1.3	Poltz and Jacob (1975)
Sylvia atricapilla	17.4	71.5	9.9	—	—	1.2	Poltz and Jacob (1975)
Sylvia curruca	8.9	76.2	14.8	—	—	0.1	Poltz and Jacob (1975)
Sylvia communis	12.7	59.0	25.9	—	—	2.4	Poltz and Jacob (1975)
Certhia familiaris	67.7	26.6	5.7	—	—	—	Jacob and Grimmer (1975b)
Certhia brachydactyla	58.9	37.5	3.6	—	—	—	Jacob and Grimmer (1975b)
Leiothris lutea	3.2	89.9	6.9	—	—	—	Jacob and Grimmer (1975b)
Mesia argentauris	20.7	79.3	—	—	—	—	Jacob and Grimmer (1975b)
Paradoxornis gularis	2.3	74.4	21.2	2.1	—	—	Jacob and Grimmer (1975b)
Paradoxornis webbianus	57.2	36.9	5.9	—	—	—	Jacob and Grimmer (1975b)
Regulus regulus	64.9	35.1	—	—	—	—	Jacob and Grimmer (1975b)
Siva cyanouroptera	13.2	74.3	12.5	—	—	—	Jacob and Grimmer (1975b)
Remiz pendulinus	78.7	20.5	0.8	—	—	—	Jacob and Grimmer (1975b)
Aegithalos caudatus	33.4	60.3	6.3	—	—	—	Jacob and Grimmer (1975b)
Panurus biarmicus	67.2	30.2	2.6	—	—	—	Jacob and Grimmer (1975b)
Parus ater	83.3	16.6	—	—	—	0.1	Poltz and Jacob (1974a)
Parus melanolophus	95.7	3.5	—	—	—	0.8	Poltz and Jacob (1974a)
Parus ater x P. melanolophus	90.5	9.4	—	—	—	0.1	Poltz and Jacob (1974a)

TABLE IV (*Continued*)

Species	Un-branched	Mono-methyl-subst.	Di-methyl-subst.	Tri-methyl-subst.	Tetra-methyl-subst.	Others	References
Parus maior	96.6	3.4	—	—	—	—	Poltz and Jacob (1974a)
Parus caeruleus	91.7	8.3	—	—	—	—	Poltz and Jacob (1974a)
Parus atricapillus montanus	84.6	15.3	—	—	—	0.1	Poltz and Jacob (1974a)
Parus atricapillus rhenanus	89.5	9.7	—	—	—	0.8	Poltz and Jacob (1974a)
Parus xanthogenys	80.4	18.8	—	—	—	0.8	Poltz and Jacob (1974a)
Emberiza schoeniclus	51.1	46.9	—	—	—	2.0	Poltz and Jacob (1974d)
Emberiza bruniceps	47.7	52.3	—	—	—	—	Poltz and Jacob (1974d)
Fringilla coelebs	67.8	28.0	—	—	—	4.2	Poltz and Jacob (1974d)
Fringilla montifringilla	70.6	26.8	—	—	—	2.6	Poltz and Jacob (1974d)
Serinus serinus	70.2	29.8	—	—	—	—	Poltz and Jacob (1974d)
Carduelis spinus	50.8	48.6	0.4	—	—	0.2	Poltz and Jacob (1974d)
Carduelis chloris	61.7	38.3	—	—	—	—	Jacob and Zeman (1971c)
Carduelis flammea	81.9	16.8	—	—	—	1.3	Poltz and Jacob (1974d)
Carduelis cannabina	81.0	19.0	—	—	—	—	Jacob and Zeman (1971c)
Loxia curvirostra	78.4	20.7	—	—	—	0.9	Poltz and Jacob (1974d)
Pyrrhula pyrrhula	34.0	66.0	—	—	—	—	Jacob and Zeman (1971c)
Passer domesticus	100.0	—	—	—	—	—	Jacob and Zeman (1970a)
Passer montanus	78.4	19.5	—	—	—	2.1	Poltz and Jacob (1974d)
Ploceus cucullatus	100.0	—	—	—	—	—	Poltz and Jacob (1973)
Ploceus subaureus	100.0	—	—	—	—	—	Poltz and Jacob (1973)
Ploceus galbula	100.0	—	—	—	—	—	Poltz and Jacob (1973)
Ploceus spec.	100.0	—	—	—	—	—	Poltz and Jacob (1973)
Quelea quelea	98.9	1.1	—	—	—	—	Poltz and Jacob (1973)
Cinclus cinclus	100.0	—	—	—	—	—	Bertelsen *et al*. (1975)

a. Hydrocarbons. Squalene is a common constituent of human sebum but has been detected in the uropygial secretion of only two bird species, both from the order Anseriformes. Odham (1967b) has shown the free flowing uropygial gland secretion of *Cairina moschata* to contain 66% squalene, the structure of which was proved by hydrogenation and MS. Edkins and Hansen (1972) detected 87% squalene and 13% monoester waxes as uropygial gland secretion constituents in *Anseranas semipalmata.*

The Podicipediformes is the only order so far investigated in which paraffins were observed as uropygial gland wax constituents (Jacob, 1977c). The preen gland secretion of *Podiceps occipitalis* contains about 50% aliphatic hydrocarbons, whereas *P. rolland* possesses 60%. The structure of the hydrocarbons of both species was very similar. Their chain length varied between C_{15}–C_{26}. Methyl branches were detected in 2-, 3-, 5-, 7-, 9-, 11-, and 13-position. Olefins in the range C_{21}–C_{25} with double bonds in 9-position as well as diens, also were identified.

b. Monoester Waxes. Primary alcohols containing monoester waxes predominate among the constituents of uropygial gland secretions. The acidic and alcoholic components of the waxes vary in chain length and the degree of substitution from order to order, or even on the familial or generic level (for details, see Section I,C,3). I will discuss the monoester waxes on the basis of the degree and kind of substitution of the fatty acids; the *main* wax components is considered as the relevant parameter. This is necessary because minor amounts of acids of various structures can be found in almost all preen waxes.

*c. Unbranched Acids (*n*-Fatty Acids).* Two species of the order Ciconiiformes possess uropygial gland waxes containing only *n*-fatty acids esterified with *n*-alkanols, the Heron *(Ardea cinerea)* (Poltz and Jacob, 1974b) and the White Stork *(Ciconia ciconia)* (Jacob, 1976f). Although unbranched acids are common in the Laridae only one Alcidae *(Fratercula arctica)* contains exclusively *n*-fatty acids and *n*-alkanols (Jacob and Grimmer, 1970a). Other Alcidae species contain less than 50% *n*-fatty acids and considerable amounts of 2-, 4-, 6-methyl- as well as 2,*x*- (2,6-, 2,8-, 2,10-), 4,*x*- (4,6-, 4,8-, 4,10-, 4,12-) dimethyl-branched and 2,*x*,*y*- (2,6,8-, 2,6,10-, 2,6,12-) and 4,*x*,*y*-trimethyl-substituted acids (4,6,10-, 4,8,12-). The 2,4,8-trimethyl-branched acids which make up about 10% in the Laridae were detected only in trace amounts in Alcidae. The alcoholic components of Alcidae species are about 50% unbranched, also both mono- and dimethyl-alkanols have been observed. Laridae species possess *n*-fatty acids, plus variable amounts of 2-methyl-, 2,6- 2,8-, and 2,10-dimethyl- and 2,4,8- and 2,6,*x*-trimethyl-branched acids. No 4-methyl-branched acids have been detected (Jacob and Zeman, 1972a; Zeman and Jacob, 1972, 1973b).

One Procellariiforme, *Garrodia nereis*, was shown to possess 16.7% unbranched acids in the preen waxes (Jacob, 1976d). Some Anseriformes species show also relatively high contents of *n*-wax acids: 93.5% in *Anseranas semipalmata* (Edkins and Hansen, 1972), 2.5–12.7% in *Anas platyrhynchos* (Odham, 1967c), 4.3% in *Somateria mollissima* (Odham, 1967b), 13.6% in *Tachyeres patachonicus*, and 34.4% in *Tachyeres pteneres* (Jacob, 1977a). Minor amounts of *n*-acids were detected among Passeriforms, except *Coloeus monedula* (Jacob and Grimmer, 1973c) and *Mesia argentauris* (Jacob and Grimmer, 1975b) where 35.7% and 47.1% *n*-fatty acids were found, respectively.

d. Branched Fatty Acids. i. 2-Methyl-branched acids. 2-Methyl-substituted fatty acids were the predominant wax constituents in some Corvidae species of the order Passeriformes (*Corvus frugilegus* 100%, *Coloeus monedula* 64.3%, *Corvus corax* 85.3%, *Pyrrhocorax graculus* 37.4%, *Pica pica* 34.2%) except *Corvus corone* and *C.c. cornix* where 4- and 6-methyl-substituted acids have been detected (Jacob and Glaser, 1970; Jacob and Grimmer, 1973c; Poltz and Jacob, 1974c). In these species 2-methyl-branched acids were accompanied by 2,*x*-dimethyl-substituted acids.

Among the wax alcohols *n*- and 2-methyl-alkanols predominate. 2-Methyl-substituted wax acids are found in some passerine birds also, e.g., *Mesia argentauris* (51.5%), *Certhia familiaris* (33.2%), *C. brachydactyla* (75.2%), *Leiothrix lutea* (73.5%) (Jacob and Grimmer, 1975b). The wax alcohols of these species also belong to the *n*- or 2-methyl series. Considerable amounts of 2-methyl-substituted acids occur in some Laridae, e.g., *Rissa tridactyla* (46.6%) (Jacob and Zeman, 1972a) and in the Limicolae species *Gallinago gallinago* (41%) (J. Jacob, unpublished results, 1975) and *Haematopus ostralegus* (60.9%) (Jacob and Poltz, 1973a; Karlsson and Odham, 1969). Moreover, this acid type is found in some species from the order Anseriformes (*Anas platyrhynchos* 43.3%, *A. strepera* 72.5%, *A. clypeata* 94.5%) esterified mainly with *n*-alkanols (Odham, 1967c; Jacob and Glaser, 1975). They also occur as main components in the genus *Falco* of the order *Falconiformes* (Jacob and Poltz, 1975a).

ii. 3-Methyl-branched acids. These acids are characteristic components of the uropygial gland secretions of the Fringillidae and Emberizidae (Jacob and Zeman, 1970a, 1971c; Poltz and Jacob, 1974d) in which they are esterified predominantly with *n*-alkanols. Methyl-branched alcohols are also observed. 3-Methyl-substituted acids occur as main constituents in the preen waxes of Piciformes (Jacob and Poltz, 1974b), Cuculiformes (Jacob and Poltz, 1972), and in the owl

Tyto alba (Jacob and Poltz, 1974a). They occur together with a great variety of other acid types in the Sphenisciformes (Jacob, 1976e), Procellariiformes (Jacob and Zeman, 1971a; Jacob, 1976e), and in Podicipediformes (Jacob, 1977c).

iii. 4- and 6-Methyl-branched acids. These acids contribute more than 20% of the total wax acids in some Procellariiformes *(Pachyptila belcheri, Garrodia nereis, Fulmarus glacialis)* (Jacob and Zeman, 1971a; Jacob, 1976d), in *Spheniscus magellanicus* (Jacob, 1976g), in the hammerhead *(Scopus umbretta)* (J. Jacob, unpublished results, 1975), and in some *Anas* species *(Anas platyrhynchos, A. strepera)* (Odham, 1967c; Jacob and Glaser, 1975). They are main wax constituents along with higher branched acids in *Corvus corone corone* and *C. c. cornix* (Jacob and Grimmer, 1973c), in *Sturnus vulgaris* and *Gracula religiosa* (J. Jacob, unpublished results, 1975), and at least in one parrot *(Agapornis fisheri)* (J. Jacob, unpublished results, 1975).

iv. Iso- and anteiso-branched acids. These acids predominate in some parrot species, e.g., *Psittacula eupatria, Loriculus* sp. (J. Jacob, unpublished results, 1975), and *Melopsittacus undulatus* (Jacob and Poltz, 1974c).

v. 2,4-Dimethyl-branched acids. The 2,4-dimethyl-substituted fatty acids are the only acidic constituents of preen waxes from *Ploceus* species, e.g., *Ploceus cucullatus, P. subaureus,* and *P. galbula* (Poltz and Jacob, 1973).

vi. 2,x-Dimethyl-branched acids. Among this type of acids (with x = even numbered) the 2,6-dimethyl-branched compounds predominate. They are distributed widely and predominate or occur in considerable amounts in the Falconiformes (Jacob and Poltz, 1975a), Gruiformes (Jacob and Zeman, 1971b; Jacob and Poltz, 1975b), Charadriidae (Jacob and Poltz, 1973a; J. Jacob, unpublished results, 1975), Laridae (Jacob and Zeman, 1972a; Zeman and Jacob, 1973a,b), Corvidae (Poltz and Jacob, 1974c), Oriolidae (J. Jacob, unpublished results, 1975), and the Bombycillidae (J. Jacob, unpublished results, 1975). They are common in the Sylviidae (Poltz and Jacob, 1975) and the closely related *Remiz pendulinus, Aegithalos caudatus,* and *Panurus biarmicus* (Jacob and Grimmer, 1975b) as well as in *Certhiidae, Paradoxornithidae,* and in *Siva cyanouroptera* (Jacob and Grimmer, 1975b). Amounts up to 36% of 2,(ω-1)- and 2,(ω-2)-dimethyl-branched acids have been detected in some Falconiformes species (Jacob and Poltz, 1975a).

vii. 3,x-Dimethyl-branched acids. These acids are generally observed in species in which 3-substituted fatty acids occur, e.g., Procellariiformes (Jacob and Zeman, 1971a; Jacob, 1976d), Podicipediformes (Jacob, 1977c), Emberizidae and Fringillidae (Jacob and

Zeman, 1970a, 1971c; Poltz and Jacob, 1974d), Piciformes (Jacob and Poltz, 1974b), and *Tyto-alba* (Jacob and Poltz, 1974a). The second methyl group is located at odd numbered C atoms, and 3,5-, 3,7-, 3,9-, 3,11-, 3,13-, and 3,15-dimethyl-branched acids have been observed.

viii. 4,x-Dimethyl-branched acids. The 4,*x*-dimethyl-branched wax acids are often associated with the 4-methyl-substituted acids and they occur in the preen waxes in minor amounts in Procellariiformes (Jacob and Zeman, 1971a; Jacob, 1976e), *Scopus umbretta* (J. Jacob, unpublished results, 1975), and in *Tachyeres* (Jacob, 1977a). They are main constituents of the uropygial waxes from some rails, e.g., *Gallinula chloropus, Porzana porzana,* and *Fulica atra* (Jacob and Zeman, 1971b; Jacob and Poltz, 1975b), Alcidae species as *Uria aalge* and *Alca torda* (Jacob and Zeman, 1973), from several passerines such as *Corvus corone corone* and *C. c. cornix* (Jacob and Grimmer, 1973c), and *Gracula religiosa* (J. Jacob, unpublished results, 1975) and from some parrots (J. Jacob, unpublished results, 1975).

ix. 2,4,x-Trimethyl-branched acids. There are two types of 2,*x*,*y*-trimethyl-branched fatty acids which differ significantly in structure, those which start with a 2,4-dimethyl substitution as 2,4,6- and 2,4,8-trimethyl-substituted acids, and those which start with a 2,6-dimethyl substitution as 2,6,8-, 2,6,10-, 2,6,12-, . . . trimethyl-substituted acids. 2,4,*x*-Branched acids have been observed in Laridae (Jacob and Zeman, 1972a; Zeman and Jacob, 1972, 1973b), in *Diomedea melanophris* (Jacob, 1976d), and in *Phoenicopterus ruber* (Bertelsen, 1970). They are main constituents of the preen waxes of Anseriformes species, especially in the genera *Dendrocygna, Coscoroba, Cygnus, Anser, Branta, Tadorna, Cairina, Aythya, Somateria, Melanitta* and *Mergus* (Odham, 1963, 1965, 1966, 1967a,b,c; Bertelsen, 1973; Edkins and Hansen, 1972; Andersson and Bertelsen, 1975a,b; Jacob and Zeman, 1970b, 1972c; Jacob and Glaser, 1975). 2,4,6-Trimethylnonanoic acid is the predominant wax acid in *Quelea quelea* and thus differs from the closely related Ploceus species (Poltz and Jacob, 1973). This acid also predominates in the uropygial waxes of the dipper *(Cinclus cinclus)* (Bertelsen *et al.*, 1975).

x. 2,x,y-Trimethyl-branched acids. Among the 2,*x*,*y*-trimethyl-substituted acids (with *x* and *y* even numbered) the 2,6,10-trimethyl-branched acids predominate although 2,6,8-, 2,6,12-, 2,6,14- . . . etc., as well as 2,8,*x*-, 2,10,*x*- . . . etc., -trimethyl-branched acids have been observed. They are common in Strigiformes (Jacob and Poltz, 1974a), Gruiformes (Jacob and Poltz, 1975b; Jacob and Zeman, 1971b), Charadriidae (Jacob and Poltz, 1973a), Podicipediformes (Jacob, 1976b) and the Laridae (Jacob and Zeman, 1972a; Zeman and Jacob, 1972, 1973b). Various passerine families also produce this type of acid as a preen wax constituent, e.g., Corvidae (Jacob and Grimmer, 1973c; Poltz and Jacob, 1974c), Oriolidae (J. Jacob, unpublished results, 1975), Sturnidae (J. Jacob, unpublished results, 1975) and Sylviidae (Jacob and Grimmer, 1975b; Poltz and Jacob, 1975). They have been detected in the Thinocoridae also (J. Jacob, unpublished results, 1975).

*xi. 2,4,*x,y*-Tetramethyl-branched acids.* 2,4,x,y-Tetramethyl-branched acids (with x and y even numbered) have been observed in species of the Anseriformes (Odham, 1963, 1966, 1967b; Jacob and Zeman, 1970b; Bertelsen, 1973; Jacob and Glaser, 1975). They are normally accompanied by small amounts of 2,4,6-trimethyl-branched acids . 2,4,6,8-Tetramethyl-branched acids have shown to be the main acidic wax constituent in the kingfisher *(Alcedo atthis)* (Jacob, 1976c).

*xii. 2,*x,y,z*-Tetramethyl-branched acids.* Among these acids 2,6,x,y- and 2,6,10,x-tetramethyl-branched acids predominate. They are often accompanied by less branched isomers such as 2,6,10-trimethyl- and 2,6-dimethyl-branched acids. They have been detected in the preen waxes of some parrots (J. Jacob, unpublished results, 1975) and in the Sylviidae and the closely related Paradoxornithidae (Poltz and Jacob, 1975; Jacob and Grimmer, 1975b).

*xiii. 4,*x,y,z*-Tetramethyl-branched acids.* This type of acid has been detected only among the Psittaciformes (J. Jacob, unpublished results, 1975).

xiv. Pentamethyl-branched acids. The 2,4,6,8,10-pentamethyl-branched acids are minor acidic components of the uropygial gland wax of the Kingfisher *(Alcedo atthis)* (Jacob, 1976c) and may occur in species of the Anseriformes (Edkins and Hansen, 1972).

xv. Ethyl-substituted acids. The 2-ethyl-substituted acids are the main preen wax constituents in the family Paridae (Poltz and Jacob, 1974a) and are also common in the order Strigiformes (Jacob and Poltz, 1974a). They occur as minor wax constituents in species of the Podicipediformes (Jacob, 1977c). Together with 4-ethyl- and 4,x-diethyl-substituted acids they occur as uropygial gland wax constituents in *Theristicus caudatus* (J. Jacob, unpublished results, 1975).

*xvi. 2-Ethyl-*x*-methyl- and 2-ethyl-*x,y,*-dimethyl-substituted acids.* These acids occur as minor preen wax constituents in Strigiformes and Podicipediformes as well as in *Theristicus caudatus* and as main constituents in the Paridae.

xvii. Propyl- and butyl-substituted acids. 2-Propyl- and 2-butyl-substituted acids as preen wax constituents have been detected in *Asio otus* (Jacob and Poltz, 1974a).

The acidic constituents of all preen waxes so far investigated are summarized and listed quantitatively in Table III.

e. Alcohols. Among the alcoholic constituents of the uropygial gland waxes unsubstituted alcohols predominate. Generally variable amounts of mono-methyl-substituted alcohols with branches at even C atoms are detected, although higher branched alkanols have been observed in various species.

The quantitative wax alcohol composition of the hitherto investigated species are summarized in Table IV. Distinction is made only among un-, mono-, di-, tri-, and higher-branched alcohols but no details regarding the branching position are included.

f. Secondary Alcohols Containing Monoester Waxes. This wax type has been detected only in one species, the Night Heron *(Nycticorax nycticorax).* It consists of alkane-2-ols esterified with *n*-, 2-methyl-, 2,4-, and 2,6-dimethyl-substituted fatty acids. In addition to this, primary alcohols containing monoester waxes (12.7%) have been observed (Jacob, 1975b).

g. Diester Waxes. Four types of diester waxes have been observed in uropygial gland secretions: 3-hydroxy acids (a), 1,2-diols (b), and *erythro*- and *threo*-2,3-diols containing (c_1 and c_2) diester waxes.

```
(a)        R1—CH—CH2—C—O—R2
              |       ||
              O       O
              |
              C=O
              |
              R3

(b)        R1—CH—CH2O—C—R2
              |        ||
              O        O
              |
              C=O
              |
              R3

(c1)       R1—CH—CH—CH3
              |   |
              O   O
              |   |
            O=C   C=O
              |   |
              R2  R3

(c2)              R3
                  |
                  C=O
                  |
                  O
                  |
           R1—CH—CH—CH3
              |
              O
              |
              C=O
              |
              R2
```

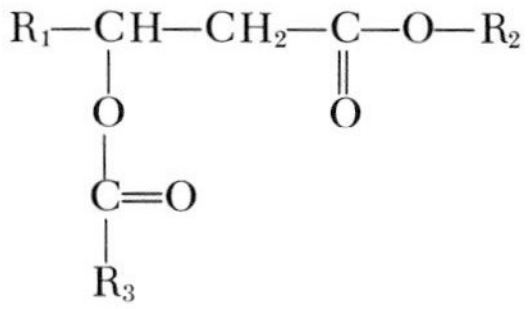

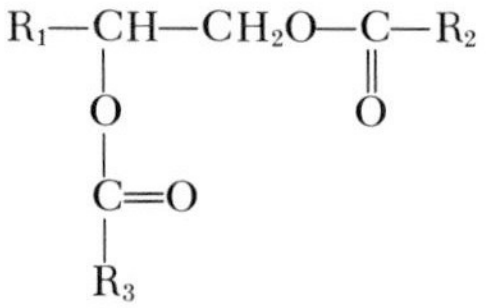

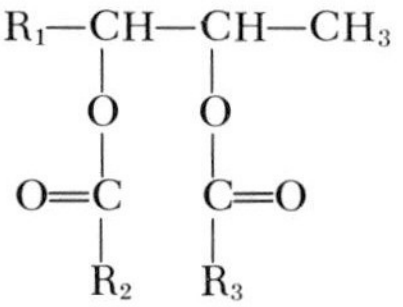

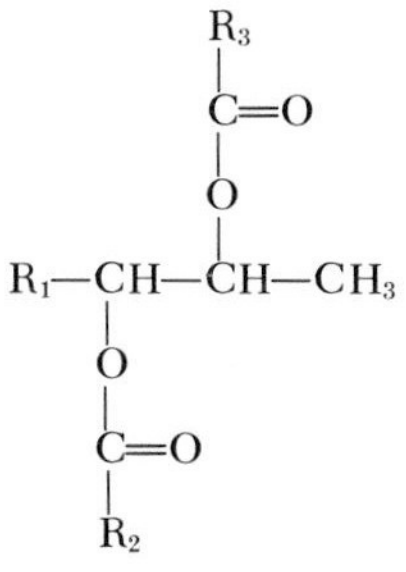

(with R_1, R_2, R_3 unbranched)

1. The 3-hydroxy acids esterified with *n*-alkanols and unbranched fatty acids have been shown to occur in the uropygial gland secretion (Jacob and Zeman, 1972b) and the feather waxes (Jacob and Grimmer, 1975a) of *Columba palumbus*. They occur in minor amounts in the preen wax of *Lanius collurio* (J. Jacob, unpublished results, 1975).

2. Alkane-1,2-diol containing waxes occur in various species of the Gruiformes, Passeriformes, and Piciformes (Saito and Gamo, 1972; Kolattukudy, 1972).

3. In the Galliformes, *erythro*- and *threo*-alkane-2,3-diols esterified with *n*-fatty acids have been identified as preen gland secretions. Both stereo isomers occur in *Gallus domesticus* (Hansen *et al.*, 1969) and *Coturnix pectoralis* (Edkins and Hansen, 1971), whereas only the *erythro*-form was detected in the *Phasianus colchicus* (Jacob and Grimmer, 1970b; Saito and Gamo, 1970; Sawaya and Kolattukudy, 1972), *Perdix perdix* (Jacob and Grimmer, 1970c), *Leipoa ocellata* (Edkins and Hansen, 1971) and *Meleagris pavo* (Hansen *et al.*, 1969). Chain lengths of the fatty acids are in the range C_6–C_{21}.

h. Triester Waxes. Two types of triester waxes have been observed in uropygial gland secretions: alkyl-hydroxy malonic acids containing lipids, and triglycerides. The first type (alkyl-hydroxy malonic acids esterified with *n*-fatty acids and *n*-alkanols) were detected in various species of the Anseriformes (Jacob and Grimmer, 1973a), Gruiformes (Jacob and Poltz, 1975b), Piciformes (Jacob and Grimmer, 1973a), and Passeriformes (Jacob and Grimmer, 1973a; Poltz and Jacob, 1974c). Triglycerides which contain saturated fatty acids only occur as true secretion constituents in the preen wax of the Heron *(Ardea cinerea)* (Poltz and Jacob, 1974b). Triglycerides occur as constituent of the gland cells in all birds. They show a very similar composition to the depot fats (see Section II).

3. *Chemotaxonomic Aspects*

Investigations on the chemical composition of the uropygial gland secretions suggest a correlation between this chemical parameter and avian classification. Specifically, this means that closely related birds possess *qualitatively* similar composition of preen waxes, which differ only *quantitatively*, in contrast to species belonging to different orders or families which are distinguished by qualitatively different wax compositions. Thus chemical composition of the uropygial gland lipids may be a useful criterion for classification of birds. Although the preen wax composition is only one of numerous genetically fixed properties it has the advantage of high constancy and it can be quantified by GLC/MS techniques. Nevertheless it should be mentioned that the quantitative composition can vary within certain limits within species. Furthermore age-dependent quantitative differences within chickens have been reported (Kolattukudy and Sawaya, 1974), so that only adult animals should be compared. Generally, significant qualitative differences in the composition of uropygial secretions are observed on the ordinal level, but they have been detected on the familial and generic levels as well, especially in the large and heterogenous orders (Gruiformes, Passeriformes). Subspecific categories cannot be distinguished as shown in races of *Corvus corone* (Jacob and Grimmer, 1973c) and *Uria aalge* (Jacob and Zeman, 1973).

In order to use the information on the wax acids and alcohols given in Tables III and IV in avian taxonomy it is useful to compare taxa by projecting the qualitatively distinct types of acids and alcohols (using the different degrees of substitution) onto a n-dimensional space matrix, where each dimension represents one type of fatty acid or alcohol. Because of the difficulty of depicting a more than three-dimensional space, the parameters should be reduced to three if possible. Thus generally only the acids are compared and among these some different types have to be combined together into one dimension. This model has been used successfully for chemotaxonomic investigations of the Anseriformes (Jacob and Glaser, 1975; Jacob, 1977a), Procellariiformes (Jacob, 1976d), and Sphenisciformes (Jacob, 1976e). A number of examples of "preen wax chemotaxonomy," are summarized here. I report these data as a basis for comparison with more traditional characteristics or other biochemical data.

1. Procellariiformes and Sphenisciformes are closely related (Jacob, 1976d,e), although both orders must be divided into at least two groups. One group is characterized by large amounts of 3-methyl-branched and considerable quantities of 4-methyl-branched acids *(Pygoscelis papua, Eudyptes crestatus, Puffinus griseus, Pelecanoides urinatrix, Pachyptila belcheri, Fulmarus glacialis)*. In the second group 2-methyl-branched together with 4-methyl-branched acids predominate *(Spheniscus magellanicus, Garrodia nereis, Diomedea melanophris)*, whereas *Procellaria aequinoctialis* falls between these two groups. The results (Fig. 6) include these plus selected additional species. The unbranched acids have been omitted in order to reduce the problem to three parameters. The preen gland secretion of *Garrodia nereis* (Table III) shows a certain relationship to the auks, *Alca, Cepphus,* and *Uria* which, on the other hand, are not far from the gulls. All possess additional unbranched wax acids. *Diomedea* obviously is the most atypical Procellariiform of all species investigated, possessing large amounts of 2,4,6- and 2,4,8-trimethyl-

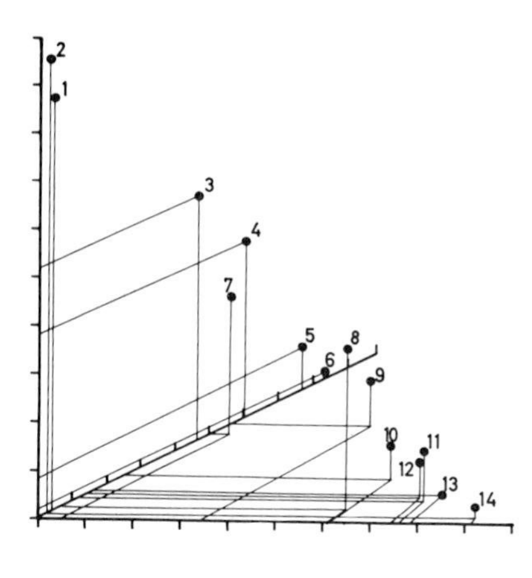

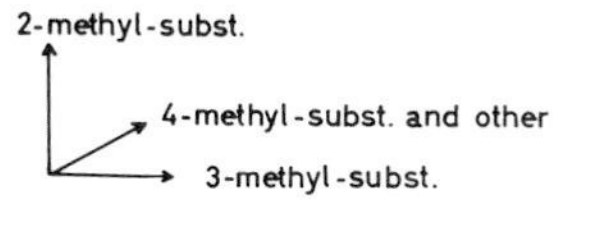

FIG. 6. Chemotaxonomic relationships between *Procellariiformes, Sphenisciformes, Phoenicopterus,* and *Alcidae,* based on the kind of methyl-substitution of the fatty acids from the uropygial gland waxes. (Nos. 3–7 contain considerable amounts of unbranched acids). (1) *Diomedea melanophris;* (2) *Phoenicopterus ruber;* (3) *Spheniscus magellanicus;* (4) *Cepphus grylle;* (5) *Alca torda;* (6) *Uria aalge;* (7) *Garrodia nereis;* (8) *Procellaria aequinoctialis;* (9) *Pachyptila belcheri;* (10) *Fulmarus glacialis;* (11) *Eudyptes crestatus;* (12) *Pygoscelis papua;* (13) *Pelecanoides urinatrix;* (14) *Puffinus griseus.*

substituted acids. Its preen wax composition is strikingly similar to that of the Flamingo *(Phoenicopterus ruber),* but it should be noted that 2,4,*x*-trimethyl-substituted wax acids also have been detected in the Anseriformes and Laridae. This distribution does not disprove the suggested relationship. The Procellariiformes and Sphenisciformes can be distinguished on the bases of the alcohol compositions.

2. The Podicipediformes show some relation to both abovementioned orders and seem most like the Phalacrocoracidae by possessing highly branched wax acids with a first substituent in 3-position (J. Jacob, unpublished results, 1975). No relation to any other order was suggested.

3. The Ciconiiformes are a heterogenous order. The two herons, *Ardea cinerea* and *Nycticorax nycticorax,* and both ibises, *Theristicus caudatus* and *Threskiornis aethiopicus,* investigated did not show chemotaxonomic similarities. The preen wax pattern of *Scopus umbretta* differs from both groups by possessing predominantly

4-methyl-branched acids (J. Jacob, unpublished results, 1975). The present material does not allow relevant conclusions on the taxonomy of this order.

4. *Phoenicopterus ruber* seems to be more like the Anseriformes than the Ciconiiformes. Flamingos have mainly 2,6-dimethyl- and 2,4,6-, as well as 2,4,8-trimethyl-branched acids.

5. The Anseriformes show a clear differentiation in the three-dimensional space matrix (Figs. 7 and 8). In Fig. 7, un-, mono-, and disubstituted acids were combined together in one dimension, the other two dimensions representing tri- and tetra-substituted acids. In Fig. 8, tri- and tetramethyl-substituted acids were combined in one dimension, un- and monosubstituted fatty acids in a second, and di-substituted acids constitute the third. Both figures demonstrate several

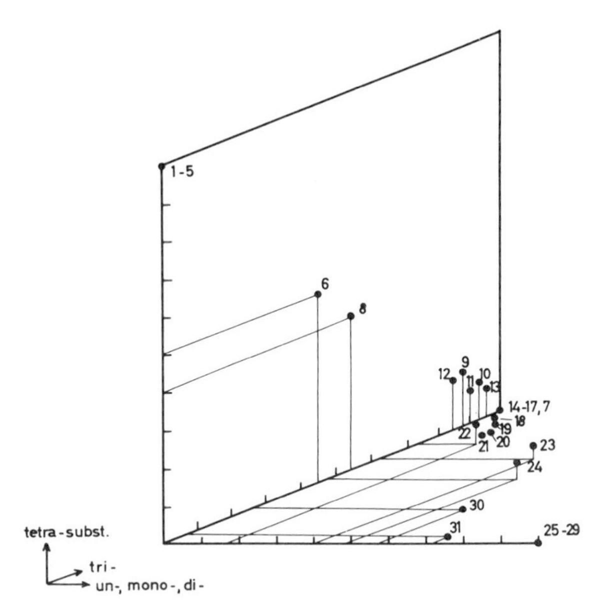

FIG. 7. Chemotaxonomic relationships among *Anseriformes* species, based on a comparison of their uropygial gland secretion constituents (unbranched, mono- and disubstituted acids are combined in one dimension). (1) *Anser anser;* (2) *A.a.f. domesticus;* (3) *A. caerulescens;* (4) *Cygnus cygnus;* (5) *C. columbianus;* (6) *C. melanocoryphus;* (7) *Coscoroba coscoroba;* (8) *Dendrocygna viduata;* (9) *Cygnus olor;* (10) *Anser fabalis;* (11) *A. indicus;* (12) *Aythya ferina;* (13) *Tadorna tadornoides;* (14) *Cereopsis novaehollandiae;* (15) *Branta leucopsis;* (16) *Cairina moschata;* (17) *Mergus serrator;* (18) *M. albellus;* (19) *Tadorna tadorna;* (20) *T. ferruginea;* (21) *Melanitta nigra;* (22) *Stictonetta naevosa,* (23) *Aythya fuligula,* (24) *Somateria mollissima;* (25) *Chenonetta jubata;* (26) *Anas strepera;* (27) *A. clypeata;* (28) *A. platyrhynchos;* (29) *A. pl. f. domesticus;* (30) *Tachyeres pteneres* (Tpt); (31) *T. patachonicus* (Tpat).

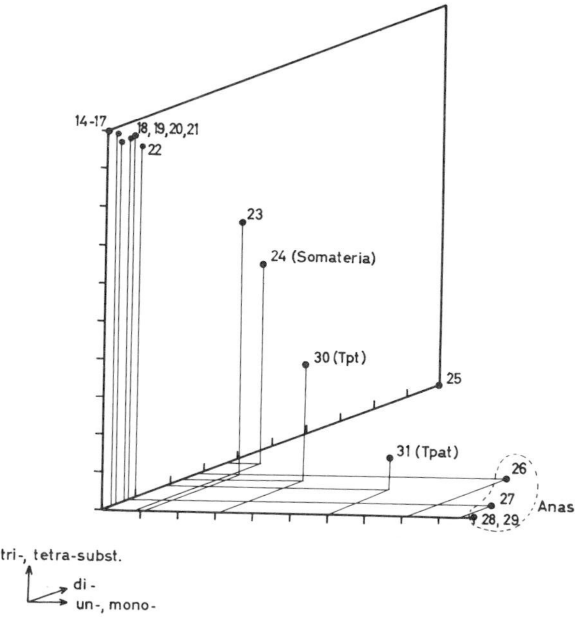

FIG. 8. Chemotaxonomic relationships among different Anseriformes species, based on the comparison of their uropygial gland secretion constituents. (Tetra- and trisubstituted as well as unbranched, mono- and disubstituted acids are combined in one dimension, respectively.) Numbers as in Fig. 7.

interesting chemotaxonomic relationships. *Anser anser* and *Anser caerulescens* together with some *Cygnus* species as, e.g., *C. cygnus* and *C. columbianus*, are closely related on the basis of possessing only tetramethyl-substituted fatty acids. This group also shows affinities with *Cygnus melanocoryphus* and *Dendrocyna viduata*. Both species are connected with group 9–18 (including 7): *Coscoroba coscoroba, Cygnus olor, Anser fabalis, Anser indicus, Aythya ferina, Tadorna tadornoides, Cereopsis novaehollandiae, Branta leucopsis, Cairina moschata, Mergus albellus* by possessing tetra- and trimethyl-branched acids. Group 19–22 *(Tadorna tadorna, Tadorna ferruginea, Melanitta nigra,* and *Stictonetta neavosa)* possess small amounts of lower branched acids, but seem related closely to group 9–18. *Aythya fuligula, Somateria mollissima, Tachyeres,* and the Anatinae are separate and distinct. The relationship among the latter is better demonstrated in Fig. 8. The Steamer Ducks *(Tachyeres)* are located equidistant between *Somateria* and *Anas*.

6. The Falconiformes appear isolated with no relationship to the Strigiformes. As was the case in other orders (Charadriiformes and

Anseriformes) an intraordinal series can be observed leading from the low-branched *Falco* species via *Accipiter* to the higher branched *Buteo*. There was a parallel increase in the degree of substitution and in chain length in alcohol components from *Falco* to *Buteo*.

7. The Galliformes are a homogenous group that do not show relationship to any order so far investigated. The occurrence of alkane-2,3-diols esterified with *n*-fatty acids is unique.

8. The Gruiformes species possess either a preen gland, or powder downs, or both. The fatty acids are characterized by a branching system with methyl branches at every fourth C atom (2,6,10- or 4,8,12-, etc.). This acid type has been reported to occur also in the Scolopacidae and in the Sylviidae.

9. Within the Charadriiformes a series of stepwise relationships exists. At one end of this series are the Alcidae with only, or predominantly, unbranched acids. At the other extreme are certain Charadriidae *(Haematopus)* with branched acids only. They are connected by birds with mixed gland compositions (Laridae). Recent work based on preen gland secretion indicates that the Chionididae (sheath bills) may be closely related to the Laridae and must be located between the Laridae and Charadriidae (Jacob, 1977b). *Thinocorus*, however, is more closely related to Charadridae (J. Jacob, unpublished results, 1975).

10. The Columbiformes are an isolated group. There are only rudimentary preen glands in these species and their contribution to the feather lipids is minimal. 3-Hydroxy fatty acids containing diester waxes are typical. This feather wax type was detected as minor preen wax constituent in *Lanius collurio* (J. Jacob, unpublished results, 1975).

11. The Psittaciformes show no relationship to the Columbiformes as often postulated. They are divided at least into two chemical groups, one possessing predominantly fatty acids with a (ω-1)- or (ω-2)-methyl branch *(Loriculus, Psittacula eupatria, Melopsittacus undulatus)* and a second possessing 4-methyl-branched acids as the main uropygial gland wax constituents *(Agapornis fisheri, Psittaculirostris desmarestii, Chalcopsitta atra)* (Jacob and Poltz, 1974c; J. Jacob, unpublished results, 1975).

12. The Cuculiformes and Piciformes are closely related and cannot be separated. Their preen gland waxes are similar to some passeriform groups *(Emberizidae, Fringillidae)*.

13. The Strigiformes have distinctive preen gland secretions, acids with larger alkyl substituents. Ethyl, propyl, and butyl derivated have been detected. Thus, they differ from all other orders, except the

Paridae where ethyl-branched acids were observed also. Within the Strigiformes and Tytonidae are clearly separated from the Strigidae by possessing 3-methyl-branched acids. *Tyto alba* shows great similarity with Cuculiformes, Piciformes, and some passerine birds.

14. Only one Coraciiforme species *(Alcedo atthis)* has been investigated. It differs from all other birds so far investigated by possessing the highest branched wax ever identified in uropygial gland secretions, composed of multi-branched acids and alcohols.

15. The Passeriformes can be divided into several groups. Fringillidae and Emberizidae including *Passer domesticus* and *Passer montanus* are clearly separated from Ploceidae. The Estrildinae and Viduinae are closely related to Ploceinae. It should be noted that *Cinclus cinclus* possesses a very similar uropygial gland wax. Corvidae, Oriolidae, and Bombycilla show similar preen waxes and seem to be closely related. The Paridae are a well-defined group within the Passeriformes and show no relationship to the other Families. *Remiz pendulinus, Aegithalos caudatus,* and *Panurus biarmicus* are not true tits but should be placed together with Regulidae, Certhiidae, Timaliidae, and Paradoxornithidae proximate to Sylviidae.

When viewed broadly, there seems to occur a progressive specialization, basically a simplification, of the uropygial gland secretion compositions associated with classification. The presumably phylogenetically older taxa such as Procellariiformes and Sphenisciformes show almost all types of fatty acids detected in preen waxes, *n*-, 2-, 3-, 4-, methyl-, and higher branched acids. This seems to be true for the Podicipediformes also. On the other hand, phylogenetically more recent groups such as Passeriformes in many cases show more simple compositions. For example, the Fringillidae and Emberizidae possess only 3-methyl-branched acids, and the Ploceinae show only one distinct wax, namely octadecyl-2,4-dimethyl heptanoate. Both the branching positions (isomerism) and chain length (homologism) become extremely reduced. The enzyme specificity associated with wax secretion production must be precise.

It is possible that the complexity of preen gland secretions may have undergone a simplification which can be related to a classificatory scheme and represents a phylogenetic sequence.

I expect that the analysis of preen gland sections will contribute further to our understanding of systematic relations of the large and heterogenous orders. In addition other chemotaxonomic methods have been applied to the class Aves. The results of the electrophoretic investigations of the egg white, eye lens, and the hemoglobin protein pattern by Sibley and co-workers and other authors are summarized elsewhere (Sibley, 1970; Sibley and Ahlquist, 1972; Sibley *et al.*, 1974). It should be noted that the results of the various biochemical analyses are not always in agreement.

4. *Biosynthesis of Uropygial Gland Waxes and their Constituents*

The data available on uropygial gland secretion constituents show that a great variety of unbranched and branched fatty acids and alcohols, hydroxy acids, and diols are synthesized in the preen gland. Nevertheless only a few papers deal with the enzymology of this gland and no enzyme from this source has been purified. At least four substrates are of importance in the biosynthesis of the acid moieties: acetate, propionate, malonate, and methylmalonate. Buckner and Kolattukudy (1975b) have shown that a partially purified fatty acid synthetase from cell-free extracts from the uropygial gland of the goose catalyzes the incorporation of malonyl-CoA and methylmalonyl-CoA into *n*- and multibranched fatty acids, respectively. Acetyl-CoA is preferred as primer for the reaction and NADPH as reductant. Noble *et al.* (1963) provide evidence for the incorporation of [3-^{14}C]-propionate into 2,4,6,8-tetramethyldecanoic and -undecanoic acid of the preen gland wax of the goose, obviously operating via methylmalonyl-CoA. Buckner and Kolattukudy (1975a) concluded that the protein-catalyzed malonyl-CoA-decarboxylase activity controls the fatty acid synthesis in the preen gland of the goose, converting malonyl-CoA to acetyl-CoA, whereas methylmalonyl-CoA is decarboxylated to only a small extent, and therefore remains as a substrate.

On the other hand, we know that the ratio of acetyl-CoA/malonyl-CoA determines the chain length (Sumper *et al.*, 1969) and the ratio acetyl-CoA/propionyl-CoA (as primers) determines the amount of even- and odd-numbered acids among the end products. Tang and Hansen (1970, 1972) demonstrated that acetyl-CoA predominates as primer in the preen gland of chicken. From this it seems reasonable that the reactions summarized in Table V probably will take part in the fatty acid biosynthesis in the preen gland, although additional pathways may be involved in this complex. In addition, chain elongation and degradation seem to occur in the uropygial gland also. A critical view of the structures of preen wax acids show that, at least in certain species, there must exist synthetase systems which accept malonyl- and methylmalonyl-CoA simultaneously. For example, in the Anseriformes in addition to 2-methyl-branched acids, 4-, and 6-methyl-, and polymethyl-branched acids were detected. The methyl-substituents of the latter were distributed randomly over the molecule, although exclusively at even-numbered C atoms (2,4,6-, 2,4,8-, 2,6,8-, 2,6,10-,

TABLE V

PROBABLE PATHWAYS IN THE BIOSYNTHESIS OF FATTY ACIDS IN THE UROPYGIAL GLAND

Primer	Substrate	Enzyme	Product
Acetyl-CoA	CO_2	Acetyl-CoA carboxylase	Malonate
Propionyl-CoA	CO_2	Propionyl-CoA carboxylase	Methylmalonate
Acetyl-CoA	Malonyl-CoA	Fatty acid synthetase	Even-numbered unbranched fatty acids
Propionyl-CoA	Malonyl-CoA	Fatty acid synthetase	Odd-numbered unbranched fatty acids
Acetyl-CoA	Methylmalonyl-CoA	Fatty acid synthetase	Even-numbered methyl-branched acids
Propionyl-CoA	Methylmalonyl-CoA	Fatty acid synthetase	Odd-numbered methyl-branched acids

and 2,6,12-trimethyl-branched acid among others). In contrast uropygial waxes with a nonrandom distribution of acid substituents were found in the Rallidae, Sylviidae, Laridae, and Charadridae. In the uropygial gland waxes of these species the methyl groups are located at every fourth C atom so that structures as 2,6,10-, 4,8,12-trimethyl-, or 2,6,10-14-tetramethyl-substituted acids result. I suggest that another enzyme system is responsible for the biosynthesis of these acids.

In some species only 2-methyl- or 4-methyl-branched acids occur. For these acids it is supposed that the involvement of methylmalonyl-CoA in the biosynthesis is restricted to the last or the penultimate step of condensation. 2- and 4-Methyl-branched acids often occur in one secretion simultaneously (Alcidae, Anseriformes) which indicates a similar biosynthetic mechanism of both acids. This, of course, is important for chemotaxonomic considerations. The 3-methyl-substituted fatty acids are widely distributed in uropygial gland secretions (Sphenisciformes, Podicipediformes *Tyto alba*, Piciformes, Cuculiformes, Passeriformes), yet nothing is known about their biosynthesis. This is also true for the 3-hydroxy acids which occur in the preen waxes of Columbiformes and Laniidae and the widely distributed alkylhydroxy malonic acids as well. Although not supported by experimental evidence, the 2-ethyl-, 2-propyl-, and 2-butyl-substituted acids probably originate from butyryl, valeroyl-, and caproyl-CoA via the alkylmalonyl-CoA compounds, ethyl-, propyl-, and butylmalonyl-CoA, respectively.

The biosynthesis of the wax alcohols is incompletely known. In some species a relationship between the structures of the alcohols and acids is obvious, e.g., in the Alcidae, Ardeidae, Charadridae, Laridae, Corvidae, Cuculiformes, Falconiformes, Coraciiformes, Strigiformes, Sylviidae, Fringillidae, and Emberizidae. The alcohols presumably are produced from the acids by reduction. The preen wax alcohols of other species, however, differ significantly from the acid moieties so that additional biosynthetic pathways must be assumed. This group includes the Anseriformes, Procellariiformes, Sphenisciformes, Ralliformes, Psittaciformes, Paridae, and Ploceidae. Some knowledge on the biosynthesis of alkane-1,2-diols and alkane-2,3-diols is available. Kolattukudy (1972) observed reduction of 2-hydroxy fatty acids to alkane-1,2-diols with a 16,000 gm supernatant of uropygial gland homogenates from *Zonotrichia leucophrys* in the presence of ATP, NADH, and NADPH. The incorporation of ^{14}C-acetate into the alkane-2,3-diols of Galliformes has been described repeatedly (Tang and Hansen, 1970, 1972; Sawaya and Kolattukudy, 1972). Following injection into the preen gland, [1-^{14}C]palmitic acid and [1-^{14}C]stearic acid, were incorporated into the alkane-2,3-diols of *Phasianus colchicus* with C-1 becoming the C-3 atom of the diols (Sawaya and Kolattukudy, 1972). However, attempts to specifically label C-1 or C-2 of the diols by incorporation of [2-^{14}C]pyruvate, [U-^{14}C]alanine, or [2-^{14}C]lactate failed. In another investigation Riley and Kolattukudy (1975) gave evidence for the conversion of [3-^{14}C]3-hydroxyoctadecane-2-one to [3-^{14}C]octadecane-2,3-diol in the preen gland of *Phasianus colchicus* after injection. The authors suppose that alkane-2,3-diols result from condensation of fatty aldehydes and hydroxymethyl thiamine pyrophosphate and subsequent reduction. Sawaya and Kolattukudy (1973) found esterifying activity in the microsomal fraction of the uropygial gland of *Phasianus colchicus* using palmitoyl-CoA as substrate. In this reaction a stepwise acylation of the diols via monoacyl compounds was observed. It can be generally said that the wax patterns of various species support the hypothesis that esterification between alcohols and acids occur randomly.

5. *Physiological Functions*

Generally sebaceous gland secretions preserve the skin against desiccation or wetting and operate as a water barrier. The uropygial gland secretion which is smeared by the bill over the bird's plumage functions as an impregnation and improves the swimming ability (Friedrich II, von Hohenstaufen 1596; Paris, 1913; Weitzel, 1951). However, experiments by Rutschke (1960) with waterfowl (Anseriformes, Ralliformes) showed that the stability of the plumage against water is independent from its impregnation by preen waxes. He believed the waxes made the feathers more flexible. The preen waxes possess optimum properties of low melting point, insolubility in water, high spreading ability, and resistance against oxygen to encourage both impermeability and flexibility of feathers. Extirpation of the uropygial gland of different species did not produce serious consequences for captive animals in species such as pigeons (Kossmann, 1871; Paris, 1913; Esther, 1938), hens (Lunghetti, 1907; Paris, 1913; Ida, 1931), ducks (Philipeaux, 1872; Paris, 1913; Ida, 1931; Elder, 1954), or passerine species (Paris, 1913; J. Jacob, unpublished results, 1975). It should be mentioned that in pigeons there is a direct lipid supply through the skin (Jacob and Grimmer, 1975a). This seems to be true for hens also (Lucas, 1968, 1970; Lucas and Stettenheim, 1972; Ishida *et al.*, 1973), so that these species might be excluded from consideration. Elder (1954) concluded that it was unlikely that ducks rendered glandless could survive in the wild. In any case, the great variability in the structure of preen waxes from order to order is difficult to understand, even if all the above-mentioned explanations for their function held. We therefore have to look for other functions of these waxes. Hou reported in a series of papers (Hou, 1928, 1929, 1930) the occurrence of provitamin D in the uropygial glands of hens. Hens suffering from rickets could be cured by UV radiation only if a preen gland was present, whereas glandectomized animals did not recover. These findings have been provocative (summarized by Elder, 1954). Even if the evidence of vitamin synthesis by the uropygial gland is proved in the future this does not provide an explanation for the variety of preen wax structures.

One acceptable hypothesis for the variety in composition might be their role in protecting the skin and feather against parasites. Many of the alkyl-substituted wax acids and alcohols show antibacterial and antimycotic properties (Stanley *et al.*, 1929; Stanley and Adams, 1932; Weitzel, 1948; Weitzel and Schraufstätter, 1950; Buu-Hoi and Cagniant, 1943; Jacob, 1974; J. Jacob, unpublished results, 1975). 3-Methyl-nonanoic, 4-methyl-heptanoic acid, and 3,7-dimethyl octanol show activity against *Microsporum canis, M. gypseum, Scopulariopsis brevicaulis, Aspergillus niger,* and *Candida* species, whereas 2-ethyl- and 2-butyl-substituted acids were active against *Staphylococcus* and *Streptomyces* species (J. Jacob, unpublished results, 1975). The structure-dependent antibacterial and hemolytic effects of branched fatty acids were reported by Breusch (1964, 1969). Moreover, the 3-hydroxy fatty acids occurring in the preen waxes of *Columba palumbus* (Jacob and Zeman, 1972b) and *Lanius collurio* (J. Jacob, unpublished results, 1975) are potent fungicides (Schildknecht and Koob, 1971). Pugh (1966, 1971, 1972) and Pugh and Evans (1970a,b) cite evidence for an effect of feather waxes on the growth of dermatophytes. They probably regulate the flora of keratophilic fungi on birds. The occurrence of another type of ectoparasite, the mallophages, is comparably species-specific as that of the fore-mentioned fungi, as shown by Timmermann (1957, 1965); but no experimental data on the effect of preen waxes on mallophagia are available.

An olfactory effect of the preen wax constituents in the individual identification of birds may be an alternate role. No evidence either to support or disprove this hypothesis is available.

D. Other Feather Lipids

Lucas (1968) established that the skin of the chicken secretes sebaceous material. Positive staining reactions with Oil Red O were found in skin tissue of other birds (duck, quail, albatross, pigeon) so that there is little doubt of the potential for lipogenesis of the skin in birds (Lucas and Stettenheim, 1972). From this, the question rises whether the skin lipids are identical with those produced in the uropygial gland. Unfortunately almost no experimental data are available on this problem. Extraction of the plumage with chloroform/methanol and comparison of the lipids extracted with those from the preen gland did not show any significant differences in case of *Puffinus griseus* (J. Jacob, unpublished results, 1975). It is therefore questionable whether in birds with well-developed preen glands the skin produces the same lipids as the gland (which seems to be unlikely) or whether the skin does not contribute considerable lipid amounts to the plumage. Lucas (1968) suggested that the waxes from the preen gland are applied primarily to the plumage, but the lipids secreted from the epidermis take care of the requirements of the skin itself. On the other hand, contamination of the plumage by the skin surface lipids is to be expected. This conclusion can be drawn from an investigation on the plumage lipids from *Columba palumbus* (Jacob and Grimmer, 1975a). Only 6.7% of the whole plumage lipids of this species originates from the diester waxes of the uropygial gland secretion (Table VI). The remainder is distributed among different lipid classes of nonuropygial gland origin and includes hydrocarbons, sterol esters, ester waxes, triglycerides, cholestanol, and free fatty acids. Among the hydrocarbons saturated odd-numbered unbranched individuals predominate, but 7-methyl-alkanes have been detected. The chain length of free and esterified alcohols corresponds to those of the fatty acids, where even-numbered compounds predominate. The main alcoholic component is 5α-cholestan-3-β-ol, which is a common metabolite in the pigeon (Subbiah *et al.*, 1971). It is supposed that the lipid produced

TABLE VI
COMPOSITION OF THE PLUMAGE LIPIDS FROM *Columba palumbus*[a]

Lipid class	Percentage
Hydrocarbons	3.3
Sterol ester + monoester waxes	10.7
Diester waxes	6.7
Triglycerides	5.0
Free alcohols + cholestanol	5.0
Free fatty acids	55.6
More polar lipids	13.7

[a] From Jacob and Grimmer, 1975b.

directly by the skin is a general mechanism at least in species with a rudimentary or no uropygial gland.

II. Depot Fats of Birds

Depot fats from birds are mixtures of triglycerides containing unbranched saturated and unsaturated fatty acids with chain length in the range of C_{10}–C_{22}. They show no relation to the uropygial gland-secreted lipids but reflect the triglyceride composition of the ingested food. For example, granivorous species possess depot fats rich in linoleic and linolenic acids (18:2 and 18:3), whereas fish-eating species have considerable amounts of polyenoic acids with 20 and 22 C atoms in the depot fat. Insect and carrion-consuming species, on the other hand, possess depot fats with relatively high contents of palmitic and oleic acid (16:0 and 18:1). The depot fat composition of species with different ingestion habits are compared in Table VII. Feeding habits range from fish-eaters (*Fratercula arctica* and *Ardea cinerea*) to herbivores (*Anser caerulescens, Phasianus colchicus*), granivores (*Columba palumbus, Pyrrhula pyrrhula*), and omnivores (*Passer domesticus*). Additional analysis of depot fats from several species, e.g., *Fulmarus glacialis, Sula bassana, Megalestris catarrhactes, Larus argentatus* (Lovern, 1938), and *Anser anser, A. fabalis, A. caerulescens, A. indicus, Aythya ferina, Mergus albellus, Anas clypeata, A. strepera* (Jacob and Glaser, 1975) have been published. Further data on the depot fat composition of the following species are available: *Dromaius novaehollandiae* (Hilditch *et al.*, 1942), *Struthio camelus* and *Phoenicopterus chilensis* (Gunstone and Russell, 1954), *Fratercula arctica* (Jacob and Grimmer, 1970a), *Ardea cinerea* (Poltz and Jacob, 1974b), *Apterix australis* (Shorland and Gass, 1961), *Phasianus colchicus* (Jacob and Grimmer 1970b), *Perdix perdix* (Jacob and Grimmer, 1970c), *Gallus domesticus* (Hilditch *et al.*, 1934), *Columba palumbus* (Jacob and Zeman, 1972b), *Pyrrhula pyrrhula* (Jacob and Zeman, 1971c), and *Passer domesticus* (Jacob and Zeman, 1970a). They do not differ significantly from the pattern.

III. Stomach Oils of Birds

The procellariiform possess strong-smelling stomach oils which are ejected against enemies in case of danger, and which serve as food for the young also. It seems that these stomach oils are not endocrine secretions but originate from undegraded or only partially degraded dietary residues (Kritzler, 1948; Lewis, 1966; Cheah and Hansen,

TABLE VII

COMPARISON OF THE DEPOT FAT COMPOSITIONS OF SPECIES WITH DIFFERENT INGESTION HABITS

Triglyceride fatty acid	*Fratercula arctica*	*Ardea cinerea*	*Anser caerulescens*	*Phasianus colchicus*	*Columba palumbus*	*Pyrrhula pyrrhula*	*Passer domesticus*
12:0	—	0.1	—	—	—	0.2	0.2
14:0	5.0	2.1	0.6	0.8	1.2	0.2	1.5
16:0	25.6	24.3	19.0	28.5	18.3	12.7	33.1
16:1	6.9	8.3	4.5	9.0	7.4	1.4	6.8
17:0	0.1	0.3	0.1	trace	trace	trace	trace
18:0	8.6	7.2	4.1	6.7	2.8	5.1	6.1
18:1	22.5	28.1	54.6	48.1	44.5	21.3	33.1
18:2	1.0	5.3	14.3	6.3	25.3	58.5	17.8
18:3	2.6	7.3	2.8	0.6	0.5	trace	1.0
20: Unsaturated	15.6	3.2	—	—	—	—	—
22: Unsaturated	11.7	10.2	—	—	—	—	—
Others	0.4	3.6	—	—	—	0.6	0.4

1970a). Wax esters as main constituents of stomach oils are reported to occur in the New Zealand Muttonbird *(Puffinus griseus)* (Smith, 1911; Carter, 1921, 1928; Carter and Malcolm, 1927). The glycerylether diesters in Leach's Storm-petrel *(Oceanodroma leucorhoa)* (Lewis, 1966) probably originate from the diet which consists of anchovy, squid, shrimp, and zooplankton. The stomach oils of the Wedge-tailed Shearwater *(Puffinus pacificus)* and the Great-Winged Petrel *(Pterodroma macroptera)* contain triglycerides, cholesterol esters, and free cholesterol (Cheah and Hansen, 1970a), but the stomach oil of the closely related Short-tailed Shearwater *(Puffinus tenuirostris)* consists of 80% ester waxes, 10% triglycerides, and 10% di- and monoglycerides (Cheah and Hansen, 1970b). Diet dependence might explain the contradictory results on the composition of stomach oils from the Fulmar *(Fulmarus glacialis)*, a bird with wide dietary requirements including plankton, whale excrements, sepia, herring, crustacees, and fish carrion. Rosenheim and Webster (1927) found ester waxes while Cheah and Hansen (1970b) reported triglycerides with a high content of C_{20}–C_{24} polyenoic acids to be the main constituents.

REFERENCES

Ackman, R. G. (1972). *J. Chromatogr. Sci.* **10**, 243–246.
Andersson, B. A., and Bertelsen, O. (1975a). *Chem. Scr.* **8**, 91–94.
Andersson, B. A., and Bertelsen, O. (1975b). *Chem. Scr.* **8**, 135–139.
Bertelsen, O. (1970). *Ark. Kemi* **32**, 17–26.
Bertelsen, O. (1973). *Chem. Scr.* **4**, 163–174.
Bertelsen, O., Eliasson, B., Odham, G., and Stenhagen, E. (1975). *Chem. Scr.* **8**, 5–7.
Breusch, F. L. (1964). *Abstr. Pap., World Fat Congr., 1st, 1964* p. 45.
Breusch, F. L. (1969). *Fortschr. Chem. Forsch.* **12**, 119–184.
Buckner, J. S., and Kolattukudy, P. E. (1975a). *Biochemistry* **14**, 1768–1773.
Buckner, J. S., and Kolattukudy, P. E. (1975b). *Biochemistry* **14**, 1774–1782.
Buu-Hoi, N. P., and Cagniant, P. (1943). *Hoppe-Seyler's Z. Physiol. Chem.* **279**, 76–86.
Carter, C. L. (1921). *J. Soc. Chem. Ind., London* **40**, 220T.
Carter, C. L. (1928). *J. Soc. Chem. Ind., London* **47**, 26T–30T.
Carter, C. L., and Malcolm, J. (1927). *Biochem. J.* **21**, 484–488.
Cater, D. B., and Lawrie, N. R. (1950). *J. Physiol. (London)* **111**, 231–243.
Cater, D. B., and Lawrie, N. R. (1951). *J. Physiol. (London)* **112**, 405–419.
Cheah, C. C., and Hansen, I. A. (1970a). *Comp. Biochem. Physiol.* **31**, 757–761.
Cheah, C. C., and Hansen, I. A. (1970b). *Int. J. Biochem.* **1**, 198–202.
Edkins, E., and Hansen, I. A. (1971). *Comp. Biochem. Physiol.* B**39**, 1–4.
Edkins, E., and Hansen, I. A. (1972). *Comp. Biochem. Physiol. B* **41**, 105–112.
Elder, W. H. (1954). *Wilson Bull.* **66**, 6–31.
Esther, K. H. (1938). *Morphol. Jahrb.* **82**, 321–383.
Friedrich II von Hohenstaufen. (1596). "De arte venandi cum avibus." Augusta vindelicorum, Augsburg.
Grimmer, G., Jacob, J., and Kimmig, J. (1971). *Z. Klin. Chem. Klin. Biochem.* **9**, 111–116.
Gunstone, F. D., and Russell, W. C. (1954). *Biochem. J.* **57**, 459–461.

Haahti, E. O. A., and Fales, H. M. (1967). *J. Lipid Res.* **8**, 131–137.
Hansen, I. A., Tang, B. K., and Edkins, E. (1969). *J. Lipid Res.* **10**, 267–270.
Hilditch, T. P., Jones, E. C., and Rhead, A. J. (1934). *Biochem. J.* **28**, 786–795.
Hilditch, T. P., Sime, I. C., and Maddison, L. (1942). *Biochem. J.* **36**, 98–109.
Hou, H. C. (1928). *Chin. J. Physiol.* **2**, 345–380.
Hou, H. C. (1929). *Chin. J. Physiol.* **3**, 171–182.
Hou, H. C. (1930). *Chin. J. Physiol.* **4**, 79–92.
Ida, Z. (1931). *Med. Fak. Pathol. Inst. Jpn. Mitt.* **23**, 1–12.
Ishida, K., Kusuhara, S., and Yamaguchi, M. (1971). *Gakkai-Ho* **42**, 544–550.
Ishida, K., Suzuki, T., Kusuhara, S., and Yamaguchi, M. (1973). *Poult. Sci.* **52**, 83–87.
Jacob, J. (1974). *Fette, Seifen, Anstrichm.* **76**, 241–244.
Jacob, J. (1975a). *J. Chromatogr. Sci.* **13**, 415–422.
Jacob, J. (1975b). *Hoppe-Seyler's Z. Physiol. Chem.* **356**, 1823–1825.
Jacob, J. (1976a). *In* "Chemistry and Biochemistry of Natural Waxes" (P. E. Kolattukudy, ed.). Elsevier, Amsterdam 1976.
Jacob, J. (1976b). *Fortschr. Chem. Org. Naturst.* **34** (in press).
Jacob, J. (1976c). *Hoppe-Seyler's Z. Physiol. Chem.* **357**, 609–611.
Jacob, J. (1976d). *Biochem. Syst.* **4**, 215–221.
Jacob, J. (1976e). *Biochem. Syst.* **4**, 209–213.
Jacob, J. (1976f). *Lipids* **11**, 816–818.
Jacob, J. (1977a). *J. Ornithol.* **118**, 52–59.
Jacob, J. (1977b). *J. Ornithol.* **118**, 189–194.
Jacob, J. (1977c). *J. Lipid. Res.* (in press).
Jacob, J., and Glaser, A. (1970). *Z. Naturforsch., Teil B* **25**, 1435–1437.
Jacob, J., and Glaser, A. (1975). *Biochem. Syst.* **2**, 215–220.
Jacob, J., and Grimmer, G. (1967). *J. Lipid Res.* **8**, 308–311.
Jacob, J., and Grimmer, G. (1970a). *Z. Naturforsch., Teil B* **25**, 54–56.
Jacob, J., and Grimmer, G. (1970b). *Z. Naturforsch., Teil B* **25**, 577–580.
Jacob, J., and Grimmer, G. (1970c). *Z. Naturforsch., Teil B* **25**, 689–692.
Jacob, J., and Grimmer, G. (1973a). *Hoppe-Seyler's Z. Physiol. Chem.* **354**, 1648–1650.
Jacob, J., and Grimmer, G. (1973b). *Z. Klin. Chem. Klin. Biochem.* **11**, 297–300.
Jacob, J., and Grimmer, G. (1973c). *Z. Naturforsch., Teil C* **28**, 75–77.
Jacob, J., and Grimmer, G. (1975a). *Z. Naturforsch., Teil C* **30**, 363–368.
Jacob, J., and Grimmer, G. (1975b). *Biochem. Syst.* **3**, 267–271.
Jacob, J., and Poltz, J. (1972). *Hoppe-Seyler's Z. Physiol. Chem.* **353**, 1657–1660.
Jacob, J., and Poltz, J. (1973a). *Biochem. Syst.* **1**, 169–172.
Jacob, J., and Poltz, J. (1973b). *Z. Naturforsch., Teil C* **28**, 449–452.
Jacob, J., and Poltz, J. (1974a). *J. Lipid Res.* **15**, 243–248.
Jacob, J., and Poltz, J. (1974b). *Z. Naturforsch., Teil C* **29**, 236–238.
Jacob, J., and Poltz, J. (1974c). *J. Ornithol.* **115**, 454–459.
Jacob, J., and Poltz, J. (1975a). *Lipids* **10**, 1–8.
Jacob, J., and Poltz, J. (1975b). *Biochem. Syst.* **3**, 263–266.
Jacob, J., and Zeman, A. (1970a). *Z. Naturforsch., Teil B* **25**, 984–988.
Jacob, J., and Zeman, A. (1970b). *Z. Naturforsch., Teil B* **25**, 1438–1447.
Jacob, J., and Zeman, A. (1971a). *Z. Naturforsch., Teil B* **26**, 33–40.
Jacob, J., and Zeman, A. (1971b). *Z. Naturforsch., Teil B* **26**, 1344–1351.
Jacob, J., and Zeman, A. (1971c). *Z. Naturforsch., Teil B* **26**, 1352–1356.
Jacob, J., and Zeman, A. (1972a). *Z. Naturforsch., Teil B* **27**, 691–695.
Jacob, J., and Zeman, A. (1972b). *Hoppe-Seyler's Z. Physiol. Chem.* **353**, 492–494.
Jacob, J., and Zeman, A. (1972c). *Z. Naturforsch., Teil B* **27**, 695–698.
Jacob, J., and Zeman, A. (1973). *Z. Naturforsch., Teil C* **28**, 78–82.

Karlsson, H., and Odham, G. (1969). *Ark. Kemi* **31,** 143–158.
Kolattukudy, P. E. (1972). *Biochem. Biophys. Res. Commun.* **49,** 1376–1383.
Kolattukudy, P. E., and Sawaya, W. N. (1974). *Lipids* **9,** 290–292.
Kossmann, R. (1871). *Z. Wiss. Zool.* **21,** 568–599.
Kritzler, H. (1948). *Condor* **50,** 5–15.
Langer, S. H., Connell, S., and Wender, I. (1958). *J. Org. Chem.* **23,** 50–58.
Lennert, K., and Weitzel, G. (1951). *Hoppe-Seyler's Z. Physiol. Chem.* **288,** 266–272.
Lennert, K., and Weitzel, G. (1952). *Z. Mikrosk.-Anat. Forsch.* **58,** 208–229.
Lewis, R. W. (1966). *Comp. Biochem. Physiol.* **19,** 363–377.
Lovern, J. A. (1938). *Biochem. J.* **32,** 2142–2144.
Lucas, A. M. (1968). *Anat. Rec.* **160,** 386–387.
Lucas, A. M. (1970). *Fed. Proc. Fed. Am. Soc. Exp. Biol.* **29,** 1641–1648.
Lucas, A. M., and Stettenheim, P. (1972). *U.S. Dep. Agric., Agric. Handb.* **362,** 613–626.
Lunghetti, B. (1907). *Arch. Mikrosk. Anat.* **69,** 264–321.
Morris, L. J. (1963). *J. Chromatogr.* **12,** 321–328.
Morrison, W. R., and Smith, L. M. (1964). *J. Lipid Res.* **5,** 600–608.
Nicolaides, N. (1971). *Lipids* **6,** 901–905.
Nicolaides, N. (1974). *Science* **186,** 19–26.
Noble, R. E., Stjernholm, R. L., Mercier, D., and Lederer, E. (1963). *Nature (London)* **199,** 600–601.
Odham, G. (1963). *Ark. Kemi* **21,** 379–393.
Odham, G. (1964). *Ark. Kemi* **22,** 417–445.
Odham, G. (1965). *Ark. Kemi* **23,** 431–451.
Odham, G. (1966). *Ark. Kemi* **25,** 543–554.
Odham, G. (1967a). *Ark. Kemi* **27,** 251–255.
Odham, G. (1967b). *Ark. Kemi* **27,** 263–288.
Odham, G. (1967c). *Ark. Kemi* **27,** 289–294.
Paris, P. (1913). *Arch. Zool. Exp. Gen.* **53,** 139–276.
Philipeaux, J. M. (1872). *C.R. Seances Soc. Biol. Ses Fil.* **24,** 49–52.
Poltz, J., and Jacob, J. (1973). *Z. Naturforsch., Teil C* **28,** 449–452.
Poltz, J., and Jacob, J. (1974a). *Biochim. Biophys. Acta* **360,** 348–356.
Poltz, J., and Jacob, J. (1974b). *J. Ornithol.* **115,** 103–105.
Poltz, J., and Jacob, J. (1974c). *Z. Naturforsch., Teil C* **29,** 239–242.
Poltz, J., and Jacob, J. (1974d). *J. Ornithol.* **115,** 119–127.
Poltz, J., and Jacob, J. (1975). *Biochem. Syst.* **3,** 57–62.
Pugh, G. J. F. (1966). *J. Indian Bot. Soc.* **45,** 296–303.
Pugh, G. J. F. (1971). *Int. Colloq. Soil Zool. Inst. Natl. Rech. Agron., Paris, 4th 1971* pp. 319–327.
Pugh, G. J. F. (1972). *Ibis* **114,** 172–177.
Pugh, G. J. F., and Evans, M. D. (1970a). *Trans. Br. Mycol. Soc.* **54,** 233–240.
Pugh, G. J. F., and Evans, M. D. (1970b). *Trans. Br. Mycol. Soc.* **54,** 241–250.
Riley, R. G., and Koluttukudy, P. E. (1975). *Arch. Biochem. Biophys.* **157,** 309–319.
Rosenheim, O., and Webster, T. A. (1927). *Biochem. J.* **21,** 111–118.
Rutschke, E. (1960). *Zool. Jahrb., Abt. Syst. Oekol. Geogr. Tiere* **87,** 441–506.
Ryhage, R., and Stenhagen, E. (1960a). *Ark. Kemi* **15,** 291–315.
Ryhage, R., and Stenhagen, E. (1960b). *Ark. Kemi* **15,** 333–351.
Ryhage, R., and Stenhagen, E. (1960c). *Ark. Kemi* **15,** 545–560.
Ryhage, R., and Stenhagen, E. (1963). In "Mass Spectrometry of Organic Ions" (F. W. McLafferty, ed.), pp. 399–414. Chap. 9 Academic Press, New York.
Saito, K., and Gamo, M. (1970). *J. Biochem. (Tokyo)* **67,** 841–849.
Saito, K., and Gamo, M. (1972). *Biochim. Biophys. Acta* **260,** 164–168.

Sawaya, W. N., and Kolattukudy, P. E. (1972). *Biochemistry* **11**, 4398–4406.
Sawaya, W. N., and Kolattukudy, P. E. (1973). *Arch. Biochem. Biophys.* **157**, 309–319.
Schildknecht, H., and Koob, K. (1971). *Angew. Chem.* **83**, 110.
Shorland, F. B., and Gass, J. P. (1961). *J. Sci. Food Agric.* **12**, 174–177.
Sibley, C. G. (1970). *Bull Peabody Mus.* **32**, 1–131.
Sibley, C. G., and Ahlquist, J. E. (1972). *Bull. Peabody Mus.* **39**, 1–276.
Sibley, C. G., Corbin, K. W., Ahlquist, J. E., and Ferguson, A. (1974). *In* "Biochemical and Immunological Taxonomy of Animals" (C. A. Wright, ed.), pp. 89–176. Chap. 2. Academic Press, New York.
Smith, L. H. (1911). *J. Soc. Chem. Ind., London* **30**, 405.
Stanley, W. M., and Adams, R. (1932). *J. Am. Chem. Soc.* **54**, 1548–1557.
Stanley, W. M., Jay, M. S., and Adams, R. (1929). *J. Am. Chem. Soc.* **51**, 1261–1266.
Stern, M. (1905). *Arch. Mikrosk. Anat.* **66**, 299–311.
Subbiah, M. T. R., Kottke, B. A., and Carlo, I. A. (1971). *Lipids* **6**, 517–519.
Sumper, M., Oesterhelt, D., Riepertinger, C., and Lynen, F. (1969). *Eur. J. Biochem.* **10**, 377–387.
Tang, B. Y., and Hansen, I. A. (1970). *Proc. Aust. Biochem. Soc.* **3**, 84.
Tang, B. Y., and Hansen, I. A. (1972). *Eur. J. Biochem.* **31**, 372–377.
Thomas, A. E., III, Sharoun, J. E., and Ralston, H. (1965). *J. Am. Oil Chem. Soc.* **42**, 789–792.
Timmermann, G. (1957). "Studien zu einer vergleichenden Parasitologie der Charadriiformes oder Regenpfeifervögel," Part I. Fischer, Jena.
Timmermann, G. (1965). "Die Federlingsfauna der Sturmvögel und die Phylogenese des procellariiformen Vogelstammes." de Gruyter, Berlin.
Weitzel, G. (1948). *Angew. Chem.* **60**, 263–267.
Weitzel, G. (1951). *Fette, Seifen, Anstrichm.* **53**, 667–671.
Weitzel, G., and Schraufstätter, E. (1950). *Hoppe-Seyler's Z. Physiol. Chem.* **285**, 172–182.
Zeman, A., and Jacob, J. (1972). *Z. Anal. Chem.* **261**, 306–309.
Zeman, A., and Jacob, J. (1973a). *Fette, Seifen, Anstrichm.* **75**, 667–674.
Zeman, A., and Jacob, J. (1973b). *Petrolio ambiente* pp. 123–128.
Zeman, A., and Scharmann, H. (1972). *Fette, Seifen, Anstrichm.* **74**, 509–519.
Zeman, A., and Scharmann, H. (1973). *Fette, Seifen, Anstrichm.* **75**, 32–44.

CHAPTER 7

Avian Endocrinology

Albert H. Meier and Blaine R. Ferrell

I. Introduction

Because it deals with the chemical integration of a multicellular organism, endocrinology is of interest to a wide spectrum of biologists. The biochemist has discovered that there must be communication between cells if their activities are to be integrated for the formation of meaningful directed organismal activity. The behaviorist and naturalist suspect that the internal organization must be appreciated if one is to understand the complexities of unpredictability. The hormonal system is composed of specialized cells and tissue that discharge their products into the extracellular internal environment from which they affect cellular activities in other parts of the body. The secretions of many of the hormones are controlled directly or indirectly by the nervous system and are thus sensitive to environmental and behavioral stimuli. The activities of these hormones, in turn, produce changes in physiological condition and behavior, and in the manner in which the organism adjusts to its environment.

With the exception of mammals, more research has been done on the endocrinology of birds than on any other vertebrate class. Avian endocrinology differs from mammalian endocrinology in that mammalian studies have concentrated on a few species whereas avian studies have been performed on a wide range of both domesticated and wild species.

There are at least 35 known hormones in birds and many act synergistically with others. This review will attempt to draw the broad outlines of endocrine control and activity by considering each endocrine gland individually. In addition, it deals with several of the major functions of the neuroendocrine system as it is involved in molting, reproduction, and migration.

II. Hormone Action

There are two known cellular mechanisms that account for most hormone activities. In one system, cyclic adenosine 3′,5′-monophosphate (cyclic AMP) acts as a second messenger in the cell to elicit the response initiated by the first messenger, a hormone (Sutherland, 1972). According to Sutherland's second messenger hypothesis, the hormone binds with its specific protein receptor associated with the plasma membrane of the target cell and does not enter the cell itself. Many polypeptide hormones, such as those of the adenohypophysis, produce their effects by way of the cyclic AMP system (Fig. 1).

As in mammalian studies, avian research on cyclic AMP has been

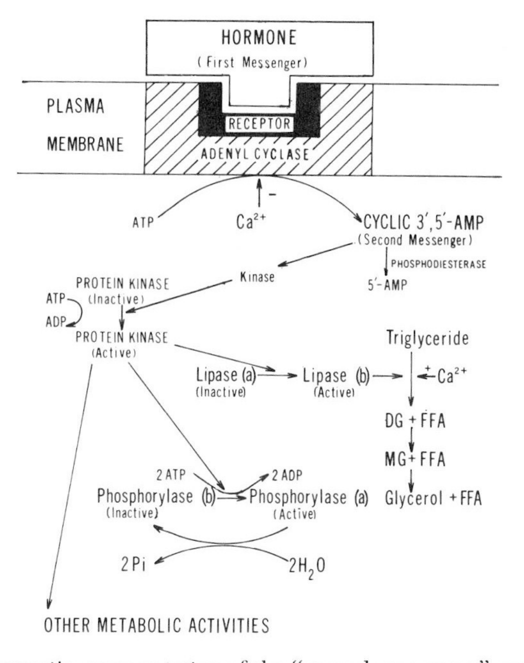

FIG. 1. Diagrammatic representation of the "second messenger" mechanism of hormone action. DG + diglyceride. MG + monoglyceride.

directed toward carbohydrate and lipid metabolism (Hazelwood, 1973; Gross and Mialhe, 1974; Boyd *et al.*, 1975; Lefebvre, 1975). For example, glucagon initiates a potent lipolytic effect in avian adipose tissue when it binds with a protein receptor associated with the adipocyte plasma membrane. This hormone–receptor complex stimulates a cascade of events involving the activation of adenylate cyclase which in turn activates protein kinase that activates a lipase leading to the breakdown of triglycerides (Rasmussen, 1970; Lefebvre, 1975). Calcium ion is necessary for cyclic AMP to have its effects (Rasmussen, 1970), and the prostaglandin, PGE_1, can inhibit the formation of cyclic AMP (Grande and Prigge, 1972; Lefebvre and Luyckx, 1974). Cyclic AMP is deactivated by phosphodiesterase.

The recent deluge of prostaglandin research in mammals points to an important role for them in mediating hormone action by way of cyclic AMP (for review Nakano, 1973; Paoletti and Puglisi, 1973; Flack, 1973). By affecting vascular permeability, prostaglandins may play a regulatory role in controlling the amount of hormones that reach the target tissues. Prostaglandins have been detected in several avian species (Karim *et al.*, 1968; McKeown *et al.*, 1974), but much more research is called for.

In a second system of hormone activity, the hormone must pass through the plasma membrane (see O'Malley and Schrader, 1976). Because of their high lipid solubilities and low molecular weights (av. 300), the steroid hormones are well suited chemically for this mechanism. Other low molecular weight hormones, such as thyroxine (Bentley, 1976), may also use this steroidal system. Specificity of the hormone for its target tissue resides in a cytoplasmic steroid-receptor protein (200,000 MW) which concentrates the steroids. Steroid-concentrating target cells have been demonstrated in chicken *(Gallus domesticus)* oviduct (O'Malley and Means, 1974; O'Malley and Schrader, 1976; Teng and Teng, 1975), duck *(Anas platorhynchos)* hypothalamus (Abel *et al.*, 1975; Martinez-Vargas *et al.*, 1975), duck salt gland (Sandor and Fazekas, 1974; Allen *et al.*, 1975b), and chicken liver (Krall and Hahn, 1975). Nontarget tissues do not concentrate the hormones. The hormone–receptor complex moves from the cytosol to the nucleus where it initiates transcription of the DNA molecule, the first step in protein synthesis.

The cytosol receptor proteins for estrogen and progesterone are different in chick oviduct tissue (O'Malley and Means, 1974). This difference offers a possible basis for the synergistic effect of these steroids wherein progesterone is ineffective unless preceded by estrogen treatment. McKnight *et al.* (1975) suggested that estrogen was necessary for the production of the receptor proteins for progesterone. Similar effects could account for many hormone synergisms and be an important regulatory mechanism.

Another possible mechanism that could influence hormone activity involves the plasma hormone-binding globulins for the steroids. Changes in plasma concentrations of the globulins could be expected to influence the physiological half-lives of the adrenal and gonadal steroids. Changes in vascular permeability to the globulins might influence an interaction between intra- and extracellular binding proteins. Sex hormones and thyroxine can elicit large increases in plasma concentrations of corticosteroid-binding globulin in mammals (Gala and Westphal, 1966; Labrie *et al.*, 1968).

III. Hypothalamus and Pituitary

A. Hypothalamus

The hypothalamus–pituitary axis performs a central role in neuroendocrine integration. Hypothalamic centers coordinate nervous imputs from external and internal stimuli and translate them into neurohormonal secretions. These secretions include two octapeptides that travel along neuronal axons directly to the pars nervosa of the neurohypophysis where they are stored until being released into the systemic circulation. Other neurohormonal secretions, or adenohypophyseal hormone releasing hormones, travel along axons to the median eminence from which they are transported via hypophyseal portal vessels to the pars distalis of the adenohypophysis. There they stimulate or inhibit the release of specific pituitary hormones (see Kobayashi and Wada, 1973) (Fig. 2).

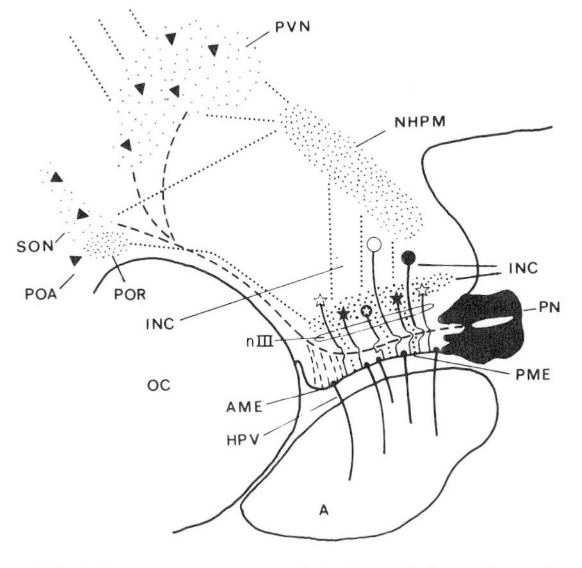

FIG. 2. A simplified diagrammatic view of the hypothalamo–hypophyseal system in birds to show the major neural and neurosecretory pathways. (1) The anterior hypothalamus. The aldehyde-fuchsin stainable neurosecretory system (triangles) arises mainly in the paraventricular (PVN) and supraoptic (SON) nuclei and terminates in the pars nervosa (PN). Some axons also end in the anterior median eminence (AME). The origin of these fibers is unclear but in passerine birds they seem to arise from the preoptic (POA) and suprachiasmatic areas. Both the SON and PVN are heavily innervated with aminergic nerve terminals (shaded areas) arising extrahypothalamically. Aminergic tracts (·········) exist to the posterior hypothalamus. (2) The tuberal hypothalamus. This region is dominated by an extensive infundibular nucleus (INC) believed to contain a number of different neurosecretory neurons (circles and stars), the axons of which form a tuberoinfundibular tract ending in the median eminence. The basal infundibular nucleus seems comparable with the nucleus tuberis. Many adrenergic terminals end in this region, and a few carry on into the median eminence itself. In the quail another nucleus (n. hypothalamicus posterior medialis, NHPM) lying in and above the infundibular nucleus, is visible in the tuberal region. It contains many aminergic nerve terminals but does not appear to be neursecretory. Fiber tracts run between the various nuclei. A: anterior pituitary (pars distalis); HPV: hypophyseal portal vessels; nIII: third ventricle; OC: optic chiasma; PME: posterior median eminence; POR: preoptic recess. From Follett and Davies, 1975.

Wingstrand's (1951) classic treatise provided the basis and much of the terminology for morphological investigations of the avian hypothalamus and pituitary. There are two major neurosecretory nuclei in the hypothalamus, the supraoptic and paraventricular nuclei. Gomori (aldehyde fuchsin)-positive neurons arise from both nuclei and are thought to transport the neurohypophyseal hormones (oxytocin and arginine vasotocin) directly to the pars nervosa. Curiously, these neuronal axons also terminate in the anterior median eminence (see also Oksche *et al.*, 1959; Farner and Oksche, 1962), but their products are released by a different mechanism (Hawkes and George, 1974). Both nuclei are supplied with monoaminergic fibers (Sharp and Follett, 1968; Oehmke *et al.*, 1969) which synapse with the neurosecretory cells (Oehmke *et al.*, 1969; Priedkalns and Oksche, 1969). Cholinergic innervation has not been demonstrated although acetylcholinesterase activity has been reported in both the pars nervosa and the median eminence (Kobayashi *et al.*, 1970).

Another neurosecretory system conveys the releasing hormones to the median eminence. The neurosecretory neurons of this system are characterized as Gomori-negative fibers. Attempts to localize these cells within the supraoptic and paraventricular nucleus and elsewhere have produced results that vary among species and between investigators (Kobayashi *et al.*, 1961; Matsui, 1966; Arai, 1963; Uemura and Kobayashi, 1963). Davies and Follett (1974b) suggested that the control of a given neurohormone such as luteinizing hormone releasing hormone may depend on a complicated pathway. The catecholamines have been implicated in the release of these neurohormones (Campbell and Wolfson, 1974).

Demonstrations of hypophysiotropic neurohormones in the hypothalamus include corticotropin releasing hormone (CRH) in pigeons *(Columba livia)* (Péczeley and Zboray, 1967), growth hormone releasing hormone in pigeons (Muller *et al.*, 1967), gonadotropin releasing hormone in Japanese Quail *(Coturnix coturnix japonica)* (Follett, 1970; Smith and Follett, 1972), and prolactin releasing hormone in several species (Nicoll and Meites, 1962; Nicoll, 1965; Gourdji and Tixier-Vidal, 1966; Chen *et al.*, 1968). There is also evidence for a prolactin inhibitory hormone in the duck (Tixier-Vidal and Gourdji, 1972). Injections of synthetic thyrotropin releasing hormone (TRH) had several effects in pigeons and chickens depending on the dose (Bolton *et al.*, 1973; Hall and Chadwick, 1974; Scanes, 1974; Chadwick and Hall, 1975; Hall *et al.*, 1975). Low doses of TRH induced the release of prolactin, and medium doses induced only the release of thyroid stimulating hormone (TSH). High doses stimulated the production and release of prolactin and also affected growth hormone release. The physiological significances of these results are uncertain, but the assumption that there is a single hypophysiotropic hormone for each of the adenohypophyseal hormones is certainly open to question. Changes in pituitary responses to various concentrations of the hypophysiotropic hormones coupled with daily rhythms in the levels of hypophysiotropic hormones (i.e., CRH: Sato and George, 1973) could have important regulatory implications.

B. Neurohypophysis

The pars nervosa of the neurohypophysis receives Gomori-positive neurosecretory axons which deliver their products around glial cells (Kobayashi *et al.*, 1961). Two octapeptides are stored in the pars nervosa, oxytocin and arginine vasotocin. Their release is probably controlled by aminergic fibers. Arginine vasotocin is thought to be an antidiuretic agent (Farner *et al.*, 1967; Follett and Farner, 1966a), and to be involved in the induction of oviposition (Munsick *et al.*, 1960). Both arginine vasotocin and oxytocin stimulate increases in blood concentrations of glucose and free fatty acids, possibly in a synergistic manner with glucagon (Kook *et al.*, 1964; Farner *et al.*, 1967; John and George, 1973; see also John *et al.*, 1974).

C. Adenohypophysis

The pars distalis of the adenohypophysis is divisible into the cephalic and caudal lobes. These lobes receive separate hypophyseal portal vessels that drain distinct portions of the median eminence (Vitums *et al.*, 1964; Dominic and Singh, 1969). The blood from the pars distalis drains into the jugulars by way of a cavernous sinus. The releasing hormones are transported along Gomori-negative neurosecretory axons to the glial cells of the median eminence.

Rahn and Painter (1941) described four cell types in the pars distalis based on staining reactions. Basophils and light staining acidophils (A_2 cells) were restricted to the cephalic lobe whereas dark staining acidophils (A_1 cells) were located in the caudal lobe. Chromophobes (nonstaining cells) were thought to be functionally inactive precursors. More recent studies have assigned eight cell types to the pars distalis, including a cell type for each one of the recognized hormones and a general precursor cell. These cells are described under the specific hormone and include Romeis' Greek letter nomenclature modified by Herlant (1964; see also Tixier-Vidal and Follett, 1973). With the availability of purified avian hormones, immunological techniques will soon be utilized to locate and verify cell types.

IV. Hormones of the Pars Distalis and Their Target Glands

A. Prolactin

Prolactin (eta) cells are concentrated in the cephalic lobe of the pars distalis (see Kobayashi and Wada, 1973; Schreibman and Holtzman, 1975). However, the boundary between the cephalic prolactin cells and the caudal somatotrophs is indistinct (Mikami *et al.*, 1973), and more refined techniques are required to clearly establish the two cell types.

Some evidence has been presented for a hypothalamic prolactin inhibitory hormone in the duck (Tixier-Vidal and Gourdji, 1972). This mechanism for prolactin control would be similar to the mammalian system. The evidence for a stimulatory hormone, however, has been more firmly established. Hypothalamic extracts stimulate prolactin production by pituitary cultures in several avian species (Meites and Nicoll, 1966; see Kobayashi and Wada, 1973). In addition, TRH has a biphasic effect on prolactin release (Hall and Chadwick, 1974; Hall *et al.*, 1975). Steroid hormones, and probably other hormones, may modify the activity of prolactin cells (Tixier-Vidal and Gourdji, 1972).

Pituitary prolactin content varies during the day in the White-throated Sparrow *(Zonotrichia albicollis)* (Meier *et al.*, 1969) and duck (Ensor, 1975). A daily rise was observed during the morning in May as well as in August in the sparrow. The time of daily release of prolactin, however, varies seasonally: during the afternoon in May and late during the night in August. These results suggest that there may be separate mechanisms for the control of prolactin production and prolactin release. The implications of a seasonal change in the daily rhythm of prolactin release are discussed in later sections. Evidences have been presented that the release of pituitary prolactin may be induced by long daily photoperiods (Alexander and Wolfson, 1970) and by dehydration (Ensor, 1975).

The first assay for prolactin was based on the cropsac response characteristic of birds belonging to the order Columbiformes (Riddle *et al.*, 1933). Investigators who use this assay should be cautioned that the response to prolactin varies greatly during the day; the greatest response occurs at 9 hours after the onset of a 12-hour daily photoperiod in pigeons (Meier *et al.*, 1971a,b). Radioimmunoassays will probably be performed soon, allowing for much greater sensitivity. Nicoll (1975) pointed out that studies using radioimmunoassays must be interpreted with caution because immunologically reactive sites are not necessarily identical with physiologically active sites. The development of assays utilizing specific tissue binding sites for prolactin would seem to offer the best hope for combining sensitivity with physiological specificity.

Since the discovery of prolactin in pigeons (Riddle *et al.*, 1932), many activities of the hormone have been reported (reviews, Riddle, 1963; Nicoll and Bern, 1972). Like growth hormone, prolactin has no specific endocrine target but affects many tissues. It synergizes with other hormones, and many of its activities vary during the day depending on the temporal relations with the daily rhythms of the adrenal corticosteroids (Meier, 1975). Synergistic influences of prolactin with other hormones may involve prolactin's effects on prostaglandins. (See Section II). The roles of prolactin in reproduction and migration are treated in later sections.

B. Growth Hormone

Growth hormone (alpha) cells resemble prolactin cells in that both have acidophilic staining characteristics. They differ in other staining and morphological criteria and are localized in the caudal lobe instead of the cephalic lobe (see Tixier-Vidal and Follett, 1973). Growth hormone secretion appears to be stimulated by a hypothalamic growth hormone releasing hormone in pigeons (Muller *et al.*, 1967) and ducks (Hawkes, 1974).

Although relatively little is known about avian growth hormone, the activities appear to be similar to those in mammals (Scanes *et al.*, 1975). Chicken growth hormone stimulated body growth, bone growth, and nitrogen retention in hypophysectomized young chickens (Glick, 1960). In addition, pituitary extracts from chickens and turkeys *(Meleagris domesticus)* had a stimulatory influence on the rat tibial epiphyseal assay although the effect was less potent than that caused by mammalian extracts (Hazelwood and Hazelwood, 1961; Hirsch, 1961). Mammalian growth hormone has little or no effect on growth and protein metabolism in birds (Glick, 1960; Nalbandov, 1966; Scanes *et al.*, 1975).

Avian growth hormone also influences lipid metabolism in birds in a manner similar to that in mammals (John and George, 1972; John *et al.*, 1973; McKeown *et al.*, 1973). Plasma free fatty acid was increased at 3 hours after intramuscular injection in pigeons indicating lipolytic activity. A daily rhythm of plasma growth hormone concentration assayed by radioimmunoassay has been reported in pigeons (McKeown *et al.*, 1973), and histological evidences have been presented for photoperiodic stimulation of the growth hormone cells (Hawkes, 1974).

C. Thyroid

Thyroid stimulating hormone (TSH) cells (delta cells) are stimulated by a hypothalamic releasing hormone in the duck (Assenmacher and Tixier-Vidal, 1964; Rosenberg *et al.*, 1967) and pigeon (Baylé *et al.*, 1966), and TSH in turn stimulates the thyroid gland. The thyroid cells are usually restricted to the caudal lobe but they may be present in both lobes of the pars distalis under certain experimental circumstances (Tixier-Vidal and Follett, 1973; Radke and Chiasson, 1974). Synthetic TRH in doses thought to be physiological stimulated TSH production and release in pigeons and chickens. Lower and higher doses of TRH had no effect on TSH but stimulated prolactin production instead (Bolton *et al.*, 1973; Hall and Chadwick, 1974; Hall *et al.*, 1975; Chadwick and Hall, 1975). There may be some autonomy of TSH production in pituitary autografts (see Tixier-Vidal and Follett, 1973). Chicken TSH is a large molecular weight glycoprotein similar to mammalian TSH (Scanes and Follett, 1972). Assays for TSH usually involve uptake or depletion (as part of thyroid hormone) of labeled iodine from the thyroid gland (Bates and Condliffe, 1966).

The avian thyroid gland is a paired structure derived by evagination from the pharyngeal floor. Its cells are fully differentiated during the first half of incubation (Assenmacher, 1973). The biosynthesis of thyroxine (tetraiodothyronine: T_4) and triiodothyronine (T_3) requires a dietary source of elemental iodine, which is concentrated in the thyroid (Astier, 1973), and the presence of a specific thyroglobulin with which the hormones are conjugated and stored in the thyroid (Assenmacher, 1973).

Birds lack a specific plasma thyroid-binding globulin, unlike mammals, but they have a thyroid-binding prealbumin which binds T_4 in most species and T_3 in the pigeon (Refetoff *et al.*, 1970) as well as a thyroid-binding albumin which binds both T_4 and T_3 in the chicken (Heninger and Newcomer, 1964). Although the total plasma concentration of thyroxine present in birds falls within the same range as that found in mammals, the amount of unbound thyroxine is greater (Refetoff *et al.*, 1970). This characteristic may account in part for the relatively short half-life (less than 18 hours) of plasma T_4 in birds (Assenmacher, 1973; Grandhi *et al.*, 1975). The major route of thyroid hormone loss in chickens is by way of the bile and feces (Hutchins and Newcomer, 1966).

Daily rhythms of ^{131}I-uptake and of plasma levels of T_4 and T_3 determined by a competitive binding assay have been reported in chickens (Newcomer, 1974). Because the peak of T_4 concentration occurred early during the daily photoperiod (16L : 8D) and the T_3 peak occurred 9 hours later, Newcomer suggested that the T_3 peak might have resulted from peripheral T_4 metabolism. A daily rhythm in the $T_3:T_4$ ratio has also been reported in chickens (Sadovsky and Bensadoun, 1971). The possible physiological significances of this ratio of thyroid hormones needs additional study (Snedecor *et al.*, 1973; Grandhi and Brown, 1975a,b).

The physiology of the thyroid is as poorly understood in birds as it is in mammals. Many of the wide range of effects are undoubtedly synergistic with or permissive for, the effects of other hormones, perhaps in part by permitting the expression of daily or circadian (daily rhythms under conditions of constant light intensity and temperature) rhythms (Meier, 1975). The role most often associated with thyroid hormones concerns their stimulatory effects on oxygen consumption. They also contribute to maturational processes such as growth (King and King, 1973) and the development of homiothermy (see Assenmacher, 1973). Injections of T_4 and T_3 increase glucose availability (Hazelwood, 1965) and depress liver glycogen levels in pigeons (Riddle and Opdyke, 1947) by increasing glucose-6-phosphatase activity (Snedecor *et al.*, 1973). Possible roles for thyroid hormones in reproduction and migration are discussed in later sections.

D. Reproductive Hormones

Despite the heroic efforts of many investigators, our understanding of the hypothalamic–pituitary pathways for the control of reproduction in birds is still rudimentary. Although there is evidence of a hypothalamic gonadotropin releasing hormone, it is still unclear whether there are separate hypothalamic releasing hormones for LH and FSH. Production of immunoreactive LH can be stimulated in Japanese Quail pituitaries *in vitro* with hypothalamic extracts (Smith and Follett, 1972; Bicknell and Follett, 1975) but similar studies for FSH stimulation await a suitable assay. Lesion studies indicate that the preoptic region of the anterior hypothalamus and the tuberal region of the posterior hypothalamus are involved in photo-induced LH production and gonadal growth in the Japanese Quail (Sharp and Follett, 1969; Davies and Follett, 1975a,b,c; Follett and Davies, 1975) and the duck (Hawkes, 1974). Adrenergic innervation has been implicated in the photo-induced release of gonadotropic hormone in the Japanese Quail (Davies and Follett, 1974a,b; Campbell and Wolfson, 1974).

The three classical methods employed to determine gonadotropin cells of the pars distalis rely on observable changes in cellular activity (1) after castration, (2) during the breeding cycle, and (3) after the administration of sex steroids. The results indicate that the FSH (beta) cells are located in the cephalic lobe and LH (gamma) cells are found in the caudal lobe of most species. The gonadotropin cells, like the TSH cells, have basophilic staining characteristics.

Purification of LH and FSH revealed that some avian gonadotropins have amino acid compositions that are similar to those found in mammalian gonadotropins except for higher glycine and alanine residues (Stockwell-Hartree and Cunningham, 1969; Godden and Scanes, 1975). Nevertheless, cross reactivity between the avian gonadotropins and antibodies to mammalian gonadotropins is poor, and successful radioimmunoassays require the formation of antibodies to purified avian gonadotropins. An LH radioimmunoassay (Follett *et al.*, 1972) has already contributed much useful information. A satisfactory FSH assay would be especially welcome.

FSH and LH act synergistically in stimulating the testes and ovaries. The specific activities of these hormones in the female are unresolved although FSH appears to have a greater role in follicular development and maturation of the ova and LH has a greater role in stimulating estrogen production and ovulation (see Section VII). In the male, FSH stimulates growth of the seminiferous tubules and meiotic divisions of primary spermatocytes to secondary spermatocytes whereas LH stimulates the production of androgens by the Leydig cells (van Tienhoven, 1961; Nalbandov, 1966). But full development of spermatozoa requires both gonadotropins. FSH probably acts on the Sertoli cells inducing the production of androgen which in turn stimulates spermatogenesis (Lofts and Murton, 1973). A high testicular growth rate until the time of spermatid maturation is associated with high secretions of both FSH and LH; a decrease in testicular growth and spermatid maturation coincides with a decrease in FSH secretion (Follett, 1975). Thus FSH may be necessary only during the initial stages of spermatogenesis (Temple, 1974). Injections of testosterone alone can maintain mature testes in the Weaver-Finch *(Quelea quelea)* past the time when they otherwise regress following reproduction; but injections of testosterone early during testicular development have an inhibitory effect, possibly by a negative feedback on FSH production (Lofts, 1962). Because prolactin can exert a potent antisteroidogenic effect (Meier, 1969a), it might allow for FSH production early during reproductive recrudescense by reducing the negative feedback by gonadal hormones. Further discussion of the roles of FSH, LH, prolactin, and other hormones in maintaining the reproductive system are presented in Section VII.

Although testosterone is the predominant hormonal product of the avian testis, progesterone and estrogen are also produced (see Lofts and Murton, 1973). Labeled testosterone is metabolized rapidly in the liver and excreted in the bile of the Japanese Quail (Fellegiová *et al.*, 1975). Gonadal steroids can be produced by embryonic chicken testes as early as day 10 of incubation (Galli and Wasserman, 1973) although, unlike in mammals, testosterone does not stimulate the development of the male gonad in birds (Adkins, 1975; Haffen *et al.*, 1975).

Witschi (1961) described the endocrine basis for secondary sexual characteristics of various avian species. The androgens are often responsible for sexual dimorphism, especially for the striking nuptial plumages of the males of many species (Lofts and Murton, 1973). Androgen administration stimulates the comb and wattles of newly hatched chicks irrespective of sex. The development of black pigmentation in the bill of photostimulated male house sparrows appears to require FSH, at least initially, as well as testosterone (Lofts *et al.*, 1973). Estrogen has been associated with the bill color change from red to nuptial straw color in the female *Quelea quẹlea* (Lofts and Murton, 1973). Hormonal influences on sexual plumages are treated later (see Section VI).

In addition to the predominant hormone, estradiol-17β, chicken and Japanese Quail ovaries can produce progesterone, androstenedione, and testosterone (Boucek and Savard, 1970; MacGregor, 1975; see also Haffen, 1975). Estrogens are required in the Japanese Quail embryo for the development of the ovary (Adkins, 1975). Estrogen is first produced at day 10 of incubation in the chicken embryo (Galli and Wasserman, 1973). The hypothalamic target receptor sites for ^{3}H-estradiol that are thought to be involved in the sex steroid feedback on the release of gonadotropin hormone releasing hormone also appear on day 10 of incubation (Martinez-Vargas *et al.*, 1975). Other target sites in the chicken brain thought to be involved in sex behavior and aggression develop later. Estrogen is the principal hormone that stimulates oviduct growth and development, but progesterone can also maintain the oviduct after the initiation by estrogen, and both hormones appear to be necessary at least in Japanese Quail to produce a functional unit (Laugier *et al.*, 1975; McKnight *et al.*, 1975; O'Malley and Schrader, 1976).

The interactions of hormones in reproduction is complicated and there are numerous exceptions to the general rule which we have tried to present. Further discussion of the hormones with respect to photoperiodism, photosensitivity, and photorefractoriness, egg laying, and incubation and brooding behavior is offered in a later section (see Section VII).

E. Melanocyte Stimulating Hormone (MSH)

Little is known concerning the control and function of avian MSH. The MSH (K) cells are thought to be restricted to cephalic lobe of the pars distalis (see Tixier-Vidal and Follett, 1973). There is no pars intermedia in adult birds, the structure where MSH is synthesized in other vertebrates. Although MSH is found in birds, the hormone differs somewhat from that found in mammals. Chickens lack the α-MSH (Shapiro *et al.*, 1972) and the β-MSH has evolved so that there is no immunological cross-reactivity between mammalian β-MSH antiserum and chicken pituitary extracts (Lowry and Scott, 1975). MSH and ACTH molecules are similar chemically.

The activity of MSH in birds is unknown. Tougard (1971) suggested that MSH might be involved in controlling feather pigmentation inasmuch as the MSH cells appeared to be more active in colored ducks and pigeons than in white birds. Considering the importance of photoreceptors in the brain, the presence of heavily pigmented membranes in the brain seems to offer a site of action for MSH that invites investigation.

F. Interrenal (Adrenal Cortical) Gland

The epsilon cells of the cephalic lobe have been correlated functionally with the production of adrenocorticotropic hormone (ACTH) (Mikami, 1958; Tixier-Vidal *et al.*, 1968; Mikami *et al.*, 1969). This finding has been corroborated using an immunocytochemical technique (Ferrand *et al.*, 1975). The presence of a hypothalamic adrenocorticotropic hormone releasing hormone (CRH) has been demonstrated for several species (see Assenmacher, 1973; Kobayashi and Wada, 1973). ACTH increases steroidogenesis in the adrenocortical tissue by mobilizing cholesterol in the glands of pigeons, Japanese Quail (Bhattacharyya *et al.*, 1975), and chickens (Freeman and Manning, 1975). The adrenocortical hormones, in turn, exert a negative feedback influence on the production of ACTH. Recently, chicken ACTH was purified and the amino acid sequence of this polypeptide determined (Lowry and Scott, 1975).

There is considerable autonomy of the interrenal gland following hypophysectomy in some species. However, the production of significant amounts of corticosterone in hypophysectomized birds may not signify an independence of ACTH stimulation inasmuch as an extrapituitary source of ACTH or ACTH-like substance has been identified in the median eminence of pigeons (Péczeley and Zboray, 1967; Frankel, 1970). Various stresses cause the production of adrenal corticosteroid hormone probably by way of ACTH release. However, neurogenic stress (restraint) promoted adrenocortical activation in hypophysectomized pigeons implanted with ectopic pituitary autografts whereas systemic stress (ether) did not (Bouillé and Baylé, 1975a,b).

Although of different embryological origin, adrenocortical tissue is intimately associated with medullary tissue in the avian interrenal gland. This relation is common to the vertebrates and may be functionally related with the stress response. The release of corticosterone during neurogenic stress causes an increase in phenylethanolamine-*N*-methyltransferase activity which promotes the conversion of norepinephrine to epinephrine in the adrenal medulla (Ghosh, 1973; Zachariasen and Newcomer, 1975). The possibility of medullary control of cortical activity invites further investigation.

The avian adrenal cortical tissue synthesizes 17-deoxycorticosteroids almost exclusively, but very small amounts of cortisol have been detected in the duck (Daniel, 1970) and chicken (Urist and Deutsch, 1960). Corticosterone is the major corticosteroid occurring in amounts 8 times greater than aldosterone. For peculiarities in the synthetic pathways of avian corticosteroids, see Assenmacher (1973).

A corticosterone binding globulin (transcortin) is present in avian plasma. Albumin also binds corticosterone but with much less affinity. The binding capacity of the plasma proteins exceeds the endogenous corticosteroid level (Steeno and DeMoor, 1966) and is higher in the rooster than in the hen (Seal and Doe, 1966; Gould and Siegel, 1974). It was suggested that transcortin-bound corticosterone is biologically inert and that the bound hormone may represent a reserve for gradual release to the tissues (Sandberg and Slaunwhite, 1962). The affinity of transcortin for corticosterone serves as the basis for a widely employed assay of plasma corticosterone (Murphy, 1967).

Daily rhythms in concentration of plasma corticosterone have been reported in the Japanese Quail (Boissin *et al.*, 1971), the White-throated Sparrow (Dusseau and Meier, 1971; Meier and Fivizzani, 1975), and the common pigeon (Joseph and Meier, 1973; Sato and George, 1973). The rhythm is entrained by the daily photoperiod; the offset of light appears to be the principal environmental cue. After 2 weeks in continuous light, the rhythm disappears in pigeons (Joseph and Meier, 1973). Although set by the daily photoperiod, the phase of the rhythm changes seasonally with respect to the photoperiod in the sparrow (Dusseau and Meier, 1971), even when the birds are kept indoors on a constant regimen of 16L:10D (Meier and Fivizzani, 1975). The significances of the corticosteroid rhythm with respect to the entrainment of other rhythms and its role in the regulation of reproduction and migration are discussed later.

The regulation of the plasma corticosteroid rhythm is generally thought to be by way of driving rhythms of CRH and ACTH. The CRH rhythm in the pigeon hypothalamus precedes the plasma corticosterone rhythm by about 4 hours (Sato and George, 1973). However, there is evidence in fish (Srivastava and Meier, 1972) and mammals (Meier, 1976) that the plasma corticosterone rhythm persists and is entrained by the daily photoperiod in hypophysectomized animals which have implanted ACTH pellets (necessary only in rat) to maintain corticosteroid production. Similar results were obtained in Japanese Quail; but because adrenocortical function was maintained by pituitary autografts instead of ACTH implants, the investigators concluded that the corticosteroid rhythm was regulated by a rhythm of CRH acting on the pituitary transplants in the kidney to produce a driving rhythm of ACTH (Assenmacher and Boissin, 1972). This topic was reviewed by Meier (1975) who concluded that the principal control of the plasma corticosteroid rhythm does not depend on a driving rhythm of ACTH from the adenohypophysis or from any other source.

A principal role of the corticosteroids in mammals involves the regulation of carbohydrate metabolism. A hyperglycemic effect of pigeon adrenal extracts was first demonstrated in birds by Riddle (1937). The increase in blood glucose results from the stimulatory effect of corticosterone on protein catabolism (Brown *et al.*, 1958) and liver gluconeogenesis (Allen *et al.*, 1975a). Allen also reported that salt stress in freshwater ducks caused an increase in plasma concentrations of both corticosterone and glucose. Curiously, however, salt stress in ducks acclimated to salt water caused an increase in corticosterone without an increase in blood glucose levels. Corticosterone is one of the few hormones that influence the otherwise biologically inert sacral glycogen body in chickens resulting in glycogen enrichment (Snedecor *et al.*, 1963).

Contrary to the mammalian system, corticosterone exerts little or no lipolytic effect on *in vitro* duck adipose tissue (Desbals, 1972). However, it does increase plasma lipid concentrations in chickens by way of a lipogenic activity in the liver (see Assenmacher, 1973). The synergistic role of corticosterone and prolactin in causing large gains or losses in body fat stores is discussed later in relation to migration.

In mammals the adrenal corticosteroids are divisible into the glucocorticoids, such as corticosterone, which influence carbohydrate and lipid metabolism, and into the mineralocorticoids, such as aldosterone, which regulate electrolytes. In some avian species, however, both aldosterone and corticosterone have important influences on electrolyte levels. As in mammals, aldosterone stimulates sodium ion (Na^+) retention in the kidney. Adrenalectomy in the duck causes a loss of Na^+ in the urine which can be reduced by aldosterone replacement (Phillips *et al.*, 1961).

Corticosterone has important roles in extrarenal responses to salt loading (Ensor, 1975; Peaker and Linzell, 1975). Salt loading causes an increase in plasma corticosterone concentration (Allen *et al.*, 1975a) that in turn may stimulate an increase in nasal gland secretions of electrolytes (Holmes *et al.*, 1961; Holmes, 1975) in ducks. The nasal salt gland has the capacity to accumulate ^3H-corticosterone (Ensor and Phillips, 1972a) and to convert it to 11-dehydroxycorticosterone which is the principal bound form of the labeled steroid (Takemoto *et al.*, 1975). Peaker and Linzell (1975) concluded that corticosterone acts in a permissive way to maintain salt secretion following neural activation by cholinergic mechanisms. Thyroid hormones (Ensor *et al.*, 1970; Phillips and Ensor, 1972), prolactin (Ensor and Phillips, 1970; Ensor, 1975), and arginine vasotocin (Ensor and Phillips, 1972b) are also thought to have stimulatory roles in regulation of salt gland secretion.

V. Peripheral Glands

A. Chromaffin (Adrenal Medullary) Tissue

As in the other vertebrates, the adrenal medullary (chromaffin) tissue of birds is closely intermingled with cortical tissue and is innervated by presynaptic sympathetic fibers. It produces both norepinephrine and epinephrine; the ratio of the catecholamines varies among the species. Cortisone injection (Ghosh, 1973) and stress-induced corticosterone elevation (Zachariasen and Newcomer, 1975) accelerate the methylation process of norepinephrine to epinephrine in pigeons and chickens.

The catecholamines are potent vasopressor agents, but their effects are limited to the peripheral circulation. Neither hormone has much effect on cardiac contraction (Carlson *et al.*, 1964), both have a hyperglycemic effect (see Assenmacher, 1973). The levels of cyclic AMP were stimulated by catecholamines in the chicken liver (Fröhlich and Marquardt, 1972). Although acute cold caused the release of norepinephrine in pigeons (Lahiri and Banerji, 1969), there are apparently no reports of significant thermogenic activity by either catecholamine.

The effects of epinephrine on lipid metabolism are variable. Epinephrine stimulated lipolysis in the adipose tissues of pigeons (Goodridge and Ball, 1965), geese *(Anser domesticus)*, and Horned Owls *(Bubo virginianus)* but to a lesser degree than that found in mammals (Prigge and Grande, 1971). The adipose tissue of the duck (Grande and Prigge, 1972) and chickens (Carlson *et al.*, 1964; Fröhlich and Marquardt, 1972) did not respond to the catecholamines.

B. Pineal Gland

The pineal organ has attracted considerable interest during the past decade as a result of its possible influence on circadian rhythms and the reproductive system. Among the lower vertebrates, this epithelial derivative from the diencephalon serves as a photoreceptor as well as a glandular tissue for the possible production of hormones. Although light can directly influence pineal metabolism in ducks (Rosner *et al.*, 1971) and chickens (Pang, 1974), there is no electrophysiological evidence of a photoreceptive role for the pineal in birds (Morita, 1966; Ralph and Dawson, 1968; Oksche and Kirschstein, 1969). Histological studies suggest an endocrine function for both mammalian and avian pineal organs (Quay, 1965; Wurtman *et al.*, 1968).

Melatonin, a derivative of serotonin, is thought to be a principal pineal hormone. Two enzymes, *N*-acetyltransferase (NAT) and hydroxyindole-*O*-methyltransferase (HIOMT), are thought to be involved in converting serotonin to melatonin; their activities are assumed to indicate synthesis of melatonin. HIOMT activity is usually assayed and may be a valid indicator of melatonin synthesis in mammals. In at least some avian species, however, there is no relation between HIOMT activity and melatonin synthesis, casting doubt on much of the early findings. NAT activity appears to be a reliable index in chickens (Binkley *et al.*, 1973; Ralph, 1975; Ralph *et al.*, 1975).

The unreliability of studies using HIOMT assays has made the literature difficult to interpret. Melatonin and NAT assays indicate that melatonin production is stimulated during darkness and inhibited during the light (Lynch, 1973; Ralph *et al.*, 1975). In addition, the melatonin rhythm is circadian in chickens in that it persists in constant light (Lynch and Ralph, 1970). These findings are consistent with the mammalian literature. As in mammals, the innervation of the avian pineal organ has been reported to be by way of sympathetic fibers that originate in the superior cervical ganglia (Hedlund, 1970). Unlike the findings in mammals, however, neither sympathectomy nor removal of the eyes disrupt photic entrainment of the NAT activity rhythm in chickens (MacBride and Ralph, 1972; MacBride, 1973; Binkley *et al.*, 1975). Thus, an extraretinal photoreceptor may entrain the pineal melatonin rhythm by way of some nonneuronal mechanism. Photoperiodic entrainment of melatonin secretion in pineal autotransplants in chickens further support this hypothesis (Pang, 1974). Whether a hormone is involved or whether light acts directly on the pineal apparently has not been ascertained.

Demonstrations that the pineal gland may have a role in regulating the reproductive system in rodents (review, Reiter, 1973) stimulated considerable interest for a similar role in birds. Extirpation of the pineal gland in Japanese Quail slightly delays ovarian maturation and egg laying, but the effects are transitory and limited to a short interval before maturity (Saylor and Wolfson, 1967). Melatonin injections in chickens produced slight progonadal effects during the first 20 days after hatching (Shellabarger, 1952) and slight antigonadal effects after 40–60 days (Shellabarger, 1952; Singh *et al.*, 1967). Pinealectomy in ducks was reported to advance the time of seasonal testicular regression (Cardinali *et al.*, 1971).

The bulk of the evidence suggests that the pineal has little, if any, physiological role in regulating avian reproduction. Pinealectomy does not influence the timing of testicular regression in the White-crowned Sparrow *(Zonotrichia leucophrys gambelii)* (Kobayashi, 1969), Harris' Sparrow *(Zonotrichia querula)* (Donham and Wilson, 1970), or the House Finch *(Carpodacus mexicanus)* (Hamner and Barfield, 1970). In addition, pinealectomy does not influence recovery from photorefractoriness in Harris' Sparrow. Extirpation or destruction of the pineal gland also had no effect on testicular recrudescense in the White-crowned Sparrow (Kobayashi, 1969; Oksche *et al.*, 1972), House Finch (Hamner and Barfield, 1970) and House Sparrow *(Passer domesticus)* (Menaker *et al.*, 1970). The often repeated assumption that the pineal gland has a central role in reproductive photoperiodism in birds lacks supportive evidence.

There is tantalizing evidence that the pineal gland may have a significant role in regulating circadian rhythms. Pinealectomy in House Sparrows has no effect on the daily rhythm of locomotor activity when birds are kept on a normal L:D schedule; however, the free-running rhythm that persists in the intact birds maintained in continuous dark (D:D) is lost following pinealectomy (Gaston and Menaker, 1968; Binkley, 1970; Binkley *et al.*, 1972) and restored by pineal autotransplants (Zimmerman and Menaker, 1975). In addition, melatonin administration in intact House Sparrows remained in continuous dark (D:D) either shortened the period of the free-running locomotor activity rhythm or induced continuous activity (Turek *et al.*, 1976). The arrhythmic activity of pinealectomized sparrows kept in DD appears similar to that observed in intact birds kept in LL. Similar findings have been reported in the White-crowned Sparrow (Gaston, 1971) and in the White-throated Sparrow (McMillan, 1972). The free-running rhythm of body temperature is also lost following pinealectomy in the House Sparrow kept in DD (Binkley *et al.*, 1971). Menaker and co-workers suggested that the pineal gland may be a circadian driving oscillator or part of a mechanism that couples driving oscillations and overt driven rhythms (Gaston and Menaker, 1968; Menaker and Oksche, 1974). Interestingly, the thyroid gland appears to have a similar effect on circadian rhythms (Meier *et al.*, 1971a,d; John *et al.*, 1972; Meier and MacGregor, 1972). The thyroidal influence has been described as a permissive effect for the expression of circadian rhythms.

Despite demonstrations that the pineal gland permits the expression of circadian rhythms in animals kept in DD, and that melatonin in chickens becomes concentrated in the brain during the dark (Pang *et al.*, 1974), it is still conjectural that the gland has a significant physiological role with regard to circadian rhythms. Given the important roles that circadian systems have in photoperiodism and other aspects of avian reproduction, it seems incongruous that the loss of an important regulator of circadian function would have little or no observable influence on reproductive indices.

C. Pancreas

The avian pancreas is located in the mesenteries of the duodenum having originated as an evagination from the primordial gut. Three endocrine cells have been identified in the Islets of Langerhans: the insulin producing (β) cells, the glucagon producing (α) cells, and the D (gastrin producing?) cells (Epple, 1968; Falkmer and Patent, 1972). In addition, another hormone, avian pancreatic polypeptide (APP), has been localized in a cell type in the exocrine parenchyma of the chicken pancreas (Larsson *et al.*, 1974).

Insulin isolated from chickens and turkeys has an amino acid sequence similar to that present in mammals (Kimmel *et al.*, 1968). However, less insulin is produced in birds, and the hormone appears to have less influence on carbohydrate metabolism (Ivy *et al.*, 1926; Mirsky *et al.*, 1941; Fröhlich and Marquardt, 1972). Insulin reduces blood glucose concentration in the penguin, *Pygocellis papua* (Farina *et al.*, 1975), and chicken (Fröhlich and Marquardt, 1972), and increases hepatic glycogenesis in ducks and chickens (Mialhe, 1958; Hazelwood *et.al.*, 1968), apparently by facilitating the cellular transport of glucose (Hazelwood, 1973). High doses of glucose, or glucagon, stimulate insulin production in several species (Mialhe, 1969; Samols *et al.*, 1969; Langslow *et al.*, 1970). Arginine stimulates insulin secretion in penguins (Farina *et al.*, 1975). Glucose and arginine have similar effects in mammals.

Avian adipose tissue, unlike that in rats, synthesizes little if any lipid *de novo* and serves primarily as a storage organ for triglycerides produced in the liver (Langslow and Hales, 1971). Thus, insulin had no effect on fatty acid synthesis and glucose uptake by adipose tissue of pigeons and passerine species (Goodridge, 1964; Goodridge and Ball, 1966). Lipogenesis is restricted to the avian liver. The hepatic lipogenic activities of insulin in the chicken appear to be similar to those found in rat liver (Goodridge, 1975). Insulin had little or no effect in counteracting the lipolytic activities of glucagon in the chicken (Goodridge, 1975) and duck (Desbals *et al.*, 1970), and did not inhibit the formation or activity of cyclic AMP in the chicken (Fröhlich and Marquardt, 1972); both of these activities occur in mammals. In fact, insulin increased cyclic AMP in blood, muscle, and liver.

Glucagon has two sources in birds, the pancreas and the gut. The glucagon extracted from the gut (6000 MW) is about twice the size of that present in the pancreas (Samols *et al.*, 1969). Glucagon has a strong hyperglycemic effect in birds, and is partly responsible for the high levels of blood glucose present even in fasted individuals (Farina *et al.*, 1975). Glucagon also has a prominent lipolytic effect in several species (Prigge and Grande, 1971; John and George, 1973; Santos and Grande, 1975; Lefebvre, 1975; Goodridge, 1975), probably through the cyclic AMP system (Boyd *et al.*, 1975; Goodridge, 1975). Growth hormone and corticosterone appear to be necessary to permit the expression of both glucagon and insulin activities (see Assenmacher, 1973). Glucagon stimulates the release of growth hormone in pigeons 5 minutes after injection (John *et al.*, 1974).

Avian pancreatic polypeptide (APP) has been purified in chickens. It has a molecular weight of 4200 and is composed of 36 amino acid residues. It has hepatic glycogenolytic and plasma hypoglycerolemic effects. In cannulation experiments of the stomach (proventriculus), APP in low dosages increased hydrogen ion concentration and release of pepsin within seconds (Hazelwood *et al.*, 1973). The regulation of APP production and release has not been ascertained.

D. Parathyroid and Ultimobranchial Glands

The regulation of calcium metabolism is the principal activity of the parathyroid and ultimobranchial glands. The paired parathyroid glands derive from the dorsal endodermal epithelium of the third and fourth pharyngeal pouches and the paired ultimobranchial glands arise from the fifth and sixth pharyngeal pouches (see Copp, 1972). However, Le Douarain and Le Lievre (1970) demonstrated in the chicken that the secretory cells of the ultimobranchial glands are actually derived from ectodermal neural crest cells which migrate to surround the branchial pouch.

Although avian parathormone (PTH) has not been isolated and no radioimmunoassay has been developed for this polypeptide, a bioassay based on the rapid (7–8 minutes: Candlish and Taylor, 1970) hypercalcemic response to parathormone in immature Japanese Quail (Dacke and Kenny, 1973; Kenny and Dacke, 1974) has proved reliable. The control of PTH secretion from the chief cells is similar to that found in mammals. Low levels of plasma calcium ions stimulate parathormone secretion by way of the cyclic AMP system whereas high levels inhibit the Ca^{2+} sensitive adenylate cyclase activity (Copp, 1972; Care and Bates, 1972). Mammalian parathormone injections cause a marked rapid increase in plasma Ca^{2+} levels in chickens (Assenmacher, 1973) and parathyroidectomy leads to a decrease in plasma calcium ions and death by tetanic contractions (Copp, 1972).

The hypercalcemic response to parathormone results from increased Ca^{2+} mobilization from bone and Ca^{2+} absorption from the intestine. Vitamin D is required for the stimulation of intestinal absorption (Taylor, 1965; Gonnerman *et al.*, 1975), probably by increasing mucosal permeability to calcium (Harrison and Harrison, 1965) and by stimulating the formation of calcium binding protein (CaBP) which aids in the transport of calcium in the mucosal cells (Wasserman and Taylor, 1966). Vitamin D is also required for parathormone-induced calcium mobilization from bone, but the hyperphosphatemic response to parathormone can occur in the presence of vitamin D deficiency (Gonnerman *et al.*, 1975). Apparently, phosphate uptake from the intestine involves a different mechanism than calcium uptake. In addition to specific receptor sites for parathormone in bone and the intestinal mucosa, parathormone receptors have also been found in chicken kidney indicating an inhibitory effect on renal calcium loss (Martin and Mosely, 1975). There is also evidence that parathormone acts on the Japanese Quail oviduct limiting eggshell calcification (Dacke, 1976).

Estrogen and parathormone act synergistically to provide the calcium required for egg formation. Estrogen stimulates the production of Ca^{2+}-binding serum phosphoproteins (Urist *et al.*, 1960) which reduce the level of free Ca^{2+} feedback on the parathyroid and greatly increase the plasma concentrations of bound calcium (Copp, 1972). Estrogen also stimulates the formation of medullary bone which serves as a Ca^{2+} storage site (Taylor, 1965), but it is questionable whether parathormone acts on this tissue (see Simkiss, 1975).

The ultimobranchial hormone, calcitonin, is secreted by the C cells. It has been isolated from the turkey, goose, pigeon, and Japanese Quail (review, Simkiss and Dacke, 1971). Calcitonin is a 32 amino acid polypeptide with a molecular weight ranging from 3000 to 5000. The amino acid sequence in birds differs markedly from mammalian calcitonin, but is similar to salmon calcitonin (Nieto *et al.*, 1973).

Hypercalcemia stimulates calcitonin secretion (Copp, 1972; Care and Bates, 1972), probably by stimulating adenyl cyclase activity in the C cells (see Care and Bates, 1972; Nieto *et al.*, 1975). Despite these clear demonstrations, the role of calcitonin in avian calcium metabolism is unresolved. Calcitonin does not depress serum Ca^{2+} in birds as it does in mammals (see Boelkins and Kenny, 1973; Simkiss, 1975). It may, however, prevent hypercalcemia (see Assenmacher, 1973) and be important in maintaining serum phosphorus homeostasis (Brown *et al.*, 1970). Calcitonin is not important in regulating calcium metabolism of the laying hen (Speers *et al.*, 1970). Further studies are required to elucidate calcitonin activity (see Chapter 3).

VI. Molting

A. INTRODUCTION

The preservation of a functional feather coat for insulation and flight is ensured through the process of molting. A diversity of environmental demands placed on the different avian species has produced a myriad of molting schedules, usually mutually exclusive of the breeding and migration seasons. This variability in molting patterns has impeded progress toward understanding the physiological bases.

Molting involves the loss of feathers and their subsequent replacement. A discussion of the development of a feather, as well as the stages and variabilities in the plumage development patterns, would be cumbersome and have been dealt with in detail elsewhere (Voitkevich, 1966; Van Tyne and Berger, 1971; Palmer, 1972; Stettenheim,

1972). This discussion will deal with those features which may be directly influenced by hormones.

B. Thyroid Hormones

The time of thyroid gland development and the initiation of biological activity vary between pigeons (nidicolous) and chickens (nidifugous), but parallel the onset of feather quill development in both species (Voitkevich, 1966). This generalization is supported by data from a variety of species in each category. Feather development has been correlated with hyperactivity of the thyroid as measured by the tadpole metamorphosis bioassay, not only with the prejuvenal molt but also with replacement by the definitive plumage and subsequent seasonal molts. On the other hand, histophysiological correlations of thyroid activity with the molting season have provided contradictory results; however, changes in thyroid histology representing increased or decreased thyroid production are not necessarily indicative of a change in the secretion rate (Voitkevich, 1966; Payne, 1972), nor of specific activity of the hormones at the feather follicle.

Thyroidectomy has been helpful in clarifying the role of the thyroid in molting. However, complete thyroidectomy is extremely difficult in many species (Voitkevich, 1966), and that might explain some of the discrepancies in the data (see Payne, 1972). Thyroidectomy generally inhibits molting. In the European Starling *(Sturnus vulgaris)*, thyroidectomies performed at different times relative to the natural molting season produced different effects on molting, the earlier the operation, the greater the inhibition (Voitkevich, 1966). Voitkevich concluded that the effects of thyroidectomy on molting involved indirect effects by way of changes in the total body physiology as well as the direct effects on the feather tissues. Although the feather germ can develop in thyroidectomized cormorants *(Phalacrocorax carbo)*, proper feather growth and pigmentation did not occur even when thyroidectomy was performed after feather development began (Voitkevich, 1966).

Thyroxine treatment usually induces molting or accelerates ongoing molts (Wagner, 1961; Voitkevich, 1966). Low doses are stimulatory in gallinaceous birds whereas high doses are required in some water birds. However, thyroxine apparently has no effect in some passerine species and some birds of prey (see Voitkevich, 1966; Payne, 1972). Because of the well-known biphasic responses to thyroxine that vary with dose, negative results must be treated with caution. Following molt there is a temporary refractory period when thyroxine cannot induce molting (Tanabe and Katsuragi, 1962; Voitkevich, 1966; Ashmole, 1968).

C. Reproductive Hormones

As molting is often associated with a hyperactive thyroid, it is also associated with hypoactive gonads in many species. Many birds undergo molt following the reproductive period when the gonads regress. Although some birds molt during the breeding season, reproduction and molt are, in general, mutually exclusive (see Voitkevich, 1966; Payne, 1972). Estrogens and androgens suppress molt in a series of species (Kobayashi, 1952; Wagner and Müller, 1963), and castration produces continuous molt in chickens and ducks or advancement of molt in pigeons (Assenmacher, 1958; Voitkevich, 1966). However, castration has no effect, or alters the timing only slightly, in many other species (see Payne, 1972). Payne proposed that sex hormones usually inhibit molting during the reproductive cycle except in those species which lack the inhibitory mechanism. Other factors such as thyroxine were thought to have stimulatory roles.

The effect of prolactin on the integument of other vertebrates (Bern and Nicoll, 1968) and its involvement in development of the brood-patch in birds (see Section VII) suggest that it might play a role in molting. Although it may have some stimulatory effects on molting as in chickens (Juhn and Harris, 1958), in other instances it has no effect or, more often, inhibitory effects (see Payne, 1972). The inhibitory effect on molting is best documented. It was first demonstrated in the chicken (Podhradsky *et al.*, 1937), later in pigeons (Kobayashi, 1953), and several wild European species (Wagner and Müller, 1963). The inhibitory influence is consonant with the roles that prolactin has in migration (see Section VIII) and during incubation and caring of the young (see Section VII) when molting is inappropriate and seldom occurs. However, prolactin can have both stimulatory and inhibitory influences on many physiological parameters including the production of gonadal hormones depending on the time of day when it is released from the pituitary or injected (see Sections VII and VIII). Our laboratory has accumulated data indicating that the time of prolactin injection may also influence molting in a similar manner under some conditions, but no published account of this possibility is available.

Although gonadal hormones are generally inhibitory to molt, low levels of these hormones are often necessary in determining the sexual dimorphism of the plumage during the breeding season. The androgens are responsible for the brilliant nuptial plumage of the male in many species (see Lofts and Murton, 1973), and even for the colorful nuptial plumage worn by female phalaropes (*Phalaropus tricolor* and *P. lobatus*) (Johns, 1964). On the other hand, estrogen has a key role in gallinaceous and anatid species, the brilliant plumage of the male being permitted by the absence of estrogen; castration produces no changes (Voitkevich, 1966). LH is responsible for the plumage change in the male Weaver, *Euplectes orix,* which formed the basis of a bioassay for LH (Witschi, 1961). Estrogen appears to inhibit the LH effect in the female weaver. In still other instances, castrations of both sexes in many sexually dimorphic passeriformes result in no changes in feather structure or color.

D. Neural Control

Although thyroid and gonadal hormones appear to have stimulatory and inhibitory roles, respectively, in the control of molting, the exceptions to the rule suggest that some other important mechanism is involved in timing molt. This conclusion is reinforced by the observations that the orderly arrangement of molting with the other events of the annual cycle may be altered by manipulating environmental conditions. For example, birds that in nature undergo a prenuptial molt before gonadal recrudescence in the spring may molt simultaneously with gonadal growth and migratory fat deposition when they are placed abruptly on long daylengths (Lesher and Kendeigh, 1941; Miller, 1954; Farner and Mewaldt, 1955; Farner, 1964). In addition, the independence of molting relative to the lipid and reproductive cycles in at least some species maintained on constant daylengths (Merkel, 1963; King, 1968) indicate a partial separation of the regulatory mechanisms. Both thyroxine and gonadal hormones are involved in too many important regulatory systems to be directly responsible for timing molting. They are more likely involved in permissive ways that modify the cycle in a manner appropriate for the behavioral and physiological strategies of each species.

Another possible mechanism might involve direct neural innervation of the dermal papillae of a feather follicle by autonomic ganglia and a network of efferent fibers around the axial artery (Stettenheim, 1972). These nerve fibers grow into the feather and then degenerate with each molt. Retention of the feathers appears to depend in some instances on motor impulses from the brain (Smith, 1964; Ostmann *et al.*, 1964), and denervation of the feather retards feather growth (Voitkevich, 1966). Voitkevich suggested that the pattern of molt may be neurally regulated by affecting the blood supply and the relative amounts of hormone each feather receives. This hypothesis is attractive because it would allow the hormones to act both at the local level of the feather as well as on the neural integrative mechanism. The total amount of the hormone present may not be critical. Thyroxine is especially well suited to influence differentiation and growth directly and to influence the autonomic nervous system (see Meier, 1975). Further discussion of thyroxine in regulating events during the annual cycle may be found in Sections VII and VIII.

VII. Reproduction

A. Introduction

Because biologists have defined the broad roles that regulatory mechanisms must fulfill, the avian system is an especially useful tool for the reproductive physiologist. This information has been particularly helpful in studying seasonality, which is virtually a universal phenomenon among birds. The evolutionary mold for seasonality was shaped by environmental factors that are more conducive for reproductive success at one season than at others. However, because reproductive maturation requires considerable time, avian biologists have concluded that the reproductive system of many species must be sensitive to predictive environmental changes that may be unlike those conditions existing at the time of reproduction. Seasonal change in daylength is the most reliable and apparently the most used predictive information among temperate zone species, and alternations of wet and dry seasons appear to be the most important environmental timer among many tropical species. Although predictive information is responsible for bringing birds into reproductive readiness, other environmental factors may be expected to modify or even block reproduction. Thus, restriction of food causes a reduction in gonadal size and function (King, 1973), and a deficiency of dietary calcium reduces egg production in some species (Scott, 1973), apparently by reducing gonadotropin production (see Chapter 3). Low temperatures may also delay reproduction (Immelmann, 1973). In some opportunistic breeders, for example, species inhabiting the dry areas of Australia, the reproductive system is maintained at a stage of near readiness. The final maturation and reproduction occur in response to rainfall which also produces the environmental factors that are conducive for successful breeding.

Inasmuch as the entire bird must be fully integrated within itself and must also be synchronized with its environment in order to carry out the important process of reproduction, the overall coordinating mechanism should be responsive to predictive environmental cues and be able to direct many behavioral and metabolic activities. It should be able to maintain reproductive events separate from other seasonal events such as migration and molt. Although this mechanism should be rigid in some respects so that predictive information has a reliable effect, it should also be sufficiently flexible to allow for modification by environmental factors existing near the reproductive period. These considerations and others have caused reproductive physiologists to concentrate their studies on the neuroendocrine system.

B. Photoperiodism

The importance of daylength in the regulation of the annual cycle of reproduction was first demonstrated experimentally in the Slate-colored Junco *(Junco hyemalis)* (Rowan, 1926). Migratory indices were also stimulated indicating that the daily photoperiod influences the mechanisms involved in migration as well (see Section VIII, B). Since Rowan's discoveries, the predictive role of daylength in reproductive maturation has been demonstrated extensively (Follett and Farner, 1966b; see also Wolfson, 1965a; Lofts and Murton, 1968). The photoperiodic effect probably involves the release of both LH and FSH (Lofts and Murton, 1968; Murton *et al.*, 1970; Gibson *et al.*, 1975; Follett *et al.*, 1975).

The realization that circadian rhythms are involved in photoperiodism in plants and animals provided considerable stimulus for studies which greatly enhanced our understanding of reproductive mechanisms. Bünning (1936, 1960) hypothesized that light acts in two capacities at different times of day to stimulate photoperiodic effects such as flowering in plants, diapause in insects, and reproductive maturation in vertebrates. The vertebrate mechanism involves a daily rhythm of reproductive photosensitivity entrained by the daily photoperiod. The daily interval of reproductive photosensitivity (photoinducible phase) occurs during the light when the photoperiod is sufficient (often called the coincidence model of photoperiodism).

Hamner (1963, 1964) was first to test Bünning's hypothesis in birds. In one experiment, Hamner demonstrated in the House Finch that cycles that are multiples of 24 hours (e.g., 6L:18D, 6L:42D) are nonstimulatory whereas cycles that are multiples of 24 hours plus 12 hours (e.g., 6L:30D, 6L:54D) are stimulatory for gonadal growth. According to the coincidence model, the photoinducible phase for gonadal stimulation is entrained so that it occurs 36–42 hours after the onset of light (30–36 hours after the offset) and every 24 hours thereafter in the absence of another photophase or until the system becomes desynchronized. Other *resonance* experiments have produced similar results in juncos (Wolfson, 1965b), crowned sparrows (Farner, 1964, 1965; Turek, 1974; Meier, 1976), and Japanese Quail (Follett and Sharp, 1969).

In another test of Bünning's hypothesis (scotophase scan or interrupted night experiments), a nonstimulatory cycle (e.g., 6L:18D) was interrupted by short intervals (15 minutes–2 hours) of light at different times during the dark. When the light interruption coincided with the photoinducible phase, gonadal growth occurred. Bünning's hypothesis was confirmed in this manner in the White-crowned Sparrow (Farner, 1965), House Finch (Hamner, 1966), House Sparrow (Menaker, 1965; Lofts and Lam, 1973), and Green Finch (*Chloris chloris*) (Murton *et al.*, 1969).

The coincidence of light during the photoinducible phase does not appear to stimulate gonadotropin release directly by way of a sensitized pathway. Once gonadal development was initiated in House Sparrows (Vaugien and Vaugien, 1961) and White-throated Sparrows (Wolfson, 1966) by long daily photoperiods, further growth occurred in sparrows transferred to continuous darkness. These results suggest that the coincidence of light during the photoinducible phase entrains, or couples, another rhythm that interacts with the photosensitive rhythm and produces a stimulatory effect on the reproductive system. In continuous darkness the two oscillators free-run and maintain the stimulatory relation. The internal coincidence model derives from a number of sources (e.g., Pittendrigh, 1960).

In a series of studies beginning in 1934 Benoit and co-workers studied the photoreceptors involved in photostimulation of the reproductive system in ducks (Benoit, 1961). Although they concluded that the eye has sensitive receptors for gonadal stimulation, they also demonstrated that photostimulation was possible in enucleated ducks. The principal area of extraoccular light perception was found in the hypothalamus.

Homma and Sakakibara (1971) induced gonadal growth in Japanese Quail by inserting probes tipped with luminescent paint in the hypothalamus as well as in specific regions of the optic and olfactory lobes. Although several reports suggested that the pineal organ may also have photoreceptors for gonadal stimulation, the general concensus is that receptors may be near the pineal in some species but they are not contained within the pineal itself (Menaker, 1968; Menaker and Oksche, 1974).

It is important to recognize that more than one kind of photoreceptor is involved in photostimulation of the reproductive system. There are many reasons to conclude that separate photoreceptors are involved in the entrainment of a daily rhythm of photosensitivity and in the photoinduction of gonadal development by light during the photoinducible phase. Menaker and Eskin (1967) offered the clearest demonstration of these two receptor systems in the House Sparrow. They maintained the sparrows on a 14L:10D schedule with light too dim (0.1 lux) to maintain testicular weights but sufficiently intense to entrain a daily rhythm of locomotor activity (and probably the rhythm of photosensitivity). A late interval (1.5 hours) of bright light during the dim light photoperiod supported the testes whereas a similar interval was ineffective early in the photoperiod.

Inasmuch as LH alone had little or no stimulatory influence on early gonadal growth in several species, the photoinduction of gonadal development probably includes stimulation of FSH production (Lofts and Murton, 1968). Assays of plasma LH in Japanese Quail (Gibson *et al.*, 1975) and in the White-crowned Sparrow (Follett *et al.*, 1975) demonstrated its release by photostimulation. Follett and co-workers found the LH was released a day after a long photoperiod was initiated and that LH release continued for at least several days after the birds were moved to continuous dark (Fig. 3).

Murton and co-workers (1969, 1970) provided evidence in the Green Finch that there may be separate photoinducible phases (and possibly separate photoreceptors) for photoinduction of FSH and LH production. In a scotophase scan experiment, they found that light interruption at 7–9 hours after the onset of a short entraining photoperiod stimulated gonadotropin production with little gonadal growth (presumably LH) whereas light interruption at 16–18 hours after the onset of the daily photoperiod stimulated gonadal growth as well (presumably FSH). Meier and Dusseau (1973) also proposed separate photoinducible phases for FSH and LH on the basis of their studies of the House Sparrow.

The daily rhythm in concentration of plasma corticosterone has been proposed as a part of the entraining system for the circadian rhythm of photosensitivity (Meier and MacGregor, 1972; Meier and Dusseau, 1973). In White-throated Sparrows maintained on a nonstimulatory daily photoperiod (6L:18D), daily injections of corticosterone elicited increases in testicular and ovarian weights when the injections were made 18 hours before the onset of light and not when the injections were made at 6 or 12 hours beforehand. When carried out for periods longer than a week, daily disturbances of handling produce similar results apparently by timing a release of endogenous corticosteroid (Meier, 1976).

Assays of plasma corticosterone concentration in White-throated Sparrows kept on nonstimulatory (10L:14D) and stimulatory (16L:8D) daily photoperiods indicate that the daily rise occurred 12 hours after the onset of light in photosensitive sparrows on either schedule (Meier and Fivizzani, 1975). On the basis of these data and those obtained

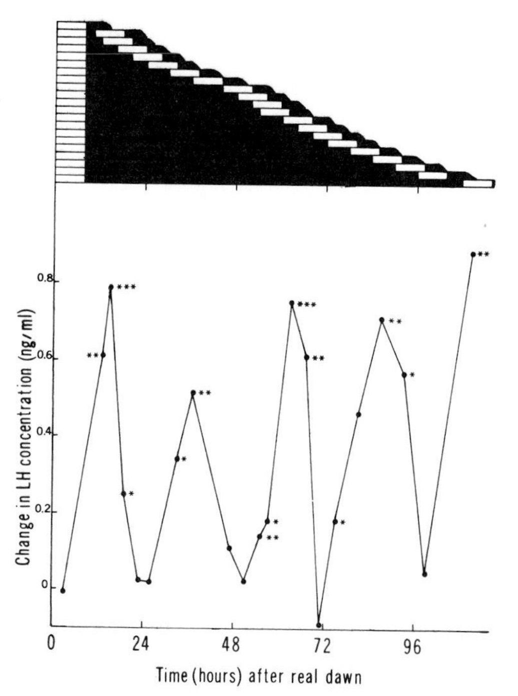

FIG. 3. The effect of an 8 hour photoperiod given at various intervals after entry to darkness on plasma LH concentration in White-crowned Sparrows. The *white* and *black* bars at the top illustrate the various treatments. Birds were previously maintained on 8L 16D and a pre-experimental blood sample was taken for all birds early in the last 8 hour light period. The postexperimental sample was taken 7–16 hours after the end of the test photoperiod. The *ordinate* shows the change in plasma LH concentration between these two samples that resulted from a particular treatment. The time after the last real dawn is shown on the *abscissa*. (From Follett *et al.*, 1974.)

previously from injection experiments, the authors proposed that the daily rise of plasma corticosterone was entrained by the offset of the daily photoperiod and that the daily increase in plasma corticosterone in turn entrains a rhythm of photosensitivity so that the photoinducible phase for FSH occurs 18–24 hours after the rise of corticosterone (30–36 hours after the exogenous entraining cue) (Fig. 4).

In house sparrows with fully developed testes and maintained on stimulatory daily photoperiods (16L:8D), handling disturbances either at the onset or at the offset of light caused a sharp reduction in testicular weights whereas handling at midday or at midnight had no such effect. The inhibitory times were distinguishable in that loss of testicular weights was accompanied by loss of bill color (index of androgenic

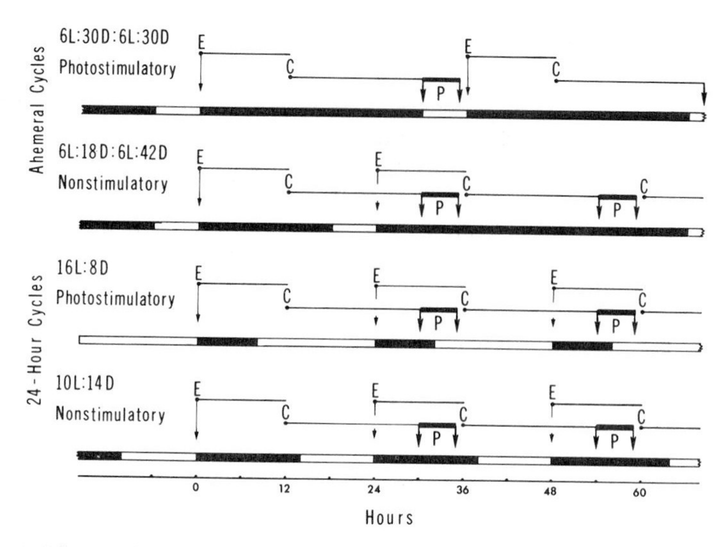

FIG. 4. Schema of the photoperiodic regulation of reproduction and migration in the White-throated Sparrow. The daily rise of plasma corticosterone concentration (C) occurs 12 hours after the entraining (E) cue of the photoperiod (offset of light). The corticosterone rhythm in turn sets the photoinducible phase (P) for hormones involved in reproduction (FSH) and migration (Prolactin). In stimulatory L:D schedules the coincidence of light during the photoinducible phase, 30–36 hours after the offset of light (18–24 hours after the daily rise of corticosterone), results in hormone production. (From Meier and Fivizzani, 1975.)

activity) in birds handled at the offset of light but not in those handled at the onset. Meier and Fivizzani (1975) suggested that handling entrains, probably by way of the adrenal cortex, daily rhythms of photosensitivity of LH and FSH so that the photoinducible phase for LH occurs during the dark as a result of handling at the offset of light and that the photoinducible phase for FSH occurs during the dark when handling is done at the onset of the daily photoperiod.

C. CIRCANNUAL SEQUENCE

Although some species, such as House Sparrows (Kirkpatrick, 1959), Japanese Quail (Wilson *et al.*, 1961), and common pigeons (Lofts and Murton, 1968) continue breeding throughout the year when daylength is sufficient, most other photoperiodic species including juncos (Wolfson, 1952), White-throated Sparrows (Shank, 1959), and White-crowned Sparrows (Farner and Mewaldt, 1955) become photorefractory after breeding during the summer when daylengths are still long. The period of refractoriness, which is characterized by unresponsiveness of the regressed reproductive apparatus to long daily photoperiods, is variable lasting from 4 weeks in some species to 4 months in others. Many become sensitive during the winter but remain sexually inactive until spring when the increasing daylength becomes stimulatory. Gonadal development, however, may occur on short photoperiods in some species (review, Lofts and Murton, 1968). The termination of photorefractoriness can be advanced in many species, including White-throated Sparrows (Shank, 1959), by exposure to short daily photoperiods (review, Wolfson, 1959b).

The importance of endogenous timing mechanisms for controlling an orderly progression of reproductive events during the annual cycle, a circannual rhythm, has long been recognized (review, Gwinner, 1975). Several species undergo appropriate seasonal changes in reproductive (as well as migratory) readiness for longer than a year while they are maintained on constant daily photoperiods. Other species, such as the White-crowned Sparrow (Farner and Lewis, 1973; King and Farner, 1974; Sansum and King, 1976), require a driving stimulus of increasing daylengths to initiate the vernal conditions; thereafter, the rest of the annual cycle proceeds under constant photoperiods. Studies of circannual periods have established that an endogenous mechanism is the primary timer of seasonal reproductive changes in many species. Changes in daylength during the annual cycle modify this mechanism. It may be premature, however, to draw too close an analogy between circadian rhythms and the endogenous seasonal timer, as "circannual rhythm" seems to imply.

A basic feature of the annual cycle and circannual sequence of reproductive changes is the alternation of photosensitive and photorefractory periods. As indicated by resonance experiments, the process of terminating the photorefractory period by short daily photoperiods, similar to the photostimulation of reproductive development by long daily photoperiods, involves circadian rhythms (Hamner, 1968; Murton *et al.*, 1970; Turek, 1972). Thus the evidence supports the concept that seasonal changes in the relations of circadian oscillators may account for seasonal changes in the interpretation of exogenous cues, at least in some species.

On the basis of a series of experiments with the White-throated Sparrow, a model involving temporal synergisms of circadian hormone rhythms has been proposed for the development and termination of photorefractoriness (see Meier and MacGregor, 1972; Meier, 1976). A principal feature of this model (Fig. 5) is a progressive change during the annual cycle in the phase relation between the daily rhythms of plasma corticosterone and prolactin concentrations. Injections of prolactin may have stimulatory or inhibitory effects on the gonads depending on the time of day when the injections are made (Meier,

1969b). The daily variation in responsiveness appears to be a circadian oscillation that is entrained by a daily rhythm of plasma corticosterone concentration. In White-throated Sparrows kept on continuous light, daily injections of prolactin given 12 hours after injections of corticosterone had a stimulatory effect on the gonads, whereas, injections given 8 hours after corticosterone injections had an inhibitory effect (Meier *et al.*, 1971c). This treatment also caused variable effects on migratory indices (see Section VIII,D). Assays of pituitary prolactin content (Meier *et al.*, 1969) and of plasma corticosterone concentration (Dusseau and Meier, 1971) in photosensitive and photorefractory White-throated Sparrows demonstrated a seasonal adjustment in the phases of the daily rhythms. The release of pituitary prolactin occurred 12 hours after the daily rise of plasma corticosterone in photosensitive birds maintained in outdoor holding aviaries in May and about 8 hours after the daily rise of corticosterone in photorefractory birds in August (Fig. 5).

The relation of corticosterone and prolactin also influences the reproductive system in the House Sparrow (Meier *et al.*, 1971c). In photorefractory sparrows maintained in continuous light during September, daily injections of prolactin provided at 4 and 8 hours after daily injections of corticosterone caused increases in testicular weights. Injections at 0, 12, 16, or 20 hours after corticosterone injections were ineffective.

The antigonadal activity of prolactin has been demonstrated in many species since the initial discovery in the Ring-neck Dove *(Streptopelia risoria)* (Riddle and Bates, 1933). Inasmuch as prolactin injections can completely suppress the oviducal response to exogenous gonadotropins in the White-throated Sparrow, at least a part of the antigonadal effect is at the gonad (Meier, 1969a). Because estrogen stimulates an increase in oviducal weight it may be concluded that prolactin exerts an antisteroidogenic effect. (See also, Section II.) Prolactin also inhibits estradiol production in ovarian tissue culture of Japanese Quail (MacGregor, 1975).

Whether the production of gonadotropins can also be inhibited by prolactin has not been ascertained. Because simultaneous injections of FSH can prevent prolactin-induced loss of testicular (Nalbandov, 1945) and ovarian (Bates *et al.*, 1935) weights in chickens, many avian physiologists have concluded that prolactin reduces FSH production. However, this effect may be an indirect one caused by the loss of gonadal hormone production.

Inasmuch as redundancy seems to be a recurrent feature of physiological mechanisms, there are probably other ways in which the corticosteroid–prolactin relation may influence gonadal development.

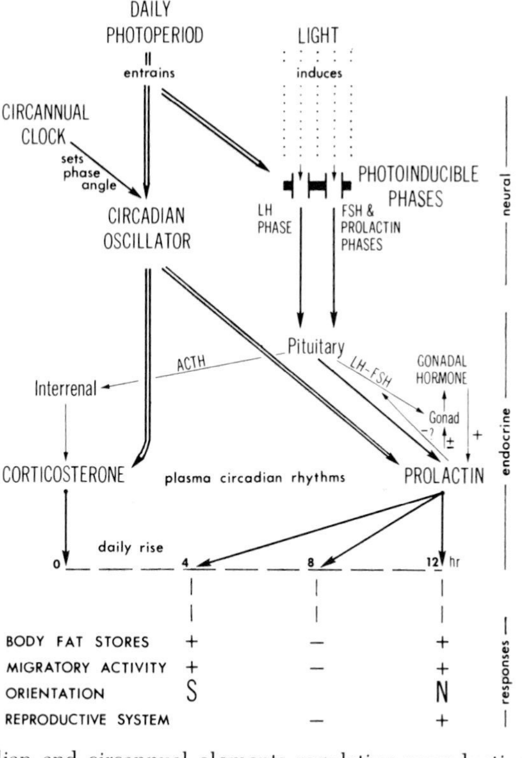

FIG. 5. Circadian and circannual elements regulating reproductive and migratory conditions in the White-throated Sparrow. (From Meier, 1976.)

Wilson and Follett (1974) described hypothalamic binding sites for the negative feedback effect of androgens and speculated that the corticosteroids might activate this mechanism. Another possibility is that the hormones influence the responsiveness of pathways involved in photoperiodic stimulation. Still another consideration is that both photosensitive and photorefractory states are dynamic conditions that may be under a series of controls. For example, early refractoriness in the White-throated Sparrow was characterized by an inhibitory relation of daily rhythms of corticosterone and prolactin (Meier and MacGregor, 1972); however, the plasma corticosterone rhythm disappeared completely at 3 weeks following the onset of refractoriness (Meier and Fivizzani, 1975). Because thyroid hormones permit the expression of response rhythms to prolactin in pigeons (John *et al.*, 1972), a reduction in thyroxine levels might account for the loss of the corticosteroid rhythm.

The role of the thyroid in the maintenance of rhythms is of special interest because thyroid hormones have also been implicated in the reproductive cycles of tropical avian species (Thapliyal and Pandha, 1967; Chandola and Thapliyal, 1973; Chandola *et al.,* 1974). Thyroxine injections inhibited ovarian follicle and oviducal development in a tropical finch, *Ploceus philippinus* (Chandola *et al.,* 1974). It was suggested that thyroxine influences the hypothalamus–pituitary–gonadal axis, and functions as a safety mechanism against unseasonal gonadal growth. It has also been reported that thyroxine decreases testosterone production and increases plasma testosterone clearance in the duck (Jallageas *et al.,* 1974).

D. Courting Behavior and Egg Laying

The male reproductive system matures before that of the female in most species. Complete reproductive readiness is induced by long days in male White-throated Sparrows (Wolfson and Kobayashi, 1962) and White-crowned Sparrows (Farner *et al.,* 1966). Only partial development was induced in females and final maturation depended on courting behavior (see also Polikarpova, 1940; Burger, 1949). The gonadal hormones are largely responsible for stimulating sexual behavior in the male (testosterone) (Guhl, 1962; Hale and Schein, 1962; Phillips and McKinney, 1962; Beach and Inman, 1965; Adkins and Nock, 1976) and in the female (estrogen) (Collias, 1962; Guhl, 1962; Adkins and Nock, 1976). Some of testosterone's activity may result from a conversion of testosterone to estrogen. Adkin and Nock (1976b) found that testosterone-induced sexual behavior in male Japanese Quail could be suppressed by a drug that blocks estrogen activity. Nest building is associated also with secretion of gonadal hormones (Eisner, 1960; Wood-Gush and Gilbert, 1975). Injections of testosterone (Noble and Wurm, 1940) and estrogen (Lehrman, 1958; Orcutt, 1965; Follett *et al.,* 1973; Hinde *et al.,* 1974; Hutchinson, 1975; Wood-Gush and Gilbert, 1975) stimulated nest building in several species.

Most of the research concerning hormone mechanisms in egg laying has been performed in chickens. Except for modifications on the general scheme, however, there is little reason to expect major species differences. Eggs are laid one a day in sequences of 2 or more days. The sequence is repeated in chickens after a day or more in which no eggs are laid. The times of oviposition (egg laying) and of ovulation are normally regulated by the daily photoperiod, but they occur at progressively later times each day of a particular sequence. Oviposition is reset to an earlier time of day at the start of a new cycle. Termination of a sequence appears to be controlled by the offset of the daily photoperiod (Fraps *et al.,* 1947; van Tienhoven *et al.,* 1954; Fraps, 1955).

LH is probably the principal ovulation inducing hormone (OIH). Hypophysectomy prevented ovulation in most hens when performed at 7.2 hours before ovulation but not at 5 hours (Rothchild and Fraps, 1949). A peak of plasma LH concentration occurred 4–6 hours before ovulation (Senior and Cunningham, 1974) and intravenous injections of 1 μg LH produced ovulation 6–7 hours later (Fraps *et al.*, 1942). The release of OIH is controlled by neural elements. Injections of Nembutal, a depressant of the central nervous system, and atropine, an anticholinergic agent, shortly before the critical period for release of pituitary OIH blocked ovulation (Fraps, 1954; van Tienhoven, 1955). The neurogenic stimulus for OIH release, in turn, is apparently a response to progesterone secretion. Progesterone injections can induce ovulation (Fraps, 1955; Wilson and Sharp, 1976a,b), and atropine injections (van Tienhoven, 1955) and destruction of hypothalamic centers (Ralph and Fraps, 1959) can block the progesterone effect.

Several hypotheses have been proposed to account for the occurrence of ovulation at a progressively later time each day during a sequence (Fraps, 1954, 1961; Bastian and Zarrow, 1955; Nalbandov, 1959). In each, daily variations constitute the basis for the explanation. Fraps (1961) proposed a daily variation in the neural threshold to progesterone. A follicular maturation cycle greater than 24 hours caused progesterone to be released later each day so that after several days it was released during a time when the neural centers are relatively insensitive, and LH release did not occur. This hypothesis, though not tested fully, is attractive because it is generally consistent with the available information and because it allows for the incorporation of the basic elements that comprise the other hypotheses as well. For example, Nalbandov (1959) proposed on the basis of earlier studies (Huston and Nalbandov, 1953) that the presence of the egg in the oviduct provided a sensory stimulus which caused the neural centers controlling OIH release to be relatively refractory. Progressive delays in the time of oviposition, then, would eventually remove the period of relative sensitivity of the neural centers to progesterone stimulation, and egg laying would be terminated for at least a day. This additional consideration obviates the need for a progressive daily delay in progesterone production and substitutes instead a progressive delay and shortening in the period of greatest neural sensitivity.

The pattern of egg laying in Japanese Quail is similar to that in chickens (Wilson and Huang, 1962; Opel, 1966). LH is the probable OIH and progesterone is the probable physiological stimulant for the release of LH. Furr *et al.* (1973) have shown that plasma progesterone concentrations increase prior to a daily increase in plasma LH which, in turn, precedes ovulation by 4–7 hours.

The pattern of egg laying appears to be a modification of a basic circadian mechanism. The ovulatory cycle is endogenous, persisting in continuous light with a periodicity somewhat greater than 24 hours in both chickens (Fraps, 1959; van Tienhoven, 1961) and Japanese Quail (Arrington *et al.*, 1962; Opel, 1966). The phases of the cycles (i.e., time of oviposition) can be set by disturbances (van Tienhoven, 1961; Arrington *et al.*, 1962) as well as by feeding (McNally, 1947; Fraps *et al.*, 1947) in animals kept in continuous light. Thus, the entraining mechanism for the ovulatory cycle is similar to that for the circadian photosensitivity rhythm in that it is sensitive to environmental cues other than the daily photoperiod.

In addition to its roles in sexual behavior and nest building, estrogen also has important physiological effects during the same period. It is primarily responsible for growth of the oviduct and attendant tissues (review, van Tienhoven, 1961). Prior to and during egg laying, estrogen produced profound changes in the blood constituents. For example, protein-bound calcium increased (review, Urist, 1959) as a result of increases in plasma phosvitin. Iron and calcium-transporting serum proteins (Greengard *et al.*, 1964; Beuving and Gruber, 1971; Planas and Frieden, 1973; Krall and Hahn, 1975), total proteins (Sturkie, 1951), total lipids (Fleischmann and Fried, 1945), vitellin (Hosoda *et al.*, 1954), biotin (Hertz *et al.*, 1949), cholesterol (Stammler *et al.*, 1955), and vitamin A (Gardiner *et al.*, 1952) increased also. These constituents were then deposited in the developing egg. Hepatic synthesis accounted for most of the increases in the lipids and proteins. The origin of calcium is discussed in Chapter 3.

E. Incubation

Estrogen contributes to the formation of the incubation patch which begins to form several weeks before to several days after egg laying is initiated. The incubation patch is usually complete early during incubation. The development involves a loss of feathers, increased vascularity, and epidermal hyperplasia so that the warmth of the parent can be applied more efficiently to the eggs. Bailey (1952) first demonstrated that estrogen injections caused the formation of the incubation patch in the White-crowned Sparrow and the Oregon Junco (*Junco oreganus*). However, estrogen alone caused only vascularization in hypophysectomized birds; prolactin injections were also necessary to complete the development. Similar results have been reported for canaries (*Serinus canarius*), European Starlings, Red-winged Blackbirds (*Agelaius phoeniceus*), House Sparrows, and chickens (Steel and Hinde, 1963; Hinde, 1967; Hutchinson *et al.*, 1967; Drent,

1973). Incubation patches usually develop in females only, even when males aid in incubation. One exception is the male phalarope which develops patches and incubates the eggs whereas the female does not participate (Johns and Pfeiffer, 1963; Jones, 1969, 1971). In phalaropes, testosterone and prolactin synergize to produce the brood patch. In the Brown-headed Cowbird *(Malothrus ater)*, an obligatory nest parasite, the tissue was unresponsive to the hormones (Selander and Kuich, 1963).

The presence of eggs provides visual and tactile information that initiates incubation behavior. Progesterone may have a contributory role in some species. Systemic injections (Lehrman, 1958) or implants in the preoptic nuclei and lateral forebrain (Komisaruk, 1967) initiated incubation behavior in reproductively experienced ring doves. However, progesterone injections did not induce incubation behavior in chickens (Riddle, 1937) or canaries (Kobayashi, 1952). Prolactin appears to be the principal hormone that regulates incubation behavior. The presence of the eggs stimulates prolactin secretion (Riddle and Dykshorn, 1932; Lehrman, 1955; Friedman and Lehrman, 1968) which maintains incubation behavior in chickens, doves, pigeons, and Bengalese Finches *(Lonchura striata)* (Riddle *et al.*, 1935; Lehrman and Brody, 1961, 1964; Slater, 1967; Eisner, 1969). Pituitary prolactin content was also high during incubation in turkeys (Breitenbach *et al.*, 1965) and White-crowned Sparrows (Meier *et al.*, 1965). The increased production of prolactin during incubation depended on prior induction by gonadal hormones, at least in male pigeons as demonstrated by castration experiments (Kaufman and Dobrowska, 1931; Schooley, 1937; Yasuda, 1953). Removal of the eggs from incubating domestic hens caused a loss of pituitary prolactin content (Saeki and Tanaba, 1955).

F. Broodiness

The role of the gonadal hormones diminishes progressively during incubation. Following hatching, injections of testosterone and estrogen interrupt brooding behavior in hens with chicks (Collias, 1950). The maintenance of low levels in the gonadal hormones during broodiness may well be a consequence of the antigonadal (antisteroidogenic) activities of prolactin (review, Lehrman, 1961). Prolactin injections elicited brooding behavior in domestic hens and cocks (Nalbandov and Card, 1945), wild turkeys *(Meleagris)* (Crispens, 1957), and male and female ring doves (Lehrman, 1955). FSH and methyltestosterone injections prevented the broodiness response to prolactin injections (Nalbandov, 1945).

In pigeons, prolactin both supported brooding behavior and stimulated hyperplasia and secretion of the mucosal layer of the cropsac (Riddle *et al.*, 1932). The cropsac of milk containing partially digested food is regurgitated by both parents as food for the young (review, Levi, 1957). The magnitude of the cropsac response to prolactin varied markedly depending on the time of day when the injections were made (Burns and Meier, 1971; Meier *et al.*, 1971a,b). Prolactin was most effective when injected at 9 hours after the onset of a 12-hour daily photoperiod but had little or no effect when administered at the onset of light. The temporal effect was demonstrable whether prolactin was administered by intramuscular injection (systemic assay) or by intradermal injection over the cropsac (local assay). The magnitude of the temporal differences in response may explain why the cropsac assay seems to have been reliable in the hands of only a few individuals during the 4 decades when it was the principal bioassay for prolactin.

Photoperiodic entrainment of the circadian rhythm of cropsac sensitivity to prolactin appears to be mediated by the adrenal corticosteroids. In pigeons kept in continuous light, corticosterone injections given systemically or locally entrained a cropsac rhythm so that the greatest sensitivity to prolactin occurred 18 hours later (Meier *et al.*, 1971a,b). When these results were related to those in which the greatest sensitivity was found at 9 hours after the onset of a 12-hour daily photoperiod, one would project that the daily rise of endogenous corticosterone should occur in the plasma about 3 hours after the onset of darkness in birds kept on a 12L:12D schedule. This prediction has been affirmed (Joseph and Meier, 1973).

The rhythm of cropsac sensitivity is circadian in the strict sense in that it persists for at least several days in pigeons kept in continuous light and constant temperature (Meier *et al.*, 1971b; John *et al.*, 1972). After 2 weeks, however, the sensitivity rhythm dampens unless the pigeons are treated with thyroid hormones. This suggests that constant light had an inhibitory effect on the thyroid and that adequate levels of thyroid hormones are necessary for the expression of free-running circadian rhythms.

The presence of a discrete daily interval of cropsac responsiveness to prolactin indicates further that the activities of prolactin during the brooding period may depend more on the timing of prolactin release rather than on the amount produced each day. Although prolactin seems to have an important role in maintaining parental behavior in birds, the pituitary content and apparent prolactin production decreases sharply following incubation (review, Lehrman, 1961).

Further, Schooley and Riddle (1938) reported that as much prolactin is produced and released in immature pigeons as in brooding adults. Because prolactin can also stimulate fattening in the pigeon at a specific time of day relative to corticosterone without significantly affecting the cropsac, it was suggested that one temporal relation of corticosterone and prolactin may be present in young pigeons and account for the relatively large amount of body fat whereas another hormonal relation may account for cropsac stimulation in the adult (Meier *et al.*, 1971a). It seems reasonable that many of the multitude of activities ascribed to prolactin must depend on temporal changes in tissue responsiveness and/or prolactin release.

VIII. Migration

A. Introduction

Migration is a major adaptive mechanism that many species employ to escape or exploit environmental changes. More than 60% of all species which breed in northern North America escape the severe winter climate by migrating south. Although the end results of migration (for example, greater survival rates during the winter; greater reproductive potential in the summer) serve to select migratory mechanisms, the environmental determinants for timing migration, as in timing reproduction, may be related only indirectly to the environment which exerts the selective force. Migration requires preparation of the internal organization so that the multiple facets of metabolism and behavior are appropriately integrated. The migratory conditions, in turn, must be organized sequentially with reproduction, molt, and other events during an annual cycle; and all the annual events must be synchronized with the environmental cycle.

There are two basic points of view with regard to migratory mechanisms. One is that migratory mechanisms differ fundamentally and that various systems must have evolved independently many times in avian species. This view is sustained by the rich diversity of migratory behavior and nourished by overly ambitious schemes that focus too narrowly on regulatory mechanisms and that fail to take into account the total biological requirements. The second is that the basic mechanisms are present in all birds, perhaps even in most vertebrates. Species differences, then, are modifications of a basic system; and sedentary behavior is an alternate life-style resulting from a genetic or environmental block of migratory behavior. Such a simple on/off mechanism is illustrated by the occurrence of both migrants and resident individuals among a single nesting of Song Sparrows (Nice, 1937). Although these two views of migration begin at opposite ends of a spectrum, they have common elements. Their significance lies in perspective and philosophical approach.

B. Timing of Migration

The timing of migration and its integration with other events during the annual cycle has long been a subject of active research. Beginning with studies by Rowan (1926) of the Slate-colored Junco increased daylength was identified as a principal exogenous cue for timing spring migration in many Temperate Zone migrants (reviews, Farner, 1955; Wolfson, 1959a; Berthold, 1975). Long daily photoperiods during the winter induce migratory restlessness and fattening in caged birds that were similar to those conditions which occur in birds under natural daylengths during the spring migratory period. Fall migration occurs as an indirect consequence of the vernal increase in daylength and is not a direct reaction to environmental changes. Thus, the timing of migration, like that of reproduction, depends on an endogenous mechanism as well as on exogenous cues.

Compared with studies of reproduction, relatively little has been done to elucidate the temporal architecture of the photoperiodic induction of migratory indices. According to an internal coincidence model, the available results (cited in Meier, 1976) are interpretable in the following manner: The photophase of a 6L:66D schedule entrains a rhythm of photosensitivity so that a photoinducible phase for nocturnal restlessness and fattening occurs 36–42 hours after the onset of the 6-hour photophase. The presence of light during the photoinducible phase sets the phase of, or couples, another circadian system so that the temporal interaction stimulates migratory conditions. Prolactin is thought to be an important mediator of the photoperiodic stimulation of migratory events (Meier, 1976).

These results indicate that more than one system is involved in photoperiodic stimulation of spring migratory conditions. This system may involve at least two separate photoreceptors and circadian mechanisms. In this respect, photostimulation of migratory indices resembles photostimulation of reproductive development. However, it has not as yet been determined whether the entraining systems for the rhythms of locomotor activity, photosensitivity of the reproductive system, and photosensitivity of the migratory complex are identical. In addition, it is not known whether the mechanisms involved in the production of reproductive and migratory hormones are parts of the same circadian mechanism.

Although previous studies indicated that daylength is interpreted in different ways during the year, the nature of the endogenous timing mechanism was more clearly defined by experiments in which birds were maintained on constant daily photoperiods for longer than a year. Changes in body weight and migratory restlessness prescribe a circannual period in some species kept on either short or long daily photoperiods (Merkel, 1963; King, 1968; Berthold, 1974a). The circannual period ranged from a low of 7 months to a high of 15 months. As many as three successive circannual cycles of body weight changes and migratory restlessness have been demonstrated. The circannual cycles of body weight and migratory restlessness may eventually become uncoupled, indicating that the control mechanisms for these two events are not completely parallel and that one is not an absolute requirement for the other (reviews, Berthold, 1974b, 1975; Gwinner, 1975). This conclusion, however, should be considered in terms of the impressive evidence that all aspects of the seasonal changes are closely integrated and synchronized under natural conditions.

C. Metabolic Support

In addition to the behavioral adjustments, metabolic changes occur during the migratory periods. The regulatory mechanisms must control both metabolism and behavior. The most conspicuous metabolic change is the deposition of large fat stores which supply much of the energy required for flights. A given quantity of fat provides about 250% of the amount of energy in the form of ATP derivable from carbohydrates or proteins. It also occupies less space and requires less water because it is not hydrated. The weight of hydration for carbohydrates and proteins makes them far less suitable for efficient energy storage (see Chapter 9).

Lipogenesis is largely restricted to the liver in all species examined (Goodridge, 1975). As measured by the incorporation in lipids of ^{3}H-acetate, hepatic lipogenesis is largely restricted to the latter half of daylight in the White-throated Sparrow during the spring migratory period (A. H. Meier, unpublished). After formation, the triglycerides are transported as complexes with serum globulins to the adipose tissues where the fat is stored.

Mobilization of fat occurs during flight and for a time thereafter. The triglycerides are converted to fatty acids and glycerol; the fatty acids are transported in the blood linked with albumin. The red muscles greatly increase fat content during the migratory period, and they are capable of utilizing lipids for sustained activity by virtue of an efficient machinery for aerobic oxidation. They are distinguishable from white muscle by having greater amounts of myoglobin and larger numbers of mitochondria (George and Berger, 1966; John and George, 1972).

Many hormones have fat mobilizing activities and may be expected to have roles in migration. In mammals ACTH, epinephrine, norepinephrine, growth hormone, and glucagon are especially important, and corticosteroids and thyroid hormones have supportive roles. This scheme is generally applicable for birds; however, the catecholomines appear to be relatively ineffective in some species (see Section V,C). Low blood sugar and various physiological and psychological stresses cause the release of fat mobilizing hormones with the possible exception of thyroid hormone. The special contribution of growth hormone in supporting avian migration was discussed by John *et al.* (1973).

Although the fat mobilizing hormones probably have significant supportive roles in migration, their activities appear to be largely reactive rather than preparatory. For preparatory roles, it may be more instructive to consider hormones that stimulate lipogenesis. This role virtually demands that the production or activity of the hormone be regulated by the nervous system because fattening is responsive to the daily photoperiod. This consideration has apparently discouraged hypotheses that would attribute an important preparatory role for insulin, a potent lipogenic hormone. Although a reduction in fat mobilizing activities could also produce increases in the amount of body fat, hypotheses that ascribe migratory fattening as a consequence of low levels of mobilizing hormones seem to miss entirely the migratory role of fat. The regulatory mechanism for fat metabolism during migration must be capable of stimulating lipogenesis and storage on the one hand and allow for maximum mobilization on the other.

The thyroid hormones have permissive effects for both lipogenesis and fat mobilization in mammals, and considerable attention has been given the thyroid with regard to avian migration. However, the results are inconsistent. Histological evidence for thyroid activity has correlated well with migration in some species (Putzig, 1937, 1938; George and Naik, 1964) but not in others (Putzig, 1937, 1938; Wilson and Farner, 1960). Similarly, thyroxine injections produced variable effects on locomotor activity and fat stores (Wagner, 1930; Merkel, 1938, 1960; Putzig, 1938; Schildmacher and Rautenberg, 1952).

A gonadal role in migration was first advocated by Rowan (1926). Since then both positive and negative results have accumulated. Gonadotropins and gonadal hormones have been reported to stimulate or depress fat stores, and castrates do or do not fatten and exhibit migratory activity in response to increasing daylength in spring (reviews, Farner, 1955; Helms, 1963; Berthold, 1971). Some of the discrepancy may be a matter of timing. Weise (1967) demonstrated that fattening and migratory restlessness could be prevented in the White-throated Sparrow in spring if castration was carried out several months previously but not if it were carried out shortly before the migratory period. Apparently gonadal hormones do not directly induce migration, but they seem to prepare the animal in some way to become responsive to increasing daylength. An interesting finding is that testosterone can promote splitting of the circadian rhythm of locomotor activity in starlings (Gwinner, 1974).

D. Temporal Hormonal Synergism

During the past 10 years, our laboratory has accumulated information on the White-throated Sparrow, and constructed an experimental model for the regulation of migration. The basic premise of the model is that the principal timing mechanism is an endogenous circannual sequence which is constructed by a changing phase relation of two circadian oscillators. The hormonal components for the two oscillators are thought to be corticosterone and prolactin (Meier *et al.*, 1971d; Meier, 1976) (see Fig. 5).

The experimental evidence for this internal coincidence model was derived from injections of corticosterone and prolactin, and from assays of the endogenous hormones. Daily injections of prolactin at midday (16L:8D) in lean photorefractory birds in summer caused rapid gains in fat stores and body weights that within 1 week of injections reached levels found naturally during migration (Meier and Davis, 1967). However, similar prolactin injections carried out at the onset of light did not induce fattening and may even have further decreased fat stores in the lean birds. In another experiment it was demonstrated that daily prolactin injections at midday and late in the photoperiod (16L:8D) stimulated nocturnal locomotor restlessness in this nocturnal migrant as well as large increases in body fat stores (Meier, 1969b). Injections made early in the day were ineffective. Stimulatory effects of prolactin injections on fattening and nocturnal locomotor activity were also observed in the White-crowned Sparrow, but temporal variations in responses to prolactin were not tested (Meier and Farner, 1964; Meier *et al.*, 1965).

Daily variations in responses to prolactin are apparently entrained by a daily rhythm in concentration of plasma corticosterone. In White-throated Sparrows maintained in continuous light in order to remove photoperiodic entrainment, daily injections of corticosterone set the phase of a circadian rhythm of fattening response to prolactin. Daily injections of prolactin at either 4 or 12 hours after the injections of corticosterone elicited large gains in the body fat stores of both photosensitive birds in the winter and of photorefractory birds in the summer. Prolactin injections at 8 hours after corticosterone, however, caused losses in fat stores. The levels of body fat in the 12- and 4-hour treated birds were equivalent to those found in spring and fall migrants, and the levels of fat in the 8-hour treated group were similar to those found in lean photorefractory birds in the summer. The 12-hour pattern of hormone treatment also stimulated gonadal development whereas the 8-hour pattern was strongly inhibitory. These results suggested that the seasonal conditions might be regulated by a temporal synergism of corticosterone and prolactin rhythms.

Further proof of this hypothesis was produced by testing the activity and orientation of White-throated Sparrows under the open night sky following treatment with corticosterone and prolactin (Meier and Martin, 1971; Martin and Meier, 1973). The 4- and 12-hour patterns of injections stimulated nocturnal locomotor restlessness whereas the 8-hour pattern was ineffective. In addition, the 12-hour pattern caused the birds to orient toward the north (toward breeding grounds) and the 4-hour pattern caused them to orient toward the south (toward the wintering quarters).

Assays of pituitary prolactin content (Meier *et al.*, 1969) and of plasma corticosterone concentration (Dusseau and Meier, 1971) in White-throated Sparrows throughout a day during spring migration and during the summer early postnuptial period have demonstrated that the release of pituitary prolactin occurs about 12 hours after the daily rise of plasma corticosterone in the spring and about 8 hours after the rise of corticosterone in the summer. The phases of circadian rhythms of both plasma corticosterone (Fivizzani and Meier, 1976) and plasma prolactin (A. J. Fivizzani and A. H. Meier, unpublished) concentration vary synchronously with seasonal changes in reproductive and migratory indices in White-throated Sparrows maintained on a constant 16-hour daily photoperiod for more than 6 months. During this period, the birds progressed through both the spring and fall migratory conditions. These results demonstrate that the circannual changes in metabolic and behavioral conditions are accompanied by, and may well be caused by, circannual changes in the phase relations of the circadian rhythms of corticosterone and prolactin.

The temporal synergism model for hormonal control of the annual cycle does not suggest a reduced role for hormones other than corticosterone and prolactin. For example, thyroxine is thought to be necessary to maintain the circadian rhythms of corticosterone concentration and the fattening response to prolactin (Meier *et al.*, 1971a,d; Meier,

1975) and gonadal hormones may be necessary to induce either the production or specific activities of prolactin (Meier and MacGregor, 1972). The multiple activities of prolactin in the various temporal relations with corticosterone may also involve alterations in production and specific activities of other hormones (see Section VII,C). The temporal synergism model is a bare framework for further investigation in the White-throated Sparrow, and a model that needs testing in other species.

IX. Concluding Remarks

Although it is profitable to study the individual endocrine organs and their secretions, it is clear that hormonal regulation of important events and conditions rarely depends on the activities of a single hormone. Rather, complexes of hormones act synergistically under the overall control of the nervous system. The complex organization that this system requires is met by circadian mechanisms.

Compared with other vertebrate taxa, our understanding of the functions of daily hormone rhythms of birds in regulating the behavioral and physiological changes during an annual cycle is advanced, primarily due to the considerable work done in establishing the circadian basis of photoperiodism. But much more needs to be accomplished especially with regard to the neural circadian and "circannual" oscillators and the manner in which they control the production and rhythms of hormones. The nature of temporal hormonal synergisms at the cellular level are equally important.

The possibility that hormone activity may be modified by local agents such as the prostaglandins has not received adequate attention. In addition, the ability to quantify specific cellular binding sites for hormones has great promise for learning the bases of hormone synergisms. One possibility to investigate is that one hormone may induce specific binding sites for another hormone. Such a mechanism might account for daily variations in tissue responsiveness to hormones, and indirectly for seasonal and developmental changes.

While concentrating efforts toward understanding the neural regulation of hormone secretion, on the one hand, and the cellular mechanisms for hormone activity, on the other, it may be folly to ignore the vascular connection between the two. Because many of the hormones are either large molecules, or are bound to large molecules in the blood, changes in capillary flow and permeability by hormone action or direct autonomic innervation could greatly alter both the secretion and activity of hormones.

ACKNOWLEDGMENTS

We thank Mrs. Avia L. Dimattia for typing the manuscript.

REFERENCES

Abel, J. H., Jr., Takemoto, D., Hoffman, D., McNeill, T., Kozlowski, G. P., Masken, J. F., and Sheridan, P. (1975). *Cell Tissue Res.* **161**, 285–291.

Adkins, E. K. (1975). *J. Comp. Physiol. Psychol.* **89**, 61–71.

Adkins, E. K., and Nock, B. (1976a). *J. Endocrinol.* **68**, 49–55.

Adkins, E. K., and Nock, B. (1976b). *Horm. Behav.* **7**, 417–429.

Alexander, B., and Wolfson, A. (1970). *Poult. Sci.* **49**, 632–640.

Allen, J. C., Abel, J. H., Jr., and Takemoto, D. J. (1975a). *Gen. Comp. Endocrinol.* **26**, 209–216.

Allen, J. C., Abel, J. H., Jr., and Takemoto, D. J. (1975b). *Gen. Comp. Endocrinol.* **26**, 217–225.

Arai, Y. (1963). *J. Fac. Sci., Univ. Tokyo, Sect. 4* **10**, 249–268.

Arrington, L. D., Abplanalp, H., and Wilson, W. O. (1962). *Br. Poult. Sci.* **3**, 105.

Ashmole, N. P. (1968). *Condor* **70**, 35–55.

Assenmacher, I. (1958). *Alauda* **26**, 242–289.

Assenmacher, I. (1973). *In* "Avian Biology" (D. S. Farner and J. R. King, eds.), Vol. III, pp. 183–286. Academic Press, New York.

Assenmacher, I., and Boissin, J. (1972). *Gen. Comp. Endocrinol., Suppl.* **3**, 489–498.

Assenmacher, I., and Tixier-Vidal, A. (1964). *Arch. Anat. Microsc. Morphol. Exp.* **53**, 83–108.

Astier, H. (1973). D.Sc. Thesis, University of Montpellier, France.

Bailey, R. E. (1952). *Condor* **54**, 121–136.

Bastian, J. W., and Zarrow, M. X. (1955). *Poult. Sci.* **34**, 776.

Bates, R. W., and Condliffe, P. G. (1966). *In* "The Pituitary Gland" (G. W. Harris and B. T. Donovan, eds.), Vol. I, pp. 374–410. Butterworth, London.

Bates, R. W., Lahr, E. L., and Riddle, O. (1935). *Am. J. Physiol.* **111**, 361–368.

Baylé, J. D., Astier, H., and Assenmacher, I. (1966). *J. Physiol. (Paris)* **58**, 459.

Beach, F. A., and Inman, N. G. (1965). *Proc. Natl. Acad. Sci. U.S.A.* **54**, 1426–1431.

Benoit, J. (1961). *Yale J. Biol. Med.* **34**, 97.

Bentley, P. J. (1976). "Comparative Vertebrate Endocrinology." Cambridge Univ. Press, London and New York.

Bern, H. A., and Nicoll, C. S. (1968). *Recent Prog. Horm. Res.* **24**, 681–720.

Berthold, P. (1971). *In* "Grundries der Vogelzugskunde" (E. Schuz, ed.), pp. 257–299. Parey, Berlin.

Berthold, P. (1974a). *J. Ornithol.* **115**, 251–272.

Berthold, P. (1974b). *In* "Circannual Clocks, Annual Biological Clocks" (E. T. Pengelley, ed.), pp. 55–94. Academic Press, New York.

Berthold, P. (1975). *In* "Avian Biology" (D. S. Farner and J. R. King, eds.), Vol. V, pp. 77–128. Academic Press, New York.

Beuving, G., and Gruber, M. (1971). *Biochim. Biophys. Acta* **232**, 529–536.

Bhattacharyya, T. K., Calas, A., and Assenmacher, I. (1975). *Gen. Comp. Endocrinol.* **26**, 115–125.

Bicknell, R. J., and Follett, B. K. (1975). *Gen. Comp. Endocrinol.* **26**, 141–152.

Binkley, S. (1970). Ph.D. Dissertation, University of Texas, Austin.

Binkley, S., Kluth, E., and Menaker, M. (1971). *Science* **174**, 311–314.

Binkley, S., Kluth, E., and Menaker, M. (1972). *J. Comp. Physiol.* **77**, 163–169.

Binkley, S., MacBride, S., Klein, D., and Ralph, C. L. (1973). *Science* **181**, 273.

Binkley, S., MacBride, S. E., Klein, D. C., and Ralph, C. L. (1975). *Endocrinology* **96**, 848–853.
Boelkins, J. N., and Kenny, A. D. (1973). *Endocrinology* **92**, 1754–1760.
Boissin, J., and Assenmacher, I. (1971). *C. R. Hebd. Seances Acad. Sci.* **273**, 1744–1747.
Boissin, J., Baylé, J. D., and Assenmacher, I. (1971). *C. R. Seances Soc. Biol. Ses Fil.* **165**, 1382.
Bolton, N., Chadwick, A., and Scanes, C. G. (1973). *J. Physiol. (London)* **238**, 78–79.
Boucek, R. J., and Savard, K. (1970). *Gen. Comp. Endocrinol.* **15**, 6–11.
Bouillé, C., and Baylé, J. D. (1975a). *Neuroendocrinology* **18**, 35–41.
Bouillé, C., and Baylé, J. D. (1975b). *Neuroendocrinology* **18**, 281–289.
Boyd, J. A., Wieser, P. B., and Fain, N. (1975). *Gen. Comp. Endocrinol.* **26**, 243–247.
Breitenbach, R. P., Nagra, C. L., and Meyer, R. K. (1965). *Anim. Behav.* **13**, 143–148.
Brown, D. M., Perey, D. Y. E., and Jowsey, J. (1970). *J. Endocrinol.* **87**, 1282–1291.
Brown, K. I., Brown, D. J., and Meyer, R. K. (1958). *Am. J. Physiol.* **192**, 43.
Bünning, E. (1936). *Ber. Dtsch. Bot. Ges.* **54**, 590.
Bünning, E. (1960). *Cold Spring Harbor Symp. Quant. Biol.* **25**, 249–256.
Burger, J. W. (1949). *Wilson Bull.* **61**, 211–230.
Burns, J. T., and Meier, A. H. (1971). *Experientia* **27**, 572–574.
Campbell, G. T., and Wolfson, A. (1974). *Gen. Comp. Endocrinol.* **23**, 302–310.
Candlish, J. K., and Taylor, T. G. (1970). *J. Endocrinol.* **48**, 143–144.
Cardinali, D. P., Cuello, A. E., Tramezzani, J. M., and Rosner, J. M. (1971). *Endocrinology* **89**, 301.
Care, A. D., and Bates, R. F. L. (1972). *Gen. Comp. Endocrinol., Suppl.* **3**, 448–458.
Carlson, L. A., Liljedahl, S. O., Verdy, M., and Wirsen, C. (1964). *Metab., Clin. Exp.* **13**, 227–231.
Chadwick, A. C., and Hall, T. R. (1975). *J. Endocrinol.* **64**, 268–278.
Chandola, A., and Thapliyal, J. P. (1973). *Gen. Comp. Endocrinol.* **21**, 305–313.
Chandola, A., Thapliyal, J. P., and Paynaskar, J. (1974). *Gen. Comp. Endocrinol.* **24**, 437–441.
Chen, C., Bixler, E. J., Weber, A., and Meites, J. (1968). *Gen. Comp. Endocrinol.* **11**, 489–494.
Collias, N. E. (1950). *In* "A Symposium on Steroid Hormones" (E. S. Gordon, ed.), pp. 277–329. Univ. of Wisconsin Press, Madison.
Collias, N. E. (1962). *In* "The Sexual Behavior of Domestic Animals" (E. S. E. Hafez, ed.), pp. 565–585. Williams & Wilkins, Baltimore, Maryland.
Copp, D. H. (1972). *Gen. Comp. Endocrinol., Suppl.* **3**, 441–447.
Crispens, C. G., Jr. (1957). *J. Wildl. Manage.* **21**, 462.
Dacke, C. G. (1976). *J. Endocrinol.* **71**, 239–243.
Dacke, C. G., and Kenny, A. D. (1973). *Endocrinology* **92**, 463–470.
Daniel, J. Y. (1970). *Ann. Endocrinol.* **31**, 209–216.
Davies, D. T., and Follett, B. K. (1974a). *J. Endocrinol.* **60**, 277–283.
Davies, D. T., and Follett, B. K. (1974b). *J. Endocrinol.* **63**, 318–328.
Davies, D. T., and Follett, B. K. (1975a). *J. Endocrinol.* **67**, 431–438.
Davies, D. T., and Follett, B. K. (1975b). *Proc. R. Soc. London, B Ser.* **191**, 285–301.
Davies, D. T., and Follett, B. K. (1975c). *Proc. R. Soc. London, B Ser.* **191**, 303–315.
Desbals, P. (1972). *D.Sc. Thesis, University of Toulouse, France.*
Desbals, P., Desbals, B., and Miakle, P. (1970). *Diabetologia* **6**, 65.
Dominic, C. J., and Singh, R. M. (1969). *Gen. Comp. Endocrinol.* **13**, 22–26.
Donham, R. S., and Wilson, F. E. (1970). *Condor* **72**, 101–102.
Drent, R. (1973). *In* "Breeding Biology of Birds" (D. S. Farner, ed.), pp. 262–311. Natl. Acad. Sci., Washington, D.C.

Dusseau, J. W., and Meier, A. H. (1971). *Gen. Comp. Endocrinol.* **16**, 399–408.
Eisner, E. (1960). *Anim. Behav.* **8**, 155–179.
Eisner, E. (1969). *Behaviour* **33**, 262–276.
Ensor, D. M. (1975). *Symp. Zool. Soc. London* **35**, 129–148.
Ensor, D. M., and Phillips, J. G. (1970). *J. Endocrinol.* **48**, 167–172.
Ensor, D. M., and Phillips, J. G. (1972a). *J. Zool.* **168**, 119–126.
Ensor, D. M., and Phillips, J. G. (1972b). *J. Zool.* **168**, 127–137.
Ensor, D. M., Thomas, D. H., and Phillips, J. G. (1970). *J. Endocrinol.* **46**, x.
Epple, A. (1968). *Endocrinol. Jpn.* **15**, 107–122.
Falkmer, S., and Patent, G. J. (1972). *Handb. Physiol., Sect. 7: Endocrinol.* **1**, 1–23.
Farina, J., Pinto, J., Basabe, J. C., and Chieri, R. A. (1975). *Gen. Comp. Endocrinol.* **27**, 209–213.
Farner, D. S. (1955). *In* "Recent Studies of Avian Biology" (A. Wolfson, ed.), pp. 198–237. Univ. of Illinois Press, Urbana.
Farner, D. S. (1964). *Am. Sci.* **52**, 137.
Farner, D. S. (1965). *In* "Circadian Clocks" (J. Aschoff, ed.), pp. 357–369. North-Holland Publ., Amsterdam.
Farner, D. S., and Lewis, R. A. (1973). *J. Reprod. Fertil., Suppl.* **19**, 19–34.
Farner, D. S., and Mewaldt, L. R. (1955). *Condor* **57**, 112–116.
Farner, D. S., and Oksche, A. (1962). *Gen. Comp. Endocrinol.* **2**, 113–147.
Farner, D. S., Follett, B. K., King, J. R., and Morton, M. L. (1966). *Biol. Bull.* **130**, 67–75.
Farner, D. S., Wilson, F. E., and Oksche, A. (1967). *In* "Neuroendocrinology" (L. Martini and W. F. Ganong, eds.), Vol. 2, pp. 529–582. Academic Press, New York.
Fellegiová, M., Adamec, O., and Kosutzky, J. (1975). *Br. Poult. Sci.* **16**, 327–8.
Ferrand, R., LeDouarin, N. M., Polak, J. M., and Pearse, A. G. E. (1975). *Experientia* **31**, 1096–1097.
Fivizzani, A. J., and Meier, A. H. (1976). *Assoc. Southeastern Biol.* **23**, 58 (abstr.).
Flack, J. O. (1973). *In* "The Prostaglandins" (P. W. Ramwell, ed.), Vol. 1 pp. 327–346. Plenum, New York.
Fleischmann, W., and Fried, I. A. (1945). *Endocrinology* **36**, 406–415.
Follett, B. K. (1970). *Gen. Comp. Endocrinol.* **15**, 165–179.
Follett, B. K. (1975). *J. Endocrinol.* **67**, 198–208.
Follett, B. K., and Davies, D. T. (1975). *Symp. Zool. Soc. London* **35**, 199–224.
Follett, B. K., and Farner, D. S. (1966a). *Gen. Comp. Endocrinol.* **7**, 111–124.
Follett, B. K., and Farner, D. S. (1966b). *Gen. Comp. Endocrinol.* **7**, 125–131.
Follett, B. K., and Sharp, P. J. (1969). *Nature (London)* **223**, 968.
Follett, B. K., Scanes, C. G., and Cunningham, F. J. (1972). *J. Endocrinol.* **52**, 359–378.
Follett, B. K., Hinde, R. A., Steel, E., and Nicholls, T. J. (1973). *J. Endocrinol.* **59**, 151–162.
Follett, B. K., Mattocks, P. W., and Farner, D. S. (1974). *Proc. Natl. Acad. Sci. U.S.A.* **71**, 1666–1669.
Follett, B. K., Farner, D. S., and Mattocks, P. W., Jr. (1975). *Gen. Comp. Endocrinol.* **26**, 126–134.
Frankel, A. I. (1970). *Poult. Sci.* **49**, 869–921.
Fraps, R. M. (1954). *Proc. Natl. Acad. Sci. U.S.A.* **40**, 348.
Fraps, R. M. (1955). *In* "Progress in the Physiology of Farm Animals" (J. Hammond, ed.), Vol. 2, pp. 661–740. Butterworth, London.
Fraps, R. M. (1959). *In* "Photoperiodism and Related Phenomena in Plants and Animals" (R. B. Withrow, ed.), Publ. No. 55, pp. 767–785. Am. Assoc. Adv. Sci., Washington, D.C.
Fraps, R. M. (1961). *In* "Control of Ovulation" (C. A. Villee, ed.), pp. 135–167. Pergamon, Oxford.

Fraps, R. M., Filey, G. M., and Olsen, M. W. (1942). *Proc. Soc. Exp. Biol. Med.* **50**, 313–317.
Fraps, R. M., Fevold, H. L., and Neher, B. H. (1947). *Anat. Rec.* **99**, 571–572.
Freeman, B. M., and Manning, A. C. (1975). *Br. Poult. Sci.* **16**, 121–129.
Friedman, M., and Lehrman, D. S. (1968). *Anim. Behav.* **16**, 233–237.
Fröhlich, A. A., and Marquardt, R. R. (1972). *Biochim. Biophys. Acta* **286**, 396–405.
Furr, B. J. A., Bonney, R. C., England, R. M., and Cunningham, F. J. (1973). *J. Endocrinol.* **57**, 159.
Gala, R. R., and Westphal, V. (1966). *Endocrinology* **79**, 67.
Galli, F. E., and Wasserman, G. F. (1973). *Gen. Comp. Endocrinol.* **21**, 77–93.
Gardiner, V. E., Phillips, W. E., Maw, W. A., and Common, R. H. (1952). *Nature (London)* **170**, 80.
Gaston, S. (1971). *In* "Biochronometry" (M. Menaker, ed.), pp. 541–548. Natl. Acad. Sci., Washington, D.C.
Gaston, S., and Menaker, M. (1968). *Science* **160**, 1125–1127.
George, J. C., and Berger, A. J. (1966). "Avian Myology." Academic Press, New York.
George, J. C., and Naik, D. V. (1964). *Pavo* **2**, 37–49.
Ghosh, A. (1973). *Proc. Indian Science Congr., 60th* Part II, pp. 1–24.
Gibson, W. R., Follett, B. K., and Gledhill, B. (1975). *J. Endocrinol.* **64**, 87–101.
Glick, B. (1960). *Poult. Sci.* **39**, 1527.
Godden, P. M., and Scanes, C. G. (1975). *J. Endocrinol.* **64**, 448–458.
Gonnerman, W. A., Ramp, W. K., and Toverud, S. U. (1975). *Endocrinology* **96**, 275–281.
Goodridge, A. G. (1964). *Comp. Biochem. Physiol.* **13**, 1–26.
Goodridge, A. G. (1975). *Fed. Proc., Fed. Am. Soc. Exp. Biol.* **34**, 117–123.
Goodridge, A. G., and Ball, E. G. (1965). *Comp. Biochem. Physiol.* **16**, 367–381.
Goodridge, A. G., and Ball, E. G. (1966). *Am. J. Physiol.* **211**, 803–808.
Gould, N. R., and Siegel, H. S. (1974). *Gen. Comp. Endocrinol.* **24**, 177–182.
Gourdji, D., and Tixier-Vidal, A. (1966). *C. R. Hebd. Seances Acad. Sci.* **263**, 162–165.
Grande, F., and Prigge, W. F. (1972). *Proc. Soc. Exp. Biol. Med.* **140**, 999–1004.
Grandhi, R. R., and Brown, R. G. (1975a). *Poult. Sci.* **54**, 488–493.
Grandhi, R. R., and Brown, R. G. (1975b). *Poult. Sci.* **54**, 499–502.
Grandhi, R. R., Brown, R. G., Reinhart, B. S., and Summers, J. D. (1975). *Poult. Sci.* **54**, 493–499.
Greengard, O., Gordon, M., Smith, M. A., and Acs, G. (1964). *J. Biol. Chem.* **239**, 2079–2082.
Gross, R., and Mialhe, P. (1974). *Diabetologia* **10**, 277.
Guhl, A. M. (1962). *In* "The Sexual Behavior of Domestic Animals" (E. S. E. Hafez, ed.), pp. 491–530. Williams & Wilkins, Baltimore, Maryland.
Gwinner, E. (1974). *Science* **185**, 72–74.
Gwinner, E. (1975). *In* "Avian Biology" (D. S. Farner and J. R. King, eds.), Vol. V, pp. 221–285. Academic Press, New York.
Haffen, K. (1975). *Am. Zool.* **15**, 257–272.
Haffen, K., Scheib, D., Guichard, A., and Cédard, L. (1975). *Gen. Comp. Endocrinol.* **26**, 70–78.
Hale, E. B., and Schein, M. W. (1962). *In* "The Sexual Behaviour of Domestic Animals" (E. S. E. Hafez, ed.), pp. 531–564. Williams & Wilkins, Baltimore, Maryland.
Hall, T. R., and Chadwick, A. (1974). *J. Endocrinol.* **63**, 458.
Hall, T. R., Chadwick, A., Bolton, N. J., and Scanes, C. G. (1975). *Gen. Comp. Endocrinol.* **25**, 198–306.
Hamner, W. M. (1963). *Science* **142**, 1294.
Hamner, W. M. (1964). *Nature (London)* **203**, 1400.
Hamner, W. M. (1966). *Gen. Comp. Endocrinol.* **7**, 224–233.

Hamner, W. M. (1968). *Ecology* **49**, 211–227.
Hamner, W. M., and Barfield, R. J. (1970). *Condor* **72**, 99–101.
Harrison, H. E., and Harrison, H. C. (1965). *Am. J. Physiol.* **208**, 370–374.
Hawkes, M. P. G. (1974). Ph.D. Dissertation, University of Guelph, Ontario.
Hawkes, M. P. G., and George, J. C. (1974). *Acta Zool.* **56**, 67–75.
Hazelwood, R. L. (1965). *In* "Avian Physiology" (P. D. Sturkie, ed.), pp. 313–371. Cornell Univ. Press (Comstock), Ithaca, New York.
Hazelwood, R. L. (1973). *Am. Zool.* **13**, 699–709.
Hazelwood, R. L., and Hazelwood, B. S. (1961). *Proc. Soc. Exp. Biol. Med.* **108**, 10–12.
Hazelwood, R. L., Kimmel, J. R., and Pollock, H. G. (1968). *Endocrinology* **83**, 1331–1336.
Hazelwood, R. L., Turner, S. D., Kimmel, J. R., and Pollock, H. G. (1973). *Gen. Comp. Endocrinol.* **21**, 485–497.
Hedlund, L. (1970). *Anat. Rec.* **166**, 406.
Helms, C. W. (1963). *Proc. Int. Ornithol. Congr., 13th, 1962* pp. 925–939.
Heninger, R. W., and Newcomer, W. S. (1964). *Proc. Soc. Exp. Biol. Med.* **114**, 624–628.
Herlant, M. (1964). *Int. Rev. Cytol.* **17**, 299–381.
Hertz, R., Dhyse, F. G., and Tullner, W. L. (1949). *Endocrinology* **45**, 451–454.
Hinde, R. A. (1967). *Proc. Int. Ornithol. Congr., 14th, 1966* pp. 135–153.
Hinde, R. A., Steel, E., and Follett, B. K. (1974). *J. Reprod. Fertil.* **40**, 383–399.
Hirsch, L. J. (1961). Ph.D. Dissertation, University of Illinois, Urbana.
Holmes, W. N. (1975). *Gen. Comp. Endocrinol.* **25**, 249–258.
Holmes, W. N., Phillips, J. G., and Butler, D. G. (1961). *Endocrinology* **69**, 483–495.
Homma, K., and Sakakibara, Y. (1971). *In* "Biochronometry" (M. Menaker, ed.), pp. 333–341. Natl. Acad. Sci., Washington, D.C.
Hosoda, T., Kaneko, T., Mogi, K., and Abe, T. (1954). *Proc. World's Poult. Congr. Expo. 10th, 1954* Part 2, pp. 134–136.
Horrobin, D. F., Matalyi, J. P., and Menaker, M. (1976). *Med. Hypoth.* **2**, 219–226.
Huston, T. M., and Nalbandov, A. V. (1953). *Endocrinology* **52**, 149–156.
Hutchins, M. O., and Newcomer, W. S. (1966). *Gen. Comp. Endocrinol.* **6**, 239–248.
Hutchinson, R. E. (1975). *J. Endocrinol.* **64**, 417–428.
Hutchinson, R. E., Hinde, R. A., and Steel, E. A. (1967). *J. Endocrinol.* **39**, 379–385.
Immelmann, K. (1973). *In* "Breeding Biology of Birds (D. S. Farner, ed.), pp. 121–147. Natl. Acad. Sci., Washington, D.C.
Ivy, A. C., Tatum, A. C., and Jung, F. T. (1926). *Am. J. Physiol.* **78**, 666.
Jallageas, M., Assenmacher, I., and Follett, B. K. (1974). *Gen. Comp. Endocrinol.* **23**, 472–475.
John, T. M., and George, J. C. (1972). *J. Interdiscip. Cycle Res.* **3**, 33–37.
John, T. M., and George, J. C. (1973). *Comp. Biochem. Physiol. A* **45**, 541–547.
John, T. M., Meier, A. H., and Bryant, E. E. (1972). *Physiol. Zool.* **45**, 34–42.
John, T. M., McKeown, B. A., and George, J. C. (1973). *Comp. Biochem. Physiol.* **46**, 497–504.
John, T. M., McKeown, B. A., and George, J. C. (1974). *Comp. Biochem. Physiol. A* **48**, 521–526.
Johns, J. E. (1964). *Condor* **66**, 449–455.
Johns, J. E., and Pfeiffer, E. W. (1963). *Science* **140**, 1225–1226.
Jones, R. E. (1969). *Gen. Comp. Endocrinol.* **13**, 1–13.
Jones, R. E. (1971). *Biol. Rev. Cambridge Philos. Soc.* **46**, 315–339.
Joseph, M. M., and Meier, A. H. (1973). *Gen. Comp. Endocrinol.* **20**, 326.
Juhn, M., and Harris, P. (1958). *Proc. Soc. Exp. Biol. Med.* **98**, 669–672.
Karim, S. M. M., Hillier, K., and Devlin, J. (1968). *J. Pharmacol.* **20**, 749.

Kaufman, L., and Dobrowska, L. (1931). *C. R. Assoc. Anat.* **26**, 295–299.
Kenny, A. D., and Dacke, C. G. (1974). *J. Endocrinol.* **62**, 15–23.
Kimmel, J. R., Pollock, H. G., and Hazelwood, R. L. (1968). *Endocrinology* **83**, 1323–1330.
King, D. B., and King, C. R. (1973). *Gen. Comp. Endocrinol.* **21**, 517–529.
King, J. R. (1968). *Comp. Biochem. Physiol.* **24**, 827–837.
King, J. R. (1973). *In* "Breeding Biology of Birds" (D. S. Farner, ed.), pp. 78–107. Natl. Acad. Sci., Washington, D.C.
King, J. R., and Farner, D. S. (1974). *In* "Chronobiology" (L. E. Scheving, F. Hallberg, and J. E. Pauly, eds.), pp. 625–629. Igaku Shoin Ltd., Tokyo.
Kirkpatrick, C. M. (1959). *In* "Photoperiodism and Related Phenomena in Plants and Animals" (R. B. Withrow, ed.), Publ. No. 55, pp. 751–758. Am. Assoc. Adv. Sci., Washington, D.C.
Kobayashi, H. (1952). *Annot. Zool. Jpn.* **25**, 371–376.
Kobayashi, H. (1953). *Jpn. J. Zool.* **11**, 21–26.
Kobayashi, H. (1969). *In* "Seminar on Hypothalamic and Endocrine Functions in Birds" (H. Kobayashi and D. S. Farner, eds.), Abstr., p. 72. Tokyo.
Kobayashi, H., and Wada, M. (1973). *In* "Avian Biology" (D. S. Farner and J. R. King) Vol. III, pp. 287–347. Academic Press, New York.
Kobayashi, H., Bern, H. A., Nishioka, R. S., and Hyodo, Y. (1961). *Gen. Comp. Endocrinol.* **1**, 545–564.
Kobayashi, H., Matsui, T., and Ishii, I. (1970). *Int. Rev. Cytol.* **29**, 281–381.
Komisaruk, B. R. (1967). *J. Comp. Physiol. Psychol.* **64**, 219–224.
Kook, Y., Cho, K. B., and Yun, L. O. (1964). *Nature (London)* **204**, 385–386.
Krall, J. F., and Hahn, W. E. (1975). *Gen. Comp. Endocrinol.* **25**, 391–393.
Labrie, F., Ho-Kim, M. A., Delgado, A., MacIntosh, B., and Fortier, C. (1968). *Ann. Endocrinol.* **29**, 29.
Lahiri, P., and Banerji, H. (1969). *Indian J. Physiol. Allied Sci.* **23**, 100–106.
Langslow, D. R., and Hales, C. N. (1971). *In* "Physiology and Biochemistry of the Domestic Fowl" (D. J. Bell and B. M. Freeman, eds.), Vol. 1, pp. 521–547. Academic Press, New York.
Langslow, D. R., Butler, E. J., Hales, C. N., and Pearson, A. W. (1970). *J. Endocrinol.* **46**, 243–260.
Larsson, L. I., Sundler, F., Hoakanson, R., Pollock, H. G., and Kimmel, J. R. (1974). *Histochemistry* **42**, 377–382.
Laugier, C., Brard, C., Sandoz, D., and Boisvieux-Ulrich, E. (1975). *Gen. Comp. Endocrinol.* **26**, 285–300.
LeDouarain, N., and Le Lievre, L. (1970). *C. R. Hebd. Seances Acad. Sci., Ser. D* **270**, 2857–2860.
Lefebvre, P. (1975). *Biochem. Pharmacol.* **24**, 1261–1266.
Lefebvre, P. J., and Luyckx, A. S. (1974). *Biochem. Pharmacol.* **23**, 2119–2125.
Lehrman, D. S. (1955). *Behaviour* **7**, 241–286.
Lehrman, D. S. (1958). *J. Comp. Physiol. Psychol.* **51**, 142–145.
Lehrman, D. S. (1961). *In* "Sex and Internal Secretions" (W. C. Young, ed.), 3rd ed., Vol. 2, pp. 1268–1382. Williams & Wilkins, Baltimore, Maryland.
Lehrman, D. S., and Brody, P. N. (1961). *J. Endocrinol.* **22**, 269–275.
Lehrman, D. S., and Brody, P. N. (1964). *J. Comp. Physiol Psychol.* **57**, 161–165.
Lesher, S. W., and Kendeigh, S. C. (1941). *Wilson Bull.* **53**, 169–180.
Levi, W. M. (1957). "The Pigeon." Levi Publ. Co., Inc., Sumter, South Carolina.
Lofts, B. (1962). *Gen. Comp. Endocrinol.* **2**, 394.
Lofts, B., and Lam, W. L. (1973). *J. Reprod. Fertil., Suppl.* **19**, 19–34.

Lofts, B., and Murton, R. K. (1968). *J. Zool.* **155,** 327–394.
Lofts, B., and Murton, R. K. (1973). *In* "Avian Biology" (D. S. Farner and J. R. King, eds.), Vol. III, pp. 1–107. Academic Press, New York.
Lofts, B., Murton, R. K., and Thearle, R. J. P. (1973). *Gen. Comp. Endocrinol.* **21,** 202–209.
Lowry, P. J., and Scott, A. P. (1975). *Gen. Comp. Endocrinol.* **26,** 16–23.
Lynch, H. (1973). *Endocrinol., Proc. Int. Symp., 4th, 1972.* Excerpta Med. Found. Int. Congr. Ser. No. 273, p. 263.
Lynch, H. J., and Ralph, C. L. (1970). *Am. Zool.* **10,** 300.
MacBride, S. E. (1973). Ph.D. Dissertation, University of Pittsburgh, Pittsburgh, Pennsylvania.
MacBride, S. E., and Ralph, C. L. (1972). *Physiologist* **15,** 204.
MacGregor, R., III. (1975). Ph.D. Dissertation, Louisiana State University, Baton Rouge.
Martin, D. D., and Meier, A. H. (1973). *Condor* **75,** 369–374.
Martin, T. J., and Moseley, J. M. (1975). *J. Endocrinol.* **65,** 28P–29P.
Martinez-Vargas, M. C., Gibson, D. B., Sar, M., and Stumpf, W. E. (1975). *Science* **190,** 1307–1308.
Matsui, T. (1966). *J. Fac. Sci., Univ. Tokyo, Sect. 4* **11,** 49–70.
McKeown, B. A., John T. M., and George, J. C. (1974). *J. Interdiscip. Cycle Res.* **4,** 221–227.
McKnight, G. S., Pennequin, P., and Schimke, R. T. (1975). *J. Biol. Chem.* **250,** 8105–8110.
McMillan, J. (1972). *J. Comp. Physiol.* **79,** 105–112.
McNally, E. H. (1947). *Poult. Sci.* **26,** 396–399.
Meier, A. H. (1969a). *Gen. Comp. Endocrinol., Suppl.* **2,** 55–62.
Meier, A. H. (1969b). *Gen. Comp. Endocrinol., Suppl.* **13,** 222–225.
Meier, A. H. (1975). *Am. Zool.* **15,** 909–916.
Meier, A. H. (1976). *Proc. Int. Ornithol. Congr., 16th, 1975* pp. 355–368.
Meier, A. H., and Davis, K. B. (1967). *Gen. Comp. Endocrinol.* **8,** 110–114.
Meier, A. H., and Dusseau, J. W. (1973). *Biol. Reprod.* **8,** 400.
Meier, A. H., and Farner, D. S. (1964). *Gen. Comp. Endocrinol.* **4,** 584–595.
Meier, A. H., and Fivizzani, A. J. (1975). *Proc. Soc. Exp. Biol. Med.* **150,** 356–362.
Meier, A. H., and MacGregor, R., III. (1972). *Am. Zool.* **12,** 257–271.
Meier, A. H., and Martin, D. (1971). *Gen. Comp. Endocrinol.* **17,** 311.
Meier, A. H., Farner, D. S., and King, J. R. (1965). *Anim. Behav.* **13,** 453–465.
Meier, A. H., Burns, J. T., and Dusseau, J. W. (1969). *Gen. Comp. Endocrinol.* **12,** 282–289.
Meier, A. H., Burns, J. T., Davis, K. B., and John, T. M. (1971a). *J. Interdiscip. Cycle Res.* **2,** 161–172.
Meier, A. H., John, T. M., and Joseph, M. M. (1971b). *Comp. Biochem. Physiol.* **40,** 459.
Meier, A. H., Martin, D. D., and MacGregor, R., III. (1971c). *Science* **173,** 1240–1242.
Meier, A. H., Trobec, T. N., Joseph, M. M., and John, T. M. (1971d). *Proc. Soc. Exp. Biol. Med.* **137,** 408.
Meites, J., and Nicoll, C. S. (1966). *Annu. Rev. Physiol.* **28,** 57–88.
Menaker, M. (1965). *In* "Circadian Clocks" (J. Aschoff, ed.), pp. 385–395. North-Holland Publ., Amsterdam.
Menaker, M. (1968). *Proc. Natl. Acad. Sci. U.S.A.* **59,** 414.
Menaker, M., and Eskin, A. (1967). *Science* **157,** 1182.
Menaker, M., and Oksche, A. (1974). *Avian Biol.* **4,** 79–118.
Menaker, M., Roberts, R., Elliott, J., and Underwood, H. (1970). *Proc. Natl. Acad. Sci. U.S.A.* **67,** 320–325.

Merkel, F. W. (1938). *Ber. Verh. Schles. Ornithol.* **25**, 1–72.

Merkel, F. W. (1960). *Umschau* **60**, 243–246.

Merkel, F. W. (1963). *Proc. Int. Ornithol. Congr., 13th, 1962* pp. 950–959.

Mialhe, P. (1958). *Acta Endocrinol. (Copenhagen), Suppl.* **36**, 1–54.

Mialhe, P. (1969). *Prog. Endocrinol. Proc. Int. Congr. Endocrinol., 3rd, 1968* Excerpta Med. Found. Int. Congr. Ser. No. 184, pp. 158–164.

Mikami, S. I. (1958). *J. Fac. Agric., Iwate Univ.* **3**, 473–545.

Mikami, S. I., Vitums, A., and Farner, D. S. (1969). *Z. Zellforsch. Mikrosk. Anat.* **97**, 1–29.

Mikami, S. I., Farner, D. S., and Lewis, R. A. (1973). *Z. Zellforsch. Mikrosk. Anat.* **138**, 455–474.

Miller, A. H. (1954). *Condor* **56**, 13–20.

Mirsky, I. A., Nelson, N., Grayman, I., and Korenberg, M. (1941). *Am. J. Physiol.* **135**, 223.

Morita, Y. (1966). *Experientia* **22**, 402.

Muller, E. E., Sawano, S., and Schally, A. V. (1967). *Gen. Comp. Endocrinol.* **9**, 349–352.

Munsick, R. A., Sawyer, W. H., and Van Dyke, H. B. (1960). *Endocrinology* **66**, 860–871.

Murphy, B. E. P. (1967). *J. Clin. Endocrinol. Metab.* **27**, 973–990.

Murton, R. K., Bagshawe, K. D., and Lofts, B. (1969). *J. Endocrinol.* **45**, 311–312.

Murton, R. K., Lofts, B., and Westwood, N. J. (1970). *Gen. Comp. Endocrinol.* **14**, 107–113.

Nakano, J. (1973). *In* "The Prostaglandins" (P. W. Ramwell, ed.), Vol. 1, pp. 239–316. Plenum, New York.

Nalbandov, A. V. (1945). *Endocrinology* **36**, 251–258.

Nalbandov, A. V. (1959). *In* "Comparative Endocrinology" (A. Gorbman, ed.), pp. 524–532. Wiley, New York.

Nalbandov, A. V. (1966). *In* "The Pituitary Gland" (G. W. Harris and B. T. Donovan, eds.), Vol. 1, pp. 295–316. Univ. of California Press, Berkeley.

Nalbandov, A. V., and Card, L. E. (1945). *J. Hered.* **36**, 34–39.

Newcomer, W. S. (1974). *Gen. Comp. Endocrinol.* **24**, 65–73.

Nice, M. M. (1937). *Trans. Linn. Soc. New York* **4**, 61–247.

Nicoll, C. S. (1965). *J. Exp. Zool.* **158**, 203–210.

Nicoll, C. S. (1975). *Am. Zool.* **15**, 881–903.

Nicoll, C. S., and Bern, H. A. (1972). *Lactogenic Horm., Ciba Found. Symp., 1971*, No. 130, p. 299.

Nicoll, C. S., and Meites, J. (1962). *Nature (London)* **195**, 606–607.

Nieto, A., Moya, F., and R-Candela, J. L. (1973). *Biochim. Biophys. Acta* **322**, 383–391.

Nieto, A., L-Fando, J. J., and R-Candela, J. L. (1975). *Gen. Comp. Endocrinol.* **25**, 259–263.

Noble, G. K., and Wurm, M. (1940). *Endocrinology* **26**, 837–850.

Oehmke, H. J., Priedkalns, J., Vaupel-von Harnack, M., and Oksche, A. (1969). *Z. Zellforsch. Mikrosk. Anat.* **95**, 109–133.

Oksche, A., and Kirschstein, H. (1969). *Z. Zellforsch. Mikrosk. Anat.* **102**, 214–241.

Oksche, A., Laws, D. F., Kamemoto, F. I., and Farner, D. S. (1959). *Z. Zellforsch. Mikrosk. Anat.* **51**, 1–42.

Oksche, A., Kirschstein, H., Kobayashi, H., and Farner, D. S. (1972). *Z. Zellforsch. Mikrosk. Anat.* **124**, 247–274.

O'Malley, B. W., and Means, A. R. (1974). *Science* **183**, 610–620.

O'Malley, B. W., and Schrader, W. T. (1976). *Sci. Am.* **234**, 32–43.

Opel, H. (1966). *Br. Poult. Sci.* **7**, 29–38.

Orcutt, F. S. (1965). *Am. Zool.* **5**, 197 (abstr).

Ostmann, O. W., Peterson, R. A., and Renger, R. K. (1964). *Poult. Sci.* **43**, 648–654.

Palmer, R. S. (1972). *In* "Avian Biology" (D. S. Farner and J. R. King, eds.), Vol. II, pp. 65–102. Academic Press, New York.
Pang, S. F. (1974). Ph.D. Dissertation, University of Pittsburgh, Pittsburgh, Pennsylvania.
Pang, S. F., Ralph, C. L., and Reilly, D. P. (1974). *Gen. Comp. Endocrinol.* **22,** 499–506.
Paoletti, R., and Puglisi, L. (1973). *In* "The Prostaglandins" (P. W. Ramwell, ed.), Vol. 1, pp. 317–326.
Payne, R. B. (1972). *Avian Biol.* **2,** 104–155.
Peaker, M., and Linzell, J. L. (1975). "Salt Glands in Birds and Reptiles," pp. 132–157. Cambridge Univ. Press, London and New York.
Péczely, P., and Zboray, G. (1967). *Acta Physiol. Acad. Sci. Hung.* **32,** 229–239.
Phillips, J. G., and Ensor, D. M. (1972). *Gen. Comp. Endocrinol., Suppl.* **3,** 393–404.
Phillips, J. G., Holmes, W. N., and Butler, D. G. (1961). *Endocrinology* **69,** 958–959.
Phillips, R. E., and McKinney, F. (1962). *Anim. Behav.* **10,** 244–246.
Pittendrigh, C. S. (1960). *Cold Spring Harbor Symp. Quant. Biol.* **25,** 159–184.
Planas, J., and Frieden, C. (1973). *Am. J. Physiol.* **225,** 423–428.
Podhradsky, J., Wodzicki, K., and Golabeck, Z. (1937). *Ann. Tschech. Akad. Landwirtsch.* **12,** 604–613.
Polikarpova, E. (1940). *Dokl. Akad. Nauk SSSR* **26,** 91–95.
Priedkalns, J., and Oksche, A. (1969). *Z. Zellforsch. Mikrosk. Anat.* **98,** 135–147.
Prigge, W. F., and Grande, F. (1971). *Comp. Biochem. Physiol.* **39,** 69–82.
Putzig, P. (1937). *Vogelzug* **8,** 116–130.
Putzig, P. (1938). *J. Ornithol.* **86,** 123–165.
Quay, W. B. (1965). *Life Sci.* **4,** 379.
Radke, W. J., and Chiasson, R. B. (1974). *J. Endocrinol.* **60,** 187–188.
Rahn, H., and Painter, B. I. (1941). *Anat. Rec.* **79,** 297–311.
Ralph, C. L. (1975). *Am. Zool.* **15,** Suppl. 1, 105–116.
Ralph, C. L., and Dawson, D. C. (1968). *Experientia* **24,** 147–148.
Ralph, C. L., and Fraps, R. M. (1959). *Endocrinology* **65,** 819.
Ralph, C. L., Binkley, S., MacBride, S. E., and Klein, D. C. (1975). *Endocrinology* **97,** 1373–1378.
Rasmussen, H. (1970). *Science* **170,** 404–412.
Refetoff, S., Robin, N. J., and Fang, V. S. (1970). *Endocrinology* **86,** 793–805.
Reiter, R. J. (1973). *Ann. Res. Physiol.* **35,** 305–328.
Riddle, O. (1937). *Cold Spring Harbor Symp. Quant. Biol.* **5,** 362–374.
Riddle, O. (1963). *J. Natl. Cancer Inst.* **31,** 1039–1110.
Riddle, O., and Bates, R. W. (1933). *Endocrinology* **17,** 689–698.
Riddle, O., and Dykshorn, S. W. (1932). *Proc. Soc. Exp. Biol. Med.* **29,** 1213–1215.
Riddle, O., and Opdyke, D. F. (1947). *Carnegie Inst. Washington Publ.* **569,** 49.
Riddle, O., Bates, R. W., and Dykshorn, S. W. (1932). *Proc. Soc. Exp. Biol. Med.* **29,** 1211–1212.
Riddle, O., Bates, R. W., and Dykshorn, S. W. (1933). *Am. J. Physiol.* **105,** 191–216.
Riddle, O., Smith, G. C., and Moran, C. S. (1935). *Proc. Soc. Exp. Biol. Med.* **32,** 1614.
Rosenberg, L. L., Astier, H., Roche, G. L., Baylé, J. D., Tixier-Vidal, A., and Assenmacher, I. (1967). *Neuroendocrinology* **2,** 113–125.
Rosner, J. M., Denari, J. H., Nagle, C. A., Cardinali, D. P., de Perez Bedes, G. D., and Orsi, L. (1971). *Life Sci.* **11,** 829–836.
Rothchild, I., and Fraps, R. M. (1949). *Endocrinology* **44,** 134.
Rowan, W. (1926). *Proc. Boston Soc. Nat. Hist.* **38,** 147.
Sadovsky, R., and Bensadoun, A. (1971). *Gen. Comp. Endocrinol.* **17,** 268–274.
Saeki, Y., and Tanabe, Y. (1955). *Poult. Sci.* **34,** 909–919.
Samols, E. J. M., Tyler, J., and Mialhe, P. (1969) *Lancet* **1,** 174–176.

Sandberg, A. A., and Slaunwhite, W. R., Jr. (1962). *J. Clin. Invest.* **41**, 1396–1397.
Sandor, T., and Fazekas, A. G. (1974). *Gen. Comp. Endocrinol.* **22**, 343–349.
Sansun, E. L., and King, J. R. (1976). *Physiol. Zool.* **49**, 407–416.
Santos, P. G., and Grande, F. (1975). *Proc. Soc. Exp. Biol. Med.* **149**, 652–655.
Sato, T., and George, J. C. (1973). *Can. J. Physiol. Pharmacol.* **51**, 743–747.
Saylor, A., and Wolfson, A. (1967). *Neuroendocrinology* **5**, 322.
Scanes, C. G. (1974). *Neuroendocrinology* **15**, 1–9.
Scanes, C. G., and Follett, B. K. (1972). *Br. Poult. Sci.* **13**, 603–610.
Scanes, C. G., Telfer, S. B., Hackett, A. F., Nightingale, R., and Sharifuddin, B. A. (1975). *Br. Poult. Sci.* **16**, 405–408.
Schildmacher, H., and Rautenberg, W. (1952). *Biol. Zentralbl.* **71**, 397–405.
Schooley, J. P. (1937). *Cold Spring Harbor Symp. Quant. Biol.* **5**, 165–177.
Schooley, J. P., and Riddle, O. (1938). *Am. J. Anat.* **63**, 313.
Schreibman, M. P., and Holtzman, S. (1975). *Am. Zool.* **15**, 867–880.
Scott, M. L. (1973). *In* "Breeding Biology of Birds" (D. S. Farner, ed.), pp. 46–59. Natl. Acad. Sci., Washington, D.C.
Seal, U. S., and Doe, R. P. (1966). *Steroid Dyn., Proc. Symp., 1965* pp. 63–90.
Selander, R. K., and Kuich, L. L. (1963). *Condor* **65**, 73–90.
Senior, B. E., and Cunningham, F. J. (1974). *J. Endocrinol.* **60**, 201–202.
Shank, M. C. (1959). *Auk* **76**, 44–54.
Shapiro, M., Nicholson, W. E., Orth, D. N., Mitchel, W. M., Island, D. P., and Liddle, G. W. (1972). *Endocrinology* **90**, 249–256.
Sharp, P. J., and Follett, B. K. (1968). *Z. Zellforsch. Mikrosk. Anat.* **90**, 245–262.
Sharp, P. J., and Follett, B. K. (1969). *Neuroendocrinology* **5**, 205–218.
Shellabarger, C. J. (1952). *Endocrinology* **51**, 152–154.
Simkiss, K. (1975). *Symp. Zool. Soc. London* **35**, 307–337.
Simkiss, K., and Dacke, C. G. (1971). *In* "Physiology and Biochemistry of the Domestic Fowl" (D. J. Bell and B. M. Freeman, eds.), Vol. 1, pp. 481–488. Academic Press, New York.
Singh, D. V., Panda, J. N., Anderson, R. R., and Turner, C. W. (1967). *Proc. Soc. Exp. Biol. Med.* **126**, 553.
Slater, P. J. B. (1967). *Anim. Behav.* **15**, 520–526.
Smith, C. J. (1964). Ph.D. Dissertation, University of Maryland, College Park.
Smith, P. M., and Follett, B. K. (1972). *J. Endocrinol.* **53**, 131–138.
Snedecor, J. G., King, D. B., and Hendrikson, R. C. (1963). *Gen. Comp. Endocrinol.* **3**, 176–183.
Snedecor, J. G., Raheja, K. L., and Freedland, R. A. (1973). *Gen. Comp. Endocrinol.* **18**, 199–209.
Speers, G. M., Perey, D. Y. E., and Brown, D. M. (1970). *Endocrinology* **87**, 1292–1297.
Srivastava, A. K., and Meier, A. H. (1972). *Science* **177**, 185.
Stammler, J., Katz, L. N., Pick, R., and Rodbard, S. (1955). *Recent Prog. Horm. Res.* **11**, 401–447.
Steel, E. A., and Hinde, R. A. (1963). *J. Endocrinol.* **26**, 11–24.
Steeno, O., and DeMoor, P. (1966). *Bull. Soc. R. Zool. Anvers.* **38**, 3–24.
Stettenheim, P. (1972). *In* "Avian Biology" (D. S. Farner and J. R. King, eds.), Vol. II, pp. 1–63. Academic Press, New York.
Stockwell-Hartree, A., and Cunningham, F. J. (1969). *J. Endocrinol.* **43**, 609–616.
Sturkie, P. D. (1951). *Endocrinology* **49**, 565–570.
Sutherland, E. W. (1972). *Science* **177**, 401–408.
Takemoto, D. J., Abel, J. H., Jr., and Allen, J. C. (1975). *Gen. Comp. Endocrinol.* **26**, 226–232.
Tanabe, Y., and Katsuragi, T. (1962). *Bull. Natl. Inst. Agric. Sci., Ser. G* **21**, 49–59.

Taylor, T. G. (1965). *Proc. Nutr. Soc.* **24**, 49–54.

Temple, S. A. (1974). *Gen. Comp. Endocrinol.* **22**, 470–479.

Teng, C. S., and Teng, C. T. (1975). *Biochem. J.* **150**, 191–194.

Thapliyal, J. P., and Pandha, S. K. (1967). *Gen. Comp. Endocrinol.* **8**, 84–93.

Tixier-Vidal, A., and Follett, B. K. (1973). *In* "Avian Biology" (D. S. Farner and J. R. King, eds.), Vol. III, pp. 109–182. Academic Press, New York.

Tixier-Vidal, A., and Gourdji, D. (1972). *Gen. Comp. Endocrinol., Suppl.* **3**, 51–64.

Tixier-Vidal, A., Follett, B. K., and Farner, D. S. (1968). *Z. Zellforsch. Mikrosk. Anat.* **92**, 610–635.

Tougard, C. (1971). *Z. Zellforsch. Mikrosk. Anat.* **116**, 375–390.

Turek, F. W. (1972). *Science* **178**, 1112–1113.

Turek, F. W. (1974). *J. Comp. Physiol.* **92**, 59–64.

Turek, F. W., McMillan, J. P., and Menaker, M. (1976). *Science* **194**, 1441–1443.

Uemura, H., and Kobayashi, H. (1963). *Gen. Comp. Endocrinol.* **3**, 253–264.

Urist, M. R (1959). *Recent Prog. Horm. Res.* **15**, 455–477.

Urist, M. R., and Deutsch, N. M. (1960). *Proc. Soc. Exp. Biol. Med.* **104**, 35–39.

Urist, M. R., Deutsch, N. M., Pomerantz, G., and McLean, F. C. (1960). *Am. J. Physiol.* **199**, 851–855.

van Tienhoven, A. (1955). *Endocrinology* **56**, 667–674.

van Tienhoven, A. (1961). *In* "Sex and Internal Secretions" (W. C. Young, ed.), 3rd ed., Vol. 2, pp. 1088–1169. Williams & Wilkins, Baltimore, Maryland.

van Tienhoven, A., Nalbandov, A. V., and Norton, H. W. (1954). *Endocrinology* **54**, 605.

Van Tyne, J., and Berger, A. J. (1971). "Fundamentals of Ornithology," pp. 70–107. Dover, New York.

Vaugien, M., and Vaugien, L. (1961). *C. R. Hebd. Seances Acad. Sci.* **253**, 2762–2764.

Vitums, A., Mikami, S. I., Oksche, A., and Farner, D. S. (1964). *Z. Zellforsch. Mikrosk. Anat.* **64**, 541–569.

Voitkevich, A. A. (1966). "The Feathers and Plumage of Birds." October House Inc., New York.

Wagner, H. O. (1930). *Z. Vergl. Physiol.* **12**, 703–724.

Wagner, H. O. (1961). *Z. Vergl. Physiol.* **44**, 565–575.

Wagner, H. O., and Müller, C. (1963). *Z. Morphol. Oekol. Tiere* **53**, 107–151.

Wasserman, R. H., and Taylor, R. H. (1966). *Science* **152**, 791–793.

Weise, C. M. (1967). *Condor* **69**, 49.

Wilson, A. C., and Farner, D. S. (1960). *Condor* **62**, 414–425.

Wilson, F. E., and Follett, B. K. (1974). *Gen. Comp. Endocrinol.* **23**, 82–93.

Wilson, S. C., and Sharp, P. J. (1976a). *Br. Poult. Sci.* **17**, 163–173.

Wilson, S. C., and Sharp, P. J. (1976b). *J. Endocrinol.* **71**, 87–98.

Wilson, W. O., and Huang, R. H. (1962). *Poult. Sci.* **41**, 1843.

Wilson, W. O., Abbott, V. L., and Abplanalp, H. (1961). *Poult. Sci.* **40**, 651–657.

Wingstrand, K. G. (1951). *CWK Gleerup Lund.*

Witschi, E. (1961). *In* "Biology and Comparative Physiology of Birds" (A. J. Marshall, ed.), Vol. 2, pp. 115–168. Academic Press, New York.

Wolfson, A. (1952). *J. Exp. Zool.* **121**, 311–326.

Wolfson, A. (1959a). *In* "Photoperiodism and Related Phenomena in Plants and Animals" (R. B. Withrow, ed.), Publ. No. 55, pp. 679–716. Am. Assoc. Adv. Sci., Washington, D.C.

Wolfson, A. (1959b). *Physiol. Zool.* **32**, 160–176.

Wolfson, A. (1965a). *Proc. Int. Congr. Endocrinol., 2nd, 1964* Excerpta Med. Found. Int. Congr. Ser. No. 83, p. 183.

Wolfson, A. (1965b). *In* "Circadian Clocks" (J. Aschoff, ed.), pp. 370–378. North-Holland Publ., Amsterdam.

Wolfson, A. (1966). *Recent Prog. Horm. Res.* **22,** 177–244.
Wolfson, A., and Kobayashi, H. (1962). *Gen. Comp. Endocrinol., Suppl.* **1,** 168–179.
Wood-Gush, D. G. M., and Gilbert, A. B. (1975). *Symp. Zool. Soc. London* **35,** 261–276.
Wurtman, R. J., Axelrod, J., and Kelly, D. E., eds. (1968). "The Pineal." Academic Press, New York.
Yasuda, M. (1953). *Proc. Jpn. Acad.* **29,** 586–594.
Zachariasen, R. D., and Newcomer, W. S. (1975). *Gen. Comp. Endocrinol.* **25,** 332–338.
Zimmerman, N., and Menaker, M. (1975). *Science* **190,** 477–479.

CHAPTER 8

Calcium Metabolism in Birds

Shmuel Hurwitz

I. Introduction

Most of the early studies on calcium were concerned with its role as the main component of bone. However, the importance of this cation in other physiological phenomena such as muscle contraction and relaxation has been also recognized for many years. More recently, it became evident that calcium was required for activation of many enzyme systems (Mahler, 1961) and for transfer of information in the cell. With regard to the latter, calcium interacts with the adenylate cyclase and guanosyl cyclase systems to control hormone secretion, cellular multiplication, and differentiation (Rasmussen *et al.*, 1975). Many of the cellular functions thus require calcium, or are sensitive to extracellular calcium concentration. It is, therefore, not surprising that calcium is regulated within a very narrow range, so that plasma calcium is considered to be one of nature's constants (McLean and Hastings, 1935).

Soft tissue contains only 1–2% of the total body calcium. Most of the soft tissue calcium appears in extracellular fluids. The extracellular calcium pool is small with regard to the turnover of this cation. If no control existed, a 10% change in calcium absorption in the rat would lead to a 50% change in the concentration of plasma calcium within one day. In the laying hen during shell formation, the entire extracellular pool equals the amount of calcium absorbed within one-half hour. This extraordinary turnover rate requires an extremely efficient regulatory mechanism in order to avoid errors in calcium concentration of the body fluids. Furthermore, shell calcification occupies about two-thirds of the 25-hour laying cycle. During this time, calcium is drained from the hen's body at a rate of about 140 mg/hour compared to only slight losses when no shell is formed. Thus, the calcium metabolism of the hen is an interrupted process, which needs turning on and off. Such an interrupted metabolic pattern represents an extreme challenge to the calcium homeostatic mechanisms. In this presentation I examine the response of the various control systems in birds to this challenge. Basic principles will be summarized rather than reviewed, and features specific to birds will be discussed. Furthermore, no attempt will be made for a thorough historical review of calcium metabolism of birds. For an excellent review of the earlier work, the reader is referred to Simkiss (1961).

II. Regulation of Calcium Metabolism

The metabolism of calcium in the bird is presented schematically in Fig. 1. In this scheme the controlled system is plasma free calcium (Ca_s) which also serves as a controlled signal. An error in Ca_s is given by ($Ca_s - U$) with U representing the reference value, or set-point. This error is detected by the various control systems: bone, kidney, and intestine which may respond directly to changes in plasma calcium. Bronner (1975) showed that bone can control plasma calcium directly, without any hormonal involvement. Similarly, the kidney may regulate calcium excretion in response to the filtered load which varies with the concentration of calcium.

Three hormonal systems are involved with the control of calcium metabolism: the parathyroid hormone (PTH), secreted by the parathyroid gland; calcitonin (CT), secreted by the ultimobranchial gland; and 1,25-dihydroxycholecalciferol (1,25-DHCC), secreted by the kidney. The first two are polypeptides while the latter may be considered a steroid hormone. Prostaglandins have been also implicated in the control of bone resorption (Klein and Raisz, 1970). To the best of my knowledge, no information exists as to the possible involvement of this group of hormones in the control of calcium metabolism in birds.

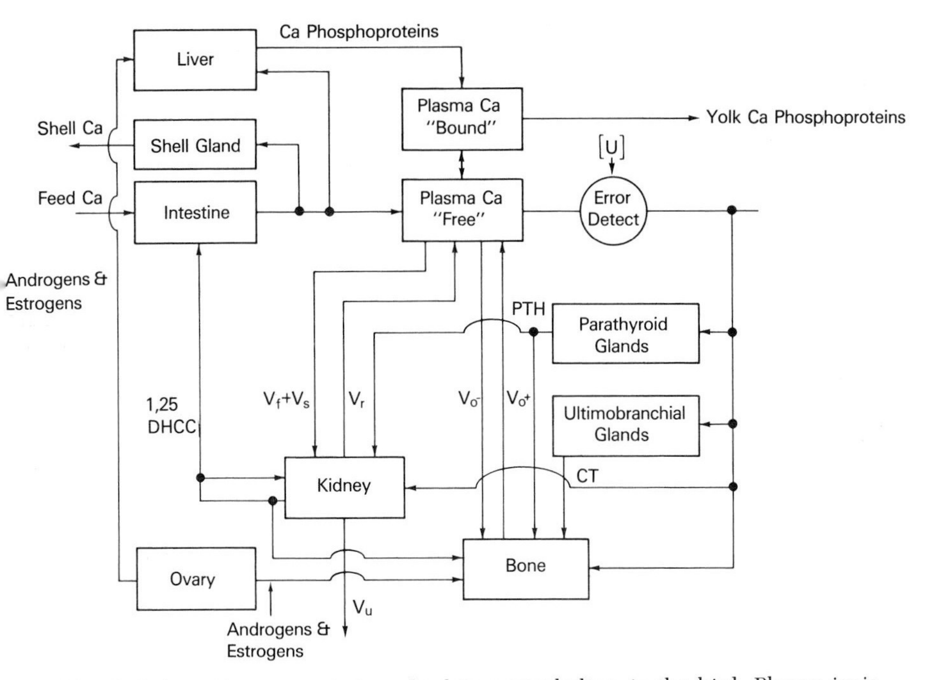

FIG. 1. Schematic representation of calcium metabolism in the bird. Plasma ionic calcium is the controlled system; its concentration serves as the controlled signal. Calcium absorbed and that secreted in the shell and egg yolk are the disturbing signals. The regulating hormones, parathyroid hormone, calcitonin, and 1,25-dihydroxycholecalciferol, are secreted by the parathyroid glands, the ultimobranchial glands, and kidney, respectively. The major controlling systems are (a) bone, with two constituent signals, formation (V_{0+}) and resorption (V_{0-}); (b) kidney, with three constituent signals, filtration (V_f), reabsorption (V_r), and secretion (V_s) of calcium; and (c) intestine, where the net flux of calcium is partially a controlling signal. The ovary, by virtue of secretion of gonadal hormones, modifies plasma calcium by induction of vitellogenin synthesis in the liver which is eventually lost from the system together with its bound calcium into the egg yolk, and induces calcium storage in bone together with medullary bone formation. For a more detailed discussion of the various feedback loops involves in the control of plasma calcium in birds, see text.

Control of plasma calcium by bone may be achieved by modulation of either of two constituent signals, bone formation (V_{0+}) or bone resorption (V_{0-}). Similarly, calcium excretion in the kidney may be regulated by changes in the amount of calcium filtered (V_f), reabsorbed (V_r), and possibly secreted (V_s). Regulation of calcium absorption by the intestine (V_i) is probably obtained primarily by changes in the calcium flux from the lumen into circulation.

Although not directly involved in the control of calcium homeostasis, gonadal hormones act in birds on the conditioning of the bone system to egg formation by inducing medullary bone formation. Another modification of calcium metabolism, in which gonadal hormones are involved, is the introduction of a steady-state disturbing signal by inducing the liver to synthesize vitellogenin. This complex protein is transferred subsequently into the circulation, raises the concentration of bound calcium and is finally transferred to the egg yolk. The net loss of calcium via the egg yolk amounts to about 25 mg/day of calcium in the domestic hen.

The most important disturbing signal, quantitatively, is the loss of calcium in the eggshell of about 2 gm/day. The significance of this transfer is realized if one considers the fact that the entire calcium content of the body is about 20 gm. Although calcium absorption can be modified by regulatory processes, it is still regarded in the scheme as a disturbing signal, since it is dependent on the supply of calcium from the diet.

In the following sections of this discussion, the various components of the control system are treated separately in order to illustrate the importance of each in calcium homeostasis in birds.

III. The Controlled Signal—Plasma Calcium

Only a fraction of the plasma calcium is in an ionized form. Due to difficulties in measuring the calcium activity a_{Ca}, various loosely defined entities have been evaluated, such as "ionized" calcium as measured by a color reaction with Muroxide. Although direct measurements using a calcium-sensitive electrode have been used, the difficulties in applying such electrodes to biological fluids limit their usefulness. The greatest body of data which approximate the "ionic" calcium concentration is derived from results of ultrafiltration through cellophane or other membranes and the measurement of ultrafilterable or diffusible calcium in the ultrafiltrates. Approximately 50–60% of the plasma calcium in mammalian species is protein bound, with albumin being the primary ligand (Moore, 1970). Typically, the plasma calcium concentration of mammals is 10 mg/100 ml, of which 5 mg/100 ml represent bound calcium. In nonlaying birds the distribution is similar (Hurwitz, 1968).

In female birds just prior to and during egg laying, plasma calcium shows a completely different pattern. The concentration of total plasma calcium increases to 20–30 mg/100 ml with almost all the increase being accounted for by bound calcium. This increase in plasma calcium, which occurs without any change in ultrafilterable calcium concentration (McDonald and Riddle, 1945), results from the appearance in plasma of a lipophosphoprotein complex which has a very high binding affinity for calcium (Urist *et al.*, 1960). This protein complex is synthesized by the liver under the influence of estrogen and is subsequently transported to the egg yolk. Of this group of proteins, two types of phosvitins, which are glycophosphoproteins, were identified from egg yolk with molecular weights of 28,000 and 34,000, respectively (Clark, 1970, 1973). α- and β-Lipovitellins have been described by Bernardi and Cook (1960) as part of this complex. More recently, Deeley *et al.* (1975) showed that actually only one precursor for all yolk proteins is synthesized by the liver under the influence of estrogen, and is the only form present in blood plasma. This protein, vitellogenin, has a molecular weight of approximately 480,000 and is made of two polypeptides, each of which contains a lipovitellin and two phosvitins. The various protein subunits are separated upon transfer into the egg yolk. The high phosphate content (approximately 30 mg P/gm protein) can account for its high calcium-binding capacity.

Thus, the increase in plasma calcium associated with egg laying is neither associated directly with calcium metabolism nor with any increased needs for shell formation. Rather, it represents the presence of yolk proteins in plasma and is associated with a very large increase in covalently bound phosphate.

The bound calcium in laying-hen plasma shows a relatively slow exchangeability with "free" calcium (Hurwitz, 1968). About 100 minutes are needed to obtain isotopic mixing between bound and free calcium in laying-hen plasma. This slow exchangeability represents a serious obstacle in conducting isotope-kinetic studies in the hen.

The regulation of plasma calcium is very efficient when diets are supplemented with vitamin D. With a normal plasma calcium of 9.5–10.5 mg/100 ml, chicks could maintain plasma calcium even when fed a diet extremely low in calcium. With normal level of dietary phosphorus (0.65–0.8%), Cipera *et al.* (1970) had to raise the calcium content from 1.36 to 3.91% in diets of growing chicks in order to raise plasma calcium from 10.4 to 14.9 mg/100 ml. With low dietary phosphate, a similar raise of plasma calcium can be obtained with much lower dietary calcium concentrations.

IV. The Disturbing Signals

A. Calcium Intake

The intake of calcium represents the only net gain of calcium from the environment. The magnitude of it depends on the availability of dietary calcium and the ability of the bird to regulate it. In nature, various types of grit can be found in bird gizzards. Those include insoluble grit, but also calcium carbonate in the form of oyster shell and other molluscan shells, or limestone grit, all depending on the environment. The larger carnivorous birds may obtain their calcium from the skeletons of their prey. Under those conditions calcium phosphate would be the major calcium source.

In farm animals, dietary calcium is supplied by a calcium phosphate supplement, usually dicalcium phosphate and by calcium carbonate–limestone, calcite grit, or oyster shell. Calcium sulfate may serve to a limited degree as a calcium supplement (Hurwitz and Rand, 1965). A diet of a growing chick typically contains about 2% dicalcium phosphate and 1% calcium carbonate. The diet of laying hens contains about 1.5% dicalcium phosphate and 7–8% calcium carbonate, due to its exceptionally high calcium requirement during shell formation.

An interesting question is whether birds have the ability to regulate their calcium intake. Griminger and Lutz (1964) reported a higher intake of calcium in laying than in nonlaying birds. The intake of this mineral was also inversely related to the dietary calcium level. Mongin and Sauveur (1974) showed that limestone grit intake followed rhythmical changes associated with the laying cycle. Peak limestone intake occurred just prior to eggshell formation. This self-regulation of calcium intake is fairly crude as a considerable variation in oyster shell intake among individual birds exists, with some birds consuming well above than required, while others consuming much less than required. The intake also varied with the type of supplement offered and was even dependent on environmental conditions (Griminger and Lutz, 1964).

B. Shell Formation

Eggshells of birds contain about 95% mineral which is almost exclusively calcite (calcium carbonate), although traces of magnesium and phosphate are found. The organic matrix of bone is made of protein similar to collagen, but also contains mucopolysaccharides (Simkiss, 1961; Simkiss and Taylor, 1971). In cross section the shell has a clearly defined anatomical structure with the mammillary core as the innermost layer giving rise to radially transverse columns of calcite (Terepka, 1963). The shell contains pores which permit a gas exchange with the environment which is important for embryological development.

From an agricultural viewpoint, the ability of the egg to withstand handling and transportation is of major concern, and depends on eggshell quality. A decline in shell quality has been also implicated in the recent years' reduction in the population of various wild bird species. The strength of the eggshell is derived from its thickness, density, and lack of faults. Many methods have been devised to measure shell quality such as resistance to pressure, deformation under pressure, shell thickness, egg density, weight per surface area, calcium content, and even measurements of β-ray backscatter. Some of these methods were compared by Tyler and Moore (1965) and by Wells (1967).

Eggshell quality is influenced by nutritional, genetic, and environmental factors. Among the dietary factors, calcium is obviously the most important. A low dietary intake of calcium results in the laying domestic fowl in a reduction in shell thickness, presumably to diminish the magnitude of calcium loss (Hurwitz and Bar, 1969a). This homeostatic response represents only partial regulation, since under those conditions the amount of calcium secreted in the eggshell exceeds the dietary intake and results in a marked drain of calcium from the skeleton (Hurwitz and Bar, 1969a). At this stage, however, egg formation is abruptly arrested probably due to disturbances in the pituitary–gonad axis (Taylor, 1965). Of the environmental factors, high ambient temperature (Tyler and Geak, 1958), and pesticide residues (Bitman *et al.*, 1969; Cecil *et al.*, 1971) were found to inhibit shell formation. The mechanism of inhibition is not clear in either case. It has been suggested that hyperventilation at high environmental temperatures may lead to a respiratory alkalosis through the loss of CO_2, which disturbs CO_2 deposition as part of the carbonate moiety of the shell (Frank and Burger, 1965). With regard to pesticides, it was suggested that DDT inhibits carbonic anhydrase (Bitman *et al.*, 1970) which was implicated previously in shell formation through its function in the hydration of CO_2 (Mueller, 1962). The factors which may affect shell formation through an influence on acid–base balance also indirectly affect the calcium transfer into the shell. This changes the magnitude of the disturbing signal.

Only sparce information exists as to the mechanism of calcium transfer across the uterine wall. Ehrenspeck *et al.* (1971) concluded, on the basis of *in vitro* experimentation, that the transport of calcium across the uterine mucosa was an active process. However, due to the very poor *in vitro* transport of calcium in their preparation, the significance of these results with regard to the *in vivo* process is still open to debate. Conversely, the electrical potential of about 10–15 mV across the uterine wall (Hurwitz *et al.*, 1970), could provide sufficient driving force to move calcium into the lumen of the shell gland.

A calcium-binding protein (CaBP), probably identical to intestinal CaBP (see Section VI,B,3), occurs in the uterine mucosa of laying domestic hens (Corradino *et al.*, 1968). This CaBP is absent in the uterus of immature birds and appears during the calcification of the first eggshell. It disappears from the uterus within days after the cessation of egg production (Bar and Hurwitz, 1973). Fluorescent antibodies against CaBP were used to localize the protein in the glandular cells of the uterus (Wasserman *et al.*, 1974). CaBP was found only in the functional uterus, and no correlation between the concentration of the protein and the intensity of calcium transport was demonstrated (Bar and Hurwitz, 1973). Further, its concentration remained unchanged during the laying cycle (Bar and Hurwitz, 1975). Thus, any possible role of CaBP in calcium translocation in the uterus still remains to be elucidated.

V. Controlling Hormones

A. Parathyroid Hormone

1. *Location and Anatomy of the Gland*

Two pairs of parathyroid glands are found on either side of the lower neck, just posterior to the thyroid and anterior to the ultimobranchial gland (Taylor, 1971). As in humans, deviations from this situation such as a deeper location of any of the glands, or the absence of one or more of the glands commonly occur. Embryological parathyroid glands originate from the third and fourth pharyngeal pouches.

The size of the gland varies with species, age, reproductive state, and calcium and vitamin D nutriture. In the young domestic chick the total weight of the glands varied from about 1–3 mg/bird in calcium-replete-vitamin D-fed chicks, to 11–14 mg in vitamin D-deficient chicks (Bar *et al.*, 1972). The glands are considerably larger in the laying domestic hen, 10–20 mg under normal mineral intake. Weights of 40–50 mg/bird were obtained in hens fed a low calcium diet for one week (Hurwitz and Griminger, 1961a). Prolonged feeding of a high calcium diet to immature pullets resulted in a pronounced degeneration of the parathyroid glands (Shane *et al.*, 1969).

The structure and ultrastructure of the parathyroid gland in various bird species (domestic chicken, pigeon, and dove) were similar to that of several mammalian glands (Stoeckel and Porte, 1973). Secretory cells from glands stimulated by *in vivo* feeding of a low calcium diet, contain well-developed organelles and many secretory granules with a pronounced margination. A prolonged inhibition by a high calcium diet resulted in the appearance of giant mitochondria, cystic cavities, and juxtaendothelial deposits of elastin-like substance.

2. *Responses to Parathyroidectomy and Exogenous Hormone*

As in mammals, parathyroidectomy resulted in decreased serum calcium and increased serum phosphate (as reviewed by Taylor, 1971). In young domestic chickens, especially the heavier strains, plasma calcium usually fell below 4 mg/100 ml. Death followed several hours after surgery (Cherian and Cipera, 1968).

Birds respond to exogenous bovine parathyroid hormone (BPTH) by an increase in plasma calcium and decrease in plasma inorganic phosphate (Mueller *et al.*, 1973). The response time is extremely rapid relative to mammals, peaking at 7–8 minutes after the intravenous hormone injection (Candlish and Taylor, 1970).

3. *Avian Parathyroid Hormone*

Avian PTH has not as yet been isolated and characterized. Consequently, no radioimmunoassay for this avian hormone is available at the present, despite the extensive use of the chick as an experimental animal to study calcium and vitamin D metabolism. The presence of a PTH-like substance in avian parathyroids was first demonstrated by Gaillard (1959), who found that avian parathyroid glands induced bone resorption *in vitro*. More recently Feinblatt *et al.* (1974) confirmed the secretion of PTH-like substance by bioassays of the media in which avian parathyroids were incubated.

Since avian parathyroid extract was found to be active in mammals and vice versa, it is not unreasonable to assume a similarity of the avian PTH to the mammalian polypeptide. This, however, awaits future evaluation.

4. *Control of Secretion*

For mammals (Sherwood *et al.*, 1966) an inverse relationship exists between the calcium concentration in the blood perfusing the parathyroid gland and PTH secretion. Suppression of PTH secretion by calcium has also been observed with chick parathyroid glands in organ culture (Feinblatt *et al.*, 1974). Thus, a clear negative feedback exists between the concentration of calcium in the extracellular fluids and PTH secretion. More recently several hormones were found to affect PTH secretion. Calcitonin and isoproterenol were found to raise cAMP concentration in bovine glands (Abe and Sherwood, 1972), and stimulate hormone secretion (Fischer *et al.*, 1971; Williams *et al.*, 1973). It will not be surprising if similar relationships will also be found in avian parathyroid glands.

The relationship between vitamin D and the parathyroid gland is of special interest. Henry and Norman (1975) demonstrated an accumulation of vitamin D metabolite 1,25-dihydroxycholecalciferol (1.25 DHCC) in chick parathyroids up to levels found in the intestinal mucosa (see Section IV,C). Furthermore, 1,25-DHCC inhibited parathyroid secretion in rats *in vivo* and in bovine glands *in vitro* (Chertow *et al.*, 1975). 1,25-DHCC and 1α-HCC but not 24,25-DHCC inhibited PTH secretion in rat parathyroid glands *in vitro* (Au and Bukowski, 1976). The mechanism by which 1,25-DHCC suppresses PTH secretion is not clear. It appears, however, that one of the first biochemical events in the gland cells is the uptake of 1,25-DHCC by a cytoplasmic protein receptor of a molecular weight of 58,000 (Brumbaugh *et al.*, 1975).

In vitamin D-deficient chicks vitamin D administration resulted in both increase in plasma calcium and retardation of parathyroid size. The reduction in parathyroid size under these conditions was believed to be due to the elevation of plasma calcium. Recently, however, it has been shown that administration of 1,25-DHCC to D-deficient chicks failed to induce a reduction of parathyroid size, despite the clear effect on plasma calcium. Such reduction in parathyroid size could be obtained either by administration of vitamin D_3 itself or by a combination of 1,25-DHCC with another vitamin D metabolite, 24,25-DHCC (Henry *et al.*, 1976). Fraser and Kodicek (1973) suggested an involvement of PTH in the control of 1,25-DHCC production by the kidney, as will be discussed in Section V,C. A feedback loop thus seems to exist between parathyroid hormone secretion and 1,25-DHCC in circulation and may be of importance in calcium homeostasis.

5. *Physiological Action*

PTH administration in ablated or normal animals produces an increase in serum calcium and a fall in serum inorganic phosphate. As postulated by Bronner (1975), this hormone is the main contributor to the integral control of plasma calcium.

The increase in plasma calcium stimulated by PTH is mainly the result of action of this hormone on bone. Classically, parathyroids were believed to induce osteoclastic bone resorption. The mechanism, however, may be too slow to account for the rapid calcium mobilization due to hormone alone, although during prolonged hyperparathyroidism an increase in the number of osteoclasts was recorded together with bone matrix breakdown. The result was an increased urinary output of hydroxyproline (Candlish, 1969). At earlier stages parathyroid hormone induces osteocytic resorption (Jande and Bélanger, 1973) which represents mostly a demineralization phenomenon without bone matrix breakdown.

The PTH-induced hypophosphatemia appears to be due to the stimulation of urinary excretion of phosphate (Ferguson and Wolbach, 1967). The action of PTH is probably localized at the proximal renal tubule where it causes a reduced phosphate reabsorption directly or as a consequence of a reduced sodium reabsorption (Aurbach and Phang, 1974). PTH has also been implicated in the induction of the hydroxylation of 25-hydroxycholecalciferol to 1,25-DHCC in the kidney (Fraser and Kodicek, 1973). However, the mode of action of the hormone and its possible interactions with calcium and phosphate ions remain unknown at the present. The suggested increase in 1,25-DHCC production by PTH may be responsible for an increase in intestinal calcium absorption attributed to PTH (Rasmussen, 1959).

The early manifestation of PTH hormone action either in bone or kidney are increases in cellular cAMP by activation of the adenylate cyclase system (Chase and Aurbach, 1967, 1970; Chase *et al.*, 1969; Shanfeld *et al.*, 1975; Dousa, 1974). Since calcitonin elicits a similar response in both tissues, it is of interest to note that Wong and Cohn (1975) were able to separate populations of cells from mouse calvaria which respond differentially to both hormones. The increase in cAMP in the renal cells results in an increased release of cAMP in the urine. This phenomenon is used to diagnose hyperparathyroidism in man.

The sequence of events precipitated by PTH binding to cell membrane and activation of the adenylate cyclase system with increased cellular cAMP, which leads eventually to phosphate excretion in the kidney, and resorption in bone, is still not understood. The induction of a protein kinase has been described for other systems, and probably operates here. In bone cells, PTH appears to cause an increase in the permeability of the cells to calcium (Dziak and Stern, 1975).

Earlier the interactions of PTH and vitamin D were described at the level of their secretion. Such interactions are also found at the level of physiological actions. Harrison *et al.* (1958) found vitamin D-deficient rats to be refractory to exogenous PTH. In chicks, Bar *et al.* (1972) found a pronounced hypocalcemia in vitamin D-deficient chicks in face of extremely hypertrophied parathyroid glands. More recently, Gonnerman *et al.* (1975) confirmed the refractiveness of D-deficient birds to PTH, with regard to elevation of plasma calcium. A marginal response to very high doses of PTH could be elicited, however, especially in birds fed a high calcium diet devoid of vitamin D. Therefore, some action of PTH on plasma calcium is still possible even in D-deficient chicks. This finding coincides with the reduction of plasma calcium in such chicks following parathyroidectomy (A. Bar and S. Hurwitz, unpublished results). It appears that the hypophosphatemic response to PTH is less affected in vitamin D deficiency and can be restored to normal level by supplementation of the vitamin D-deficient diet with calcium (Gonnerman *et al.*, 1975).

B. Calcitonin

1. *The Ultimobranchial Glands*

Experiments carried out by Copp *et al.* (1962) indicated that parathyroid hormone was not the only hormone concerned with calcium metabolism. Perfusion of the thyroid–parathyroid complex with hypercalcemic blood suggested a release of a substance, calcitonin (CT), which lowered plasma calcium earlier than could be expected from a suppression of parathyroid hormone release. Hirsch *et al.* (1963) demonstrated the thyroid origin of this substance in mammals. Foster *et al.* (1964) identified the "C" cells of the thyroid as the source of CT. Pearse and Carvarheira (1967) demonstrated the ultimobranchial origin of those cells in mammals, which led to the demonstration of calcitonin in the ultimobranchial gland of birds (Tauber, 1967). For a more comprehensive review of earlier work, see Bélanger (1971).

In chickens, the ultimobranchial glands are situated in the lower neck region usually in contact with the caudal parathyroid gland. The glands arise from the last branchial pouch of the embryological neural crest. The organ contains follicles which are lined with microvilli but devoid of granulation (Isler, 1973). Other cells which do not border the lumen are heavily granulated. These have been identified as the source of calcitonin.

2. *Chemistry of Calcitonin*

Avian calcitonin has recently been isolated and characterized (Nieto *et al.*, 1973). It is a polypeptide with 32 amino acid residues with half-cystine at the N-terminal and proline at the C-terminal. This peptide is similar to salmon calcitonin except for differences in four amino acid residues. When incubated with rat plasma *in vivo* or *in vitro*, both avian and salmon calcitonin were inactivated at a much lower rate than porcine calcitonin (Copp *et al.*, 1972). Avian CT also reacts with antibodies produced against salmon calcitonin, permitting the use of the latter in radioimmunoassay of avian calcitonin (Cutler *et al.*, 1974).

3. *Control of Synthesis and Secretion*

Copp *et al.* (1972) observed in the turkey an increase in plasma calcitonin in response to perfusion of the parathyroid–ultimobranchial complex with hypercalcemic blood. The gland hypertrophies in chicks fed a high calcium diet and atrophies when fed a low calcium diet (Chan *et al.*, 1969; Cipera *et al.*, 1970). Thus, a feedback relationship between plasma calcium and calcitonin production–secretion exists in birds. Evidence in mammals suggests that various hormones associated with the gastrointestinal tract such as pentagastrin, glucagon, and pancreozymin can stimulate the secretion of calcitonin (Care *et al.*, 1971). It has been further suggested that cAMP in the ultimobranchial cells is responsible for control of secretion (Ziegler *et al.*, 1970).

4. *Mode of Action*

In mammals, hypocalcemia produced by calcitonin was accompanied by hypophosphatemia (Hirsch *et al.*, 1964) which suggests that calcitonin inhibited the mobilization of bone mineral. In effect, bone resorption was found to be inhibited by calcitonin *in vivo* (Bélanger and Copp, 1972), and in bone cultures *in vitro* (Copp *et al.*, 1972). The rapidity of the calcitonin response suggested that the hormone does not act by inhibiting osteoclast action. Indeed, in chicks chronically injected with calcitonin a reduced osteocyte osteolysis occurred (Copp *et al.*, 1972).

Calcitonin stimulates cyclic 3′,5′-AMP in bone cells (Chase and Aurbach, 1970) which in turn was found to inhibit bone resorption in bone cultures. CT stimulated adenylate cyclase prepared from renal cortex *in vitro* (Heersche *et al.*, 1974). However, the physiological action of calcitonin in this organ has not been clearly elucidated.

5. *Importance of Calcitonin in Birds*

The role of any substance as a hormone is generally established by the following criteria: (a) ablation—metabolism of the controlled substance in the absence of a source of the hormone, and reversed by replacement therapy; (b) the establishment of a feedback mechanism between the controlled signal and the hormone; and (c) responses of target tissue(s) to the hormone.

Brown *et al.* (1970) found no significant differences in plasma calcium and bone calcification between controls and chicks ultimobranchiectomized at the age of 1 day. Neither of the constituent signals of bone metabolism, accretion or resorption, were affected by ablation, although some differences between the two groups of chicks were observed with regard to the response to parathyroid hormone. These results were similar to those of Sammon *et al.* (1969) in rats. These data suggest that the presence of a source of calcitonin is not essential for the maintenance of a normal calcium homeostasis under laboratory conditions.

Exogenous calcitonin, prepared from ultimobranchial glands failed to cause hypocalcemia in intact birds (Kraintz and Intscher, 1969; Gonnerman *et al.*, 1972). This lack of response of normal birds to exogenous calcitonin may be explained on the basis of the very rapid response of birds to parathyroid hormone (Candlish and Taylor, 1970), since a hypocalcemic response could be elicited in partially parathyroidectomized birds (Kraintz and Intscher, 1969).

Some evidence exists as to the feedback relationship between plasma calcium and calcitonin in birds. Chan *et al.* (1969) and Cipera *et al.* (1970) reported hypertrophy of the ultimobranchial glands in response to a high calcium diet. Under the same conditions, Mueller *et al.* (1970) found an increase in CT in the gland. Copp *et al.* (1972) measured a significant increase in plasma calcitonin in response to infusion with calcium in the turkey, and Ziegler *et al.* (1970) found a positive relationship between CT secretion rate as a function of the calcium concentration in the fluid perfusing the gland *in situ*. Nieto *et al.* (1975) found a linear response between medium calcium concentration and CT secretion by the ultimobranchial gland *in vitro*.

Thus, (a) calcitonin is produced and released from the ultimobranchial glands of birds, (b) under conditions of parathyroid deficiency, plasma calcium level in birds responds to exogenous calcitonin, and (c) target organs such as bone respond to calcitonin. Although the ultimobranchial glands are not essential for calcium homeostasis in birds, the improved homeostasis of calcium under an acute calcium load (Copp *et al.*, 1972) may point to some role of calcitonin in wild birds. One condition may be where the supply of calcium is not as constant as under laboratory conditions. Furthermore, various control mechanisms protecting the laying bird from wide fluctuations in plasma calcium during the laying cyclic may be affected by calcitonin. With the possibility of measuring calcitonin level in birds by radioimmunoassays (Cutler *et al.*, 1974) any possible role of calcitonin can be better evaluated in the future.

C. Vitamin D

The antirachitic activity of vitamin D in chicks was documented over 50 years ago, and many theories advanced as to its mode of action. It has also been established that cholecalciferol (vitamin D_3) can be formed in the skin from a natural precursor, 7-dehydrocholesterol, under exposure to ultraviolet irradiation. The discovery of vitamin D and its availability as a feed supplement could be regarded as the most important factor in the intensification of poultry husbandry which could be achieved by switching from outdoor rearing to total confinement of the birds.

A lag period of 4–8 hours exists between vitamin D administration to rachitic animals and the response of either plasma calcium or calcium absorption. This lag period suggested either that the vitamin had to undergo metabolic modifications in order to become active, or that the vitamin initiated a chain of biochemical events leading to the final expression of its activity. Kodicek (1963) attempted to study the metabolism of vitamin D by administration of radioactively-labeled vitamin D. Unfortunately, the labeled preparation of vitamin D used was of a specific activity too low for use at physiological doses. DeLuca and associates (1968) were able to synthesize cholecalciferol with a high specific activity. When this was administered to rats the first metabolite was characterized as 25-hydroxycholecalciferol (25-HCC) (Blunt *et al.*, 1968).

Cholecalciferol, whether of body or dietary origin, undergoes a rapid hydroxylation at position 25 in the liver. The activity of this hydroxylase appears not to be linked to calcium metabolism but is regulated by product inhibition (Omdahl and DeLuca, 1973). 25-HCC is taken up by a carrier protein (Peterson, 1971) in blood plasma and transported throughout the body. It accumulates at high concentrations in muscle, where it may be involved in synthesis of some proteins important for muscle function (Birge and Haddad, 1975). In the kidney, 25-HCC undergoes further hydroxylations. The most important compound produced is 1,25-DHCC which was identified as the active form of vitamin D (Fraser and Kodicek, 1970; Norman *et al.*, 1971; Holick *et al.*, 1971).

Kidney hydroxylation results in additional products of vitamin D such as 24,25-dihydroxycholecalciferol and 1,24,25-trihydroxycalciferol. These were much less active than 1,25-DHCC and their possible biological importance has not been clearly elucidated. The activity of the enzyme responsible for 1,25-DHCC production, 25-hydroxycholecalciferol-1-hydroxylase appears to be linked to calcium metabolism. The enzyme level is exceptionally high in rachitic birds and is fairly low in normal animals fed calcium- and phosphate-sufficient diets. However, the activity of the enzyme increased considerably under conditions of calcium but not phosphate deficiency (Henry *et al.*, 1974) when more 1,25-DHCC was found in the intestinal mucosa and kidney (Edelstein *et al.*, 1975).

The factor associated with calcium metabolism which is responsible for the regulation of the kidney 1-hydroxylase has not been identified. Parathyroidectomy caused a reduction of the activity of the enzyme implicating PTH in its control (Fraser and Kodicek, 1973). However, parathyroidectomy did not abolish 1-hydroxylase completely. Also, animals adjusted to a low calcium diet by synthesizing more 1,25-

DHCC even in the absence of the parathyroids (Galante *et al.*, 1974). Therefore, parathyroid hormone may be only one of the factors controlling 1,25-DHCC production. Other factors such as calcium concentration are involved in a rather complex mode of regulation of this hydroxylation process (Henry *et al.*, 1974).

The synthesis of 24,25-DHCC by the kidney hydroxylase system appears to be related reciprocally to that of 1,25-DHCC (Omdahl and DeLuca, 1973). 24,25-DHCC has a significant vitamin D activity, although much lower than that of 1,25-DHCC (Holick *et al.*, 1976). Due to its reciprocal relationship with 1,25-DHCC, it was hypothesized that 24,25-DHCC could be a product of catabolism (Omdahl and DeLuca, 1973). Although hardly any 24,25-DHCC can be found in the intestinal mucosa, a relatively large concentration was found in other target organs such as bone (Edelstein *et al.*, 1975), and suggests possible importance of this compound in bone metabolism.

Several vitamin D compounds have been synthesized recently. One of practical interest is 1α-hydroxycholecalciferol. This compound was as potent as 1,25-DHCC in promoting calcium absorption. Since it is already hydroxylated at position 1, its use in animals bypasses the step of kidney hydroxylation. This is similar to the conditions for 1,25-DHCC. Therefore, 1α-hydroxycholecalciferol may prove useful in treating calcium disturbances occurring with kidney diseases in which the hydroxylase system is defective. Although this situation is relatively uncommon in farm birds, a defective metabolism of vitamin D has been suggested as the cause of a widespread turkey rickets (Hurwitz *et al.*, 1973b). Binding studies suggested a much lower capacity of 1α-HCC as compared to 1,25-DHCC to bind with chromatin receptors in the intestine (Proscal *et al.*, 1975). Two studies in the rat, one conducted *in vivo* (Holick *et al.*, 1975) and one with a perfused liver system (Fukushima *et al.*, 1975) showed that 1α-HCC is hydroxylated in the liver to 1,25-DHCC, and hence its relatively high biopotency.

Early investigators of vitamin D found that the ultraviolet irradiation of ergosterol, a plant sterol, produced a compound with a high vitamin D activity which was termed vitamin D_2. Chemically this compound was similar to cholecalciferol except for the presence of an additional methyl group attached to the carbon-24 in the side chain region. The early studies also revealed that in birds and some mammalian species, vitamin D_2 had a relatively poor potency to cure rickets as compared to vitamin D_3, although both forms had a similar potency in curing rickets in most mammals.

The reason for the relative refractiveness of birds to vitamin D_2 still remains unclear. Hurwitz *et al.* (1967a) tested for the accumulation of vitamin D activity in various tissues from rachitic chicks given a single dose of vitamin D_2 or D_3 by a rat bioassay. No difference in vitamin D activity in liver, intestinal mucosa, and blood could be detected between the groups receiving either vitamin D compound. Jones *et al.* (1976) measured the ability of the chick liver and kidney to hydroxylate vitamin D_2 at position 25, or to hydroxylate 25-hydroxyergocalciferol at position 1, respectively. Their results obtained *in vitro* indicated no difference in hydroxylation of either vitamin D_2 or D_3 in the avian system. The results of those studies indicate no difference between vitamin D_3 and D_2 in their utilization from dietary origin, their metabolism, and accumulation in target tissues. MacIntyre *et al.* (1975) reported a lower affinity for vitamin D_2 than D_3 to kidney cytosol fractions. Alternately it is possible that the expression of the vitamin D activity in the cells of the target tissues is different for the two forms of the vitamin.

VI. Controlling Systems

A. Bone

1. *Function of Bone in Calcium Homeostasis*

The skeleton, which contains about 99% of the total body calcium, serves as the mechanical support of the body. In addition to this function, the skeleton is one of the major systems controlling calcium homeostasis. In the normal rat it was estimated that about 90% of the signals disturbing plasma calcium are handled by the skeleton (Bronner, 1975). This remarkable regulatory capacity depends on a constant turnover of skeletal calcium through processes of accretion and resorption. It is well recognized that neither of the processes is uniform. Bone resorption includes both osteocytic and osteoclastic resorption (demonstrated in avian medullary bone by Taylor and Bélanger, 1969). However, both accretion and resorption have been defined as single entities in studies using isotope kinetics which attempted to evaluate each separately. Using a combination of ^{45}Ca kinetics and chemical calcium balance, Bronner (1975) found that bone formation rate (V_{o+}) hardly changed in rats fed a wide range of calcium intakes. Conversely, bone resorption (V_{0-}) decreased linearly with increasing amounts of calcium absorbed from the intestine. Thus, of the two constituent signals of bone turnover, bone resorption appeared to be the main signal associated with calcium homeostasis. This is in accord with the mode of action of the two principal calcium-regulating hormones, calcitonin and parathyroid hormone, which are known to influence bone resorption.

Although PTH and possibly calcitonin regulate bone resorption, both accretion and resorption of bone continue in thyroparathyroidectomized animals, but at a reduced intensity, allowing for regulation of plasma calcium (Bronner, 1975). Under those conditions, however, errors in plasma calcium concentration which are proportional to the disturbing signal (calcium absorption) were observed. Thus, the skeleton functions in calcium homeostasis both dependently and independently of the "calcium hormones."

2. *Bone in Avian Species*

The skeleton of birds is functionally adapted to flight. Other than general features to conform with the aerodynamic requirement, the bird's skeleton is characterized by a highly developed sternum which provides support to the major mass of muscles. Also the presence of pneumatic bones is typical in birds. Few long bones are hollow by virtue of the extension of the lateral air sacs into them.

The avian skeleton contains the usual cellular elements such as bone forming cells—osteoblasts, bone cells—osteocytes; bone resorbing cells—osteoclasts, and various cartilage elements. The morphology of the various cell types, as well as their anatomical and physiological functions are not different than in mammals, and will not be further discussed.

In female birds just prior to the onset of egg laying, a unique bony structure, medullary bone, appears in the lumen of the long bones. In the domestic hen *(Gallus domesticus)* medullary bone appears about 2 weeks prior to egg laying (Hurwitz, 1965a), concomitant with the development of the ovary and its secretory activity. Medullary bone formation can be induced also in male birds by estrogen administration. The maintenance of medullary bone is dependent upon a continued presence of estrogens and androgens, since it rapidly transgresses with the decline in ovarian activity (Simkiss, 1961).

Elements of medullary bone are mixed with other elements of bone marrow. It has no interconnecting structural framework, but is made of separate aggregates which appear as fine powder upon separation and drying. Relative to cortical bone, it has a large population of both osteocytes and osteoclasts (Simkiss, 1961; Taylor *et al.*, 1971). The mineral content of medullary bone is mostly amorphous calcium phosphate rather than hydroxyapatite (Schraer *et al.*, 1967; Tannenbaum *et al.*, 1974; Miller and Schraer, 1975). This type of mineral is characteristic of young bone elements, has a very poor crystallinity, and is therefore less stabile than hydroxyapatite. Moreover, the collagen fibrils appear to be more loosely packed than in cortical bone (Stringer and Taylor, 1961). The large cell population, the vast blood supply, and the relative instability of the mineral undoubtedly contribute to the high lability of medullary bone. In effect, ^{45}Ca studies suggested a half-life of 2–4 days of medullary bone calcium in comparison to several months in structural bone (Hurwitz, 1965b).

The appearance of medullary bone, just prior to egg laying, occurs simultaneously with an increase in the retention of calcium and phosphate (Common, 1932). Therefore, medullary bone was considered to be the storage site for calcium to supply any needs during shell formation. The progressive disappearance of medullary bone in the pigeon during the laying clutch (Riddle *et al.*, 1944) was also cited as evidence for the role of medullary bone as a storage site for calcium.

The concept of medullary bone as a storage element is an oversimplification. First, Benoit and Clavert (1945) found that medullary bone was formed under the influence of estrogens in birds fed on a calcium-free diet, by necessity at the expense of structural bone. Direct utilization of structural bone calcium for medullary bone formation was also observed by Hurwitz (1965a), using early ^{45}Ca-labeling of the skeleton. Second, when laying birds are faced with a challenge of a low-calcium-diet medullary bone persists although it is reduced in quantity despite the extensive decalcification of structural bone (Hurwitz and Bar, 1966a). Third, the storage of calcium in medullary bone during the prelaying increase of calcium retention accounts for only a small fraction of the increase in bone calcium which occurs during this period (Hurwitz and Bar, 1971). Finally, the amount of medullary bone does not vary during the laying cycle (Bloom *et al.*, 1958; Hurwitz, 1964; Candlish, 1971), despite changes in cellular activity.

The above evidence clearly indicates that medullary bone does not act as a simple calcium reservoir. It is well accepted that the higher the turnover rate of any controlling system, the greater its ability to correct errors in the controlled system. It is conceivable that the presence of such a highly labile bone fraction, with a high parathyroid hormone sensitivity, may provide for a more accurate calcium homeostasis in response to acute changes in the supply of calcium during the increased demands imposed by eggshell formation.

As mentioned previously, the skeleton of birds serves as an important reservoir for calcium which can be utilized for the purpose of shell formation. In the young immature pullet, there is little net bone formation for about 2 months prior to sexual maturity. With the development of the ovary and subsequent hormone secretion, retention of calcium and phosphate occurs during the 2-week period prior to egg laying. Most of the stored mineral appears in the spongy bones such as diaphysis of the long bones in which the mineral content almost doubles. Some increase also occurs in the cortical fraction of the shafts (Hurwitz and Bar, 1971). It can be estimated that the calcium content of the chicken's body increases by about 30–50% during this prelaying storage period. Since adaptation of both calcium intake and absorption during the first few weeks of egg production is incomplete the hens are in negative calcium balance (Morgan and Mitchell, 1938), and the reserves of calcium accumulated during the prelaying storage period are utilized (Hurwitz, 1964).

When laying hens are fed a diet devoid of any inorganic calcium, egg production continues for 1–2 weeks. Although the shell progressively declines in thickness, calcified shells will be produced. Four to nine eggs are laid under such a regime depending on the calcium stores of the body, until egg laying completely ceases (Hurwitz and Bar, 1969a). During this period the chicken can lose up to 40% of her total body calcium. Essentially all bones are depleted, with the ribs and sternum suffering the greatest losses (Taylor and Moore, 1954). In the long bones the diaphysis suffers the greatest loss, although considerable losses are also evident in the cortex of the shafts and in medullary bone (Hurwitz and Bar, 1966a).

B. The Intestine

1. *Methodology*

Of the various systems involved in calcium homeostasis in birds, the intestine has received unusual attention due to its well-established relationship to vitamin D activity, and to the possibilities of a direct experimental approach to the active cellular layer, the mucosal epithelium. *In vitro* and *in situ* studies have been used widely to study the mechanisms of calcium transport at the cellular or organ level. Actually, fewer studies have been concerned with *in vivo* calcium absorption by the intact bird due to the difficulties in separating the excreta into the fecal and urine components. Thus, early information on calcium absorption *in vivo* was derived from balance studies in which the excreta was not partitioned. Retention was taken to represent intestinal absorption. Although nonabsorbed calcium represented the bulk of the calcium which appeared in the excreta of birds, the individual importance of the intestine and kidney to the homeostasis of calcium in birds could not be assessed. Fussel (1960) and Hurwitz and Griminger (1961b), attempted to partition calcium excretion in birds by colostomy. Both studies showed that urinary calcium excretion comprised about 10% of the total calcium in excreta. However, due to the technical and surgical difficulties of using colostomized birds for detailed studies, this system appears to have been abandoned.

Many of the difficulties of the balance methods were overcome by the use of nonabsorbed reference substances. Nonabsorbed substances as chromic oxide had been in extensive use in evaluation of metabolizable energy for poultry. ^{91}-Yttrium, a lantanoid with β-emission, was applied by Hurwitz and Bar (1965, 1966b) to measure calcium absorption in birds. The method can determine the contribution of each intestinal segment to the overall absorption and can estimate changes of absorption that occur within a few hours. Since the method determines the net or apparent absorption, an attempt was made to measure *in vivo* unidirectional fluxes of calcium through constant feeding of ^{91}-yttrium and intravenous injection of ^{45}Ca (Hurwitz and Bar, 1972). The method appeared to be only partially successful because of isotope mixing but was an improvement over the classical determination of endogenous calcium excretion (Aubert *et al.*, 1963).

2. Mechanism of Calcium Absorption

Schachter and associates (1960) used inverted rat duodenal sacs *in vitro* to demonstrate a vitamin D-dependent transport of calcium against a concentration gradient. Furthermore, if the transmural electrical potential (with serosal side positive) were considered, these findings suggested an active transport of calcium according to Rosenberg's (1954) criterion. Simultaneously, Harrison and Harrison (1960) were able to show a passive concentration-dependent component of calcium absorption in rat intestine which was also sensitive to vitamin D. Attempts to use the gut-sac technique in chicken duodenum failed due to an impermeability of this tissue *in vitro,* despite its ability to absorb calcium *in situ* (Hurwitz *et al.*, 1967b).

By measuring calcium fluxes across ligated chick duodenum *in situ*, Wasserman (1963) showed two components of calcium transfer. One exhibited saturation kinetics, typical of a carrier-mediated transport, and the other was nonsaturable, apparently following diffusion rules. It is of interest that both components were vitamin D-sensitive, e.g., reduced in vitamin D deficiency. Hurwitz and Bar (1972) measured the *in vivo* fluxes and used kinetic analysis to suggest the presence of a vitamin D-dependent active transport of calcium transfer in the duodenum, but in the jejunum vitamin D appeared mainly to increase the permeability to calcium. Another, yet unexplained observation suggested the presence of a vitamin D-sensitive active component of calcium absorption in the ileum *in vivo*. In this segment, Hurwitz *et al.* (1967b) also demonstrated an *in vitro* transport of calcium against a concentration gradient.

These findings suggest the possibility of both active and passive modes of transport of calcium. An important question concerns the contribution of each type of transport *in vivo*. Specifically, the question may be raised as to a possible need for active transport of calcium in the intact animal similar to that observed for sodium. Some insight into this question was gained by the measurement of the driving forces for calcium across the intestinal epithelium. The activity of calcium was measured in the intestinal contents and blood plasma with a calcium-sensitive electrode, and the electrical potential was measured through KCl-agar bridges inserted on both sides of the intestine *in vivo* to study the electrochemical potential of Ca (Hurwitz and Bar, 1968, 1969b). The results showed a positive relationship between intestinal calcium activity and calcium intake. A positive relationship existed also between the electrochemical potential difference of calcium and calcium absorption. With a sufficient calcium intake, the electrochemical potential difference of calcium was always positive on the mucosal side, suggesting no need for active transport under these conditions. On the other hand, with a low intake of calcium, the electrochemical potential difference of calcium was negative in chick intestine. Since such levels could still support a net absorption of calcium, the results suggested active calcium absorption under conditions of dietary calcium restriction.

3. *Biochemical Aspects of Calcium Absorption*

Wasserman and Taylor (1966) demonstrated the presence of a protein fraction in supernatants of chick duodenum with a high affinity for calcium. This protein fraction termed calcium-binding protein (CaBP) was found to be vitamin D-dependent. It was isolated and found to have a molecular weight of about 28,000. The protein did not contain any phosphate but had a high content of acidic amino acids. It contained no lipid, carbohydrate, or hexosamine. Thus, the site active in calcium binding had to be associated with the amino acid residues (Wasserman and Corradino, 1973). The same, or at least a similar protein fraction was observed in the intestine of other avian species such as the turkey (Hurwitz *et al.*, 1973b), the Japanese Quail (*Coturnix Coturnix japonica;* Hurwitz *et al.*, 1976) and goose (A. Bar and S. Hurwitz, unpublished). CaBP was also found in the uterus (shell gland) (Corradino *et al.*, 1968), the kidney (Taylor and Wasserman,

1972), and brain of the chicken (Taylor, 1974). The similarity of the protein fraction in these species and organs was established by acrylamide gel electrophoresis, with or without sodium duodecylsulfate, and by cross reaction with antisera produced against the intestinal protein.

The concentration of this protein in the chick duodenum was related linearly to calcium absorption capacity (Taylor and Wasserman, 1969). This relationship was taken to suggest that CaBP played an important role in intestinal calcium translocation (Wasserman and Corradino, 1973). Taylor and Wasserman (1970) localized CaBP in the brush border of the intestinal mucosa with fluorescent antibodies.

It may be speculated that CaBP could serve as the calcium carrier in the active transport process, in analogy to the bacterial permeases, or to a factor which increases the permeability of the cell membrane to calcium. Although Corradino (1975) was able to promote calcium absorption in embryonic chick intestine in tissue culture by addition of CaBP to the medium, direct supporting evidence of a role of the protein in calcium absorption is still in need.

Ca–Mg ATPase in the brush border of the chick intestine was found to be sensitive to vitamin D (Haussler *et al.*, 1970). The time course of the appearance of this enzyme in vitamin D-deficient animals following a vitamin D dose correlated well with that of calcium absorption. Those findings were the basis for implicating this enzyme in calcium transport (Omdahl and DeLuca, 1973). However, evidence supporting this contention is rather meager.

The studies on vitamin D metabolism (Section V,C) and the discovery of CaBP can provide for a construction of a sequence of cellular events leading to the regulation of calcium absorption.

A signal from the calcium homeostatic mechanisms, PTH, or a decrease in plasma calcium, brings about an increase in the activity of 25-HCC-1-hydroxylase in the kidney resulting in greater production of 1,25-DHCC. This hormone is transported to the intestinal mucosa where it comprises over 80% of the total vitamin D metabolites (Edelstein *et al.*, 1975). Upon entry into the mucosal cell 1,25-DHCC is taken up by a cytoplasmic receptor protein and transported to the nucleus where it associates with a finite number of chromatin receptor sites (Proscal *et al.*, 1975). These receptor sites have a relatively high affinity for 1,25-DHCC, some affinity for 25-HCC and 1α-HCC, but little or no affinity for vitamin D_3 itself. Thus, the hydroxyls in both positions 1 and 25 are main determinants in the binding affinity.

The attachment of 1,25-DHCC to chromatin confirms the older thesis obtained by use of actinomycin D that the action of vitamin D involves protein synthesis (Zull *et al.*, 1956). Indeed, the interaction of 1,25-DHCC with the chromatin receptors results in an increased production of a specific mRNA (Lawson and Emtage, 1974) and the interaction of this mRNA with ribosomes results in the synthesis of calcium-binding protein (CaBP). This protein, in turn, concentrates at the brush border of the intestinal mucosa (Taylor and Wasserman, 1970) where it presumably acts to increase the capacity of the membrane to transport calcium.

4. Calcium and Phosphate Intake and Calcium Absorption

Rats (Nicolaysen *et al.*, 1953), dogs, and humans (Hegsted *et al.*, 1952; Gershoff *et al.*, 1958) increased the efficiency of calcium absorption when subjected to low calcium diets. The same phenomenon was observed in laying hens (Hurwitz and Bar, 1966a). Morrissey and Wasserman (1971) showed that prior treatment with low calcium diets increased the efficiency of ^{45}Ca absorption in duodenal loops of chicks *in situ* together with the stimulation of duodenal CaBP. Treatment with a low phosphate diet also resulted in an increase in calcium absorption and CaBP.

The increase in calcium absorption in chicks fed low phosphate diets is definitely an antihomeostatic response since calcium cannot be deposited in bone due to the deficiency of phosphate. This produces a marked hypercalcemia (Bar *et al.*, 1972, 1975) and a marked increase in urinary calcium concentration (S. Hurwitz and A. Bar, unpublished data). The response of calcium absorption to low calcium treatment was confined to the duodenum, whereas the jejunum responded in the same way to a low phosphate treatment (Hurwitz *et al.*, 1973a; Bar *et al.*, 1972). Furthermore, ^{32}P-phosphate absorption increased only slightly under the same conditions. The absorption of ^{45}Ca and CaBP in the duodenum but not the jejunum were related linearly to tibia ash.

These regulatory responses appear now to be associated with the metabolism of vitamin D. A very large increase in the concentration of 1,25-DHCC in the mucosal cells of chicks followed feeding either the low calcium or the low phosphate diets (Edelstein *et al.*, 1975). However, the 25-hydroxycholecalciferol-1-hydroxylase level increased under conditions of calcium but not phosphate deficiency (Henry *et al.*, 1974). The essentiality of the hydroxylation of vitamin D in position 1 to the adaptation of calcium absorption to low calcium intakes was demonstrated by maintaining chicks from the age of 1 day on a diet supplemented with 1α-hydroxycholecalciferol (Bar *et al.*, 1975) (see Section V,C). When chicks on 1α-hydroxycholecalciferol were fed a low calcium diet, neither calcium absorption nor CaBP increased. However, phosphate deficiency resulted in some increase in CaBP and calcium absorption in chicks fed this metabolite. A similar observation was made previously by Bar and Wasserman (1973) in chicks fed dehydrotachysterol$_3$, another synthetic vitamin D compound.

Thus, it appears that the stimulation of calcium absorption by feeding a low calcium diet follows a specific sequence of events. These are increased production of 1,25-DHCC, increased accumulation of 1,25-DHCC in the intestinal mucosa, and increased CaBP synthesis. However, phosphate may influence calcium absorption capacity independently of production of 1,25-DHCC in the kidney.

5. *Egg Laying and Regulation of Calcium Absorption*

a. Onset of Egg Production. As the pullet matures calcium absorption increases in two stages: (a) at the onset of gonadal hormone secretion (about 2 weeks prior to onset of production) (Common, 1932), which results in the "prelaying storage" of calcium, and (b) at the onset of egg production, during the calcification of the first eggshell. Both stages are associated with increases in duodenal calcium absorption and CaBP (Hurwitz *et al.*, 1973a; Bar and Hurwitz, 1973), 1,25-DHCC accumulation in the duodenal mucosa (S. Hurwitz, A. Bar, and S. Edelstein, unpublished results) and the activity of the kidney 25-HCC-1-hydroxylase (Montecuccoli *et al.*, 1977). These regulatory responses of calcium absorption appear to follow a pattern similar to that observed in chicks in response to calcium restriction (Section VI,B,4).

b. Eggshell Formation. Using ^{91}yttrium as a nonabsorbed reference substance, Hurwitz and Bar (1965) showed that calcium absorption during eggshell formation was almost double that during the time when no shell was formed. An even more pronounced effect was observed in the Japanese Quail (Hurwitz *et al.*, 1976). Results of these studies and that of Hurwitz *et al.* (1973a) showed that the increase in calcium absorption occurred throughout the small intestine, but the most important site of this regulatory change was the upper jejunum.

Kinetic treatment of the results suggested that the higher calcium absorption during shell formation was the result of the greater permeability of the intestine to calcium during this time.

Increased calcium absorption occurred 2–3 hours after the start of shell calcification and even earlier in the domestic hen or quail. Furthermore, the mechanism is present already 7–8 days after the onset of egg production. Unlike the response of calcium absorption to a low calcium diet, the increase in calcium absorption due to shell formation appears to be independent of CaBP, since no significant changes in the concentration of CaBP occurred during the laying cycle (Bar and Hurwitz, 1972, 1975).

Changes in the 25-HCC-1-hydroxylase in the kidney during the laying cycle were followed by Kenny (1975) who concluded that peak activity occurred at time of ovulation and the activity was reduced when no egg was present in the oviduct. However, in their experiments with the Japanese Quail there was no reduction in the level of the enzyme 2 hours following oviposition whereas we showed that by that time calcium absorption had already decreased to about one-third of the rate during shell formation. A decrease in the activity of the enzyme did occur 20 hours after oviposition (as confirmed by Montecuccoli *et al.*, 1977) when no consecutive ovulation had occurred. Thus, it appears that modulation of vitamin D activity cannot account for the changes in calcium absorption as effected by the laying cycle. The reduction in the 1-hydroxylase enzyme level 20 hours after oviposition may reflect the "recuperation" of the system from the calcium stress imposed by shell formation since, as postulated above, the long-term adaptation to egg laying is linked to 1,25-DHCC production. More evidence that dissociates the regulation of calcium absorption from the kidney hydroxylase during the laying cycle was obtained by feeding Japanese Quail a diet with 1α-hydroxycholecalciferol as the only source of vitamin D activity, in the absence of ultraviolet irradiation (Hurwitz *et al.*, 1976). As explained in Section V,C, the use of this compound bypasses the regulation of calcium metabolism through modulation of kidney 25-HCC-1-hydroxylase in the kidney. The results showed calcium absorption was higher during shell formation in quails fed both 1α-HCC and cholecalciferol than when no shell was formed and confirmed the noninvolvement of the kidney hydroxylase system in triggering the control of calcium absorption during shell formation.

The response of calcium absorption to shell formation thus remains unexplained. The rapidity of this regulatory process is more typical of one promoted by a membrane-active hormone, which usually elicits its response through the cyclic nucleotides. Corradino (1975) demonstrated a role of cyclic AMP in promoting calcium absorption in organ cultures of embryonic chick intestine. It will not be surprising if the agent responsible for modulation of calcium absorption in the laying hen were found to belong to this class of hormones.

C. The Kidney

The role of the kidney in regulating plasma calcium is of variable importance in the various animal classes. Bronner (1975) concluded that only 1–5% of the disturbing signal in the rat is handled by the kidney. In the normal growing rat which absorbs about 50 mg calcium/day, renal excretion of calcium is only about 1 mg/day. Similarly in ruminants, the rate of urinary calcium excretion is insignificant in relation to its turnover. Conversely, the human kidney can handle over 50% of the disturbing signal, as urinary calcium excretion may be over 50% of the calcium intake. Most of the calcium filtered through the glomerulus is reabsorbed, probably in the proximal renal tubule. In humans, calcium excretion is a linear function of plasma calcium above a certain level, e.g., the filtered load of this mineral (Nordin *et al.*, 1975). This evidence indicates only a simple involvement of the kidney in calcium homeostasis.

The role of the reabsorption process in regulation of plasma calcium is still not clear. Some evidence suggests that vitamin D and PTH enhance the reabsorption of calcium resulting in an excretion of calcium lower than expected on the basis of the plasma calcium concentration (Nordin *et al.*, 1975).

Very few studies are available on the involvement of the kidney in the regulation of plasma calcium in birds. Results of Fussel (1960) and Hurwitz and Griminger (1961b) showed that in colostomized birds only a fraction of 5–10% of the absorbed calcium was eliminated in the urine. Although calcium excretion varied with calcium intake, the magnitude of these changes was too narrow to be assigned any important role in calcium homeostasis. Similar observations were made recently in the young turkey (S. Hurwitz, D. Dubrov, A. Bar, and U. Eisner, unpublished results).

The kidney may assume an important role when either bone or intestine is incapable of handling the disturbance. In the vitamin D-deficient rat, urinary calcium reached 10 mg/day as compared to a normal excretion of 1 mg/day (Hurwitz *et al.*, 1969). In chicks fed a calcium-rich phosphate-deficient diet, the efficiency of calcium absorption increased (Section VI,B,3), bone mineral was not deposited due to the lack of phosphate, and the excess calcium was excreted by the kidney. Under such conditions, urinary calcium concentration increased from 3 *mM* to 17 *mM* (Bar *et al.*, 1975).

A calcium-binding protein, similar to the intestinal one, was found in the kidney of chicks (Wasserman *et al.*, 1974). It was localized in the distal tubule by a fluorescent antibody technique and assigned a role in tubular reabsorption of calcium (Wasserman *et al.*, 1974). However, Bar *et al.* (1975) reported only a small change in kidney CaBP under conditions of calcium deficiency, during which urinary calcium excretion decreased. Conversely a major increase in CaBP occurred under conditions of phosphate deficiency, when urinary calcium output was largely elevated. Therefore, with sufficient vitamin D intake, kidney CaBP was correlated with calcium excretion, rather than with reabsorption. Thus, the control of urinary calcium excretion in avian species, and the possible role of CaBP therein, await future elucidation.

VII. Concluding Remarks

The response of the various control systems in birds to the two main disturbing signals of calcium homeostasis, changes in calcium intake and eggshell formation, have been evaluated in this discussion. As in several other multicontrol systems of vertebrates, the responses to calcium intake had two time courses. A short-term response that lasted minutes to hours was elicited by two membrane-active polypeptide hormones, parathyroid (PTH) and calcitonin (CT). The long-term response lasted hours to days and was stimulated by a steroid hormone, 1,25-dihydroxycholecalciferol (1,25-DHCC).

The secretion of the two antagonistic polypeptide hormones, presumably controlled by cellular cyclic nucleotides, was related directly to plasma calcium, but modified by other hormones. Both hormones acted to modulate bone resorption, starting with osteocytic and continuing with osteoclastic resorption, and resulting in the appropriate changes in the supply of calcium to circulation.

An increase in circulating PTH level and the reduced plasma calcium due to calcium restriction stimulated the synthesis of the steroid hormone 1,25-DHCC in the kidney, and was instrumental in initiating the long-term responses. The 1,25-DHCC produced was transported via circulation to the intestine and taken up in the mucosa first by a cytosolic and later by chromatin receptors. There, 1,25-DHCC induced the synthesis of specific mRNA which promoted the synthesis of calcium-binding protein (CaBP) and lead to an increase in the rate of calcium absorption. This change in absorption may be considered as the long-term response. With increased absorption during calcium restriction there was less need for bone breakdown and PTH. Inhibition of PTH release by 1,25-DHCC appears to be the response to this reduced need. Another long-term response to changing calcium needs was an increase in dietary calcium intake when a free choice is possible. The events leading to this response are completely unknown.

The special demands for calcium during egg laying are accompanied by changes in both major controlling systems, bone and intestine. In the 2–3 weeks preceding the onset of egg production, calcium retention increased in the immature pullet due to action of gonadal hormones and resulted in an accumulation of calcium in the skeleton. Major increases were found in the spongy bones, but some increases were also noted in the epiphyses of the long bones.

A new bone fraction, medullary bone, with an extraordinary high turnover rate, appeared in the marrow cavities of the long bones. This bone fraction probably provides for a rapid control of plasma calcium through the action of PTH in response to acute changes in the supply of calcium during shell formation.

At the level of the intestine, CaBP and calcium absorption increased during the prelaying period, followed by another increase at the onset of egg production. These changes which occur in the duodenum appear to be due to an increased uptake of 1,25-DHCC which results from an increase in the activity of the kidney 25-HCC-1-hydroxylase. The response is similar to that of feeding a low calcium diet.

Another control mechanism, manifested in increased calcium absorption along the entire intestine during shell formation, operates during the laying cycle. This short-term increase in calcium results from an increased permeability of the intestine to calcium but did not involve changes in CaBP concentration or in the production of 1,25-DHCC. The mechanism triggering this response has not been elucidated. It has been speculated that a cell membrane active agent and cAMP may be involved.

Restriction of calcium intake in the laying hen results in a reduction in the secretion of calcium into the egg shell, an increase in the calcium absorption capacity of the intestine, and a depletion of bone mineral. When the latter reached a critical level, egg production ceased possibly by an interference in the pituitary–gonadal axis. The site of this interference has not been identified.

The kidney may also be of importance in the regulation of plasma calcium. Neither the significance of this control, nor its mechanism have been elucidated.

ACKNOWLEDGMENTS

The author gratefully thanks Mrs. L. Perry for the most efficient secretarial help in preparation of this manuscript.

REFERENCES

Abe, M., and Sherwood, L. M. (1972). *Biochem. Biophys. Res. Commun.* **48,** 396–401.
Au, W. Y. W., and Bukowski, A. (1976). *Fed. Proc., Fed. Am. Soc. Exp. Biol.* **35,** 301 (abstr.).
Aubert, J.-P., Bronner, F., and Richelle, L. J. (1963). *J. Clin. Invest.* **42,** 885–897.
Aurbach, G. D., and Phang, J. M. (1974). *In* "Medical Physiology" (J. Mountcastle, ed.), 13th ed., Vol. 2, pp. 1655–1695. Mosby, St. Louis, Missouri.
Bar, A., and Hurwitz, S. (1972). *Comp. Biochem. Physiol. B* **41,** 735–745.
Bar, A., and Hurwitz, S. (1973). *Comp. Biochem. Physiol. A* **45,** 579–586.
Bar, A., and Hurwitz, S. (1975). *Poult. Sci.* **54,** 1325–1327.
Bar, A., and Wasserman, R. H. (1973). *Biochem. Biophys. Res. Commun.* **54,** 191–196.
Bar, A., Hurwitz, S., and Cohen, I. (1972). *Comp. Biochem. Physiol. A* **43,** 519–526.
Bar, A., Hurwitz, S., and Edelstein, S. (1975). *Biochim. Biophys. Acta* **385,** 438–442.
Bélanger, L. F. (1971). *J. Exp. Zool.* **178,** 125–137.
Bélanger, L. F., and Copp, D. H. (1972). *In* "Calcium, Parathyroid Hormone and the Calcitonins" (R. V. Talmage and P. L. Munson, eds.), Int. Congr. Ser. No. 243, pp. 41–50. Excerpta Med. Found., Amsterdam.
Benoit, J., and Clavert, J. (1945). *C. R. Seances Soc. Biol. Ses Fil.* **139,** 737–740.
Bernardi, G., and Cook, W. H. (1960). *Biochim. Biophys. Acta* **210,** 174–187.
Birge, S. J., and Haddad, J. C. (1975). *J. Clin. Invest.* **56,** 1100–1107.
Bitman, J. C., Cecil, H. C., Harris, S. J., and Fries, G. F. (1969). *Nature (London)* **224,** 44–46.
Bitman, J. C., Cecil, H. C., and Fries, G. F. (1970). *Science* **168,** 594–596.
Bloom, M. A., Domm, L. V., Nalbandov, A. V., and Bloom, W. (1958). *Am. J. Anat.* **102,** 411–444.
Blunt, J. W., DeLuca, H. F., and Schnoes, H. K. (1968). *Biochemistry* **7,** 3317–3322.
Bronner, F. (1975). *In* "Calcium Metabolism, Bone and Metabolic Bone Diseases" (F. Kuhlencordt and H. P. Druse, eds.), pp. 14–24. Springer-Verlag, Berlin and New York.
Brown, D. M., Perey, D. Y. E., and Jowsey, J. (1970). *Endocrinology* **87,** 1282–1291.
Brumbaugh, P. F., Hughes, M. R., and Haussler, M. R. (1975). *Proc. Natl. Acad. Sci. U.S.A.* **72,** 4871–4875.
Candlish, J. K. (1969). *Comp. Biochem. Physiol.* **32,** 703–707.
Candlish, J. K. (1971). *Br. Poult. Sci.* **12,** 119–127.
Candlish, J. K., and Taylor, T. G. (1970). *J. Endocrinol.* **48,** 143–144.
Care, A. D., Bruce, J. B., Boelkins, J. N., Kenny, A. D., Conaway, H., and Anast, C. S. (1971). *Endocrinology* **89,** 262–271.
Cecil, H. C., Bitman, J., and Harris, S. J. (1971). *Poult. Sci.* **50,** 657–659.
Chan, A. S., Cipera, J. D., and Bélanger, L. F. (1969). *Rev. Can. Biol.* **27,** 19–31.
Chase, L. R., and Aurbach, G. D. (1967). *Proc. Natl. Acad. Sci. U.S.A.* **58,** 518–525.
Chase, L. R., and Aurbach, G. D. (1970). *J. Biol. Chem.* **245,** 1520–1526.
Chase, L. R., Fedak, S. A., and Aurbach, G. D. (1969). *Endocrinology* **84,** 761–768.
Cherian, A. G., and Cipera, J. D. (1968). *Poult. Sci.* **47,** 76–82.
Chertow, B. S., Baylink, D. J., Wergedal, J. E., Su, M. H. H., and Norman, A. W. (1975). *J. Clin. Invest.* **56,** 668–678.
Cipera, J. D., Chen, A. S., and Bélanger, L. F. (1970). *Calcitonin, Proc. Int. Symp., 2nd, 1969* pp. 320–326.
Clark, R. C. (1970). *Biochem. J.* **118,** 537–542.
Clark, R. C. (1973). *Biochim. Biophys. Acta* **310,** 174–187.
Common, R. H. (1932). *J. Agric. Sci.* **22,** 576–594.

Copp, D. H., Cameron, E. C., Cheney, B. A., Davidson, A. G. F., and Henze, K. G. (1962). *Endocrinology* **70,** 638–649.

Copp, D. H., Bayfield, G. H., Kerr, C. R., Newsome, F., Walker, V., and Watts, E. G. (1972). *In* "Calcium, Parathyroid Hormone and the Calcitonins" (R. V. Talmage and P. L. Munson, eds.), Int. Congr. Ser. No. 243, pp. 12–20. Excerpta Med Found., Amsterdam.

Corradino, R. A. (1975). *In* "Calcium Regulating Hormones" (R. V. Talmage, M. Owen, and J. A. Parsons, eds.), pp. 346–361. Excerpta Med. Found., Amsterdam.

Corradino, R. A., Wasserman, R. H., Publos, M. H., and Chang, S. I. (1968). *Arch. Biochem. Biophys.* **125,** 378–380.

Cutler, G. B., Jr., Habener, J. F., and Potts, J. T., Jr. (1974). *Gen. Comp. Endocrinol.* **24,** 183–190.

Deely, R. G., Mullinix, K. P., Wetekam, W., Kronenberg, H. H., Meyers, M., Eldridge, J. D., and Goldberger, R. F. (1975). *J. Biol. Chem.* **250,** 9060–9066.

DeLuca, H. F., Weller, M., Blunt, J. W., and Neville, P. F. (1968). *Arch. Biochem. Biophys.* **124,** 122–128.

Dousa, T. P. (1974). *Am. J. Physiol.* **226,** 1193–1197.

Dziak, R., and Stern, P. H. (1975). *Endocrinology* **97,** 1281–1287.

Edelstein, S., Harell, A., Bar, A., and Hurwitz, S. (1975). *Biochim. Biophys. Acta* **385,** 438–442.

Ehrenspeck, G., Schraer, H., and Schraer, R. (1971). *Am. J. Physiol.* **220,** 962–972.

Feinblatt, J. D., Tai, L.-R., and Kenny, A. D. (1974). *Endocrinology* **96,** 282–288.

Ferguson, R. K., and Wolbach, R. A. (1967). *Am. J. Physiol.* **212,** 1123–1130.

Fischer, J. A., Oldham, S. B., Sizemore, G. W., and Arnaud, C. D. (1971). *Horm. Metab. Res.* **3,** 223–224.

Foster, G. V., Bagldiantz, A., Kumar, M. A., Slack, E., Soliman, H. A., and MacIntyre, I. (1964). *Nature (London)* **202,** 1303–1305.

Frank, F. R., and Burger, R. E. (1965). *Poult. Sci.* **44,** 1604–1606.

Fraser, D. R., and Kodicek, E. (1970). *Nature (London)* **228,** 764–766.

Fraser, D. R., and Kodicek, E. (1973). *Nature (London), New Biol.* **241,** 163–166.

Fukushima, M., Syzuki, Y., Tahira, Y., Matsunaga, I., Ochi, V., Nagoino, H., Nishii, Y., and Suda, T. (1975). *Biochem. Biophys. Res. Commun.* **66,** 632–638.

Fussel, M. H. (1960). Ph.D. Thesis, University of Cambridge.

Gaillard, P. J. (1959). *Dev. Biol.* **1,** 152–181.

Galante, L. S., Colston, K. W., Evans, I. M. A., Larkins, R. G., McAuley, S. J., and MacIntyre, I. (1974). *Clin. Sci. Mol. Med.* **46,** 9P-10P.

Gershoff, S. M., Legg, M. A., and Hegsted, D. M. (1958). *J. Nutr.* **64,** 303–312.

Gonnerman, W. A., Breitenbach, R. P., Erfling, W. F., and Anast, C. S. (1972). *Endocrinology* **91,** 1423–1429.

Gonnerman, W. A., Ramp, W. K., and Toverud, S. U. (1975). *Endocrinology* **96,** 275–281.

Griminger, P., and Lutz, H. (1964). *Poult. Sci.* **43,** 710–716.

Harrison, H. C., Harrison, H. E., and Park, E. A. (1958). *Am. J. Physiol.* **192,** 432–436.

Harrison, H. E., and Harrison, H. C. (1960). *Am. J. Physiol.* **199,** 265–271.

Haussler, M. R., Nagode, L. A., and Rasmussen, H. (1970). *Nature (London)* **228,** 1199–1201.

Heersche, J. N. M., Marcus, R., and Aurbach, G. D. (1974). *Endocrinology* **94,** 241–247.

Hegsted, D. M., Moscoso, I., and Collazas, C. (1952). *J. Nutr.* **46,** 181–201.

Henry, H. L., and Norman, A. W. (1975). *Biochem. Biophys. Res. Commun.* **62,** 781–788.

Henry, H. L., Midgett, R. J., and Norman, A. W. (1974). *J. Biol. Chem.* **249,** 7584–7592.

Henry, H. L., Taylor, A. N., and Norman, A. W. (1976). *Fed. Proc., Fed. Am. Soc. Exp. Biol.* **35,** 340 (abstr.).

Hirsch, P. F., Gauthier, G. F., and Munson, P. L. (1963). *Endocrinology* **73,** 244–252.
Hirsch, P. F., Voelkel, E. F., and Munson, P. L. (1964). *Science* **146,** 412–413.
Holick, M. F., Schnoes, H. K., and DeLuca, H. F. (1971). *Proc. Natl. Acad. Sci. U.S.A.* **68,** 803–804.
Holick, M. F., Holick, S. A., Tavela, T. E., Gallagher, B., Schnoes, H. K., and DeLuca, H. F. (1975). *Science* **190,** 576–578.
Holick, M. F., Baxter, L. A., Schraufrogel, P. K., Tavela, T. E., and DeLuca, H. F. (1976). *J. Biol. Chem.* **251,** 397–402.
Hurwitz, S. (1964). *Am. J. Physiol.* **206,** 198–204.
Hurwitz, S. (1965a). *Poult. Sci.* **43,** 1462–1472.
Hurwitz, S. (1965b). *Am. J. Physiol.* **208,** 203–207.
Hurwitz, S. (1968). *Biochim. Biophys. Acta* **156,** 389–393.
Hurwitz, S., and Bar, A. (1965). *J. Nutr.* **86,** 433–438.
Hurwitz, S., and Bar, A. (1966a). *Poult. Sci.* **45,** 345–352.
Hurwitz, S., and Bar, A. (1966b). *J. Nutr.* **89,** 311–316.
Hurwitz, S., and Bar, A. (1968). *J. Nutr.* **95,** 647–654.
Hurwitz, S., and Bar, A. (1969a). *Poult. Sci.* **48,** 1391–1396.
Hurwitz, S., and Bar, A. (1969b). *J. Nutr.* **99,** 217–224.
Hurwitz, S., and Bar, A. (1971). *Poult. Sci.* **50,** 1044–1055.
Hurwitz, S., and Bar, A. (1972). *Am. J. Physiol.* **222,** 761–767.
Hurwitz, S., and Griminger, P. (1961a). *J. Nutr.* **73,** 177–185.
Hurwitz, S., and Griminger, P. (1961b). *Nature (London)* **189,** 759–760.
Hurwitz, S., and Rand, N. T. (1965). *Poult. Sci.* **44,** 177–183.
Hurwitz, S., Harrison, H. C., and Harrison, H. E. (1967a). *J. Nutr.* **91,** 208–212.
Hurwitz, S., Harrison, H. C., and Harrison, H. E. (1967b). *J. Nutr.* **91,** 319–323.
Hurwitz, S., Stacey, R. E., and Bronner, F. (1969). *Am. J. Physiol.* **216,** 254–262.
Hurwitz, S., Cohen, I., and Bar, A. (1970). *Comp. Biochem. Physiol.* **35,** 873–878.
Hurwitz, S., Bar, A., and Cohen, I. (1973a). *Am. J. Physiol.* **225,** 150–154.
Hurwitz, S., Bar, A., and Meshorer, A. (1973b). *Poult. Sci.* **52,** 1370–1374.
Hurwitz, S., Bar, A., Edelstein, S., and Montecuccoli, G. (1976). *Fed. Proc., Fed. Am. Soc. Exp. Biol.* **35,** 772 (abstr.).
Isler, H. (1973). *Anat. Rec.* **177,** 441–460.
Jande, S. S., and Bélanger, L. F. (1973). *Clin. Orthop. Relat. Res.* **94,** 281–305.
Jones, G., Schnoes, H. K., and DeLuca, H. F. (1976). *J. Biol. Chem.* **251,** 24–28.
Kenny, A.D. (1975). *In* "Calcium Regulating Hormones" (R. V. Talmage, M. Owen, and J. H. Parsons, eds.), pp. 408–410. Excerpta Med. Found., Amsterdam.
Klein, D. C., and Raisz, L. G. (1970). *Endocrinology* **86,** 1436–1440.
Kodicek, E. (1963). *In* "The Transfer of Calcium and Strontium Across Biological Membranes" (R. H. Wasserman, ed.), pp. 185–196. Academic Press, New York.
Kraintz, L., and Intscher, K. (1969). *Can. J. Physiol. Pharmacol.* **47,** 313–315.
Lawson, D. E. M., and Emtage, J. S. (1974). *Vitam. Horm. (N.Y.)* **32,** 277–298.
MacIntyre, I., Galante, L. S., Colston, K. W., Evans, I. M. A., Larkins, R. G., McAuley, S. J., Hillyard, C. J., Greenberg, P. B., Matthews, E. W., and Byfield, P. G. H. (1975). *In* "Calcium Regulating Hormones" (R. V. Talmage, M. Owen, and J. H. Parsons, eds.), pp. 396–404. Excerpta Med. Found., Amsterdam.
McDonald, M. R., and Riddle, O. (1945). *J. Biol. Chem.* **159,** 445–464.
McLean, F. C., and Hastings, A. B. (1935). *Am. J. Med. Sci.* **189,** 601–613.
Mahler, H. R. (1961). *In* "Mineral Metabolism" (C. L. Comar and F. Bronner, eds.), Vol. 1, Part 1B, pp. 843–879. Academic Press, New York.
Miller, A. L., and Schraer, H. (1975). *Calcif. Tissue Res.* **18,** 311–324.
Mongin, P., and Sauveur, B. (1974). *Br. Poult. Sci.* **15,** 349–359.

Montecuccoli, G., Hurwitz, S., Cohen, A., and Bar, A. (1977). *Comp. Biochem. Physiol.* **57A**, 335–339.

Moore, E. W. (1970). *J. Clin. Invest.* **49**, 318–334.

Morgan, C. L., and Mitchell, J. H. (1938). *Poult. Sci.* **17**, 99–104.

Morrissey, R. L., and Wasserman, R. H. (1971). *Am. J. Physiol.* **220**, 1509–1515.

Mueller, G. L., Anast, C. S., and Breitenbach, R. P. (1970). *Am. J. Physiol.* **218**, 1718–1722.

Mueller, W. J. (1962). *Poult. Sci.* **41**, 1792–1796.

Mueller, W. J., Hall, K. L., Maurer, C. A., Jr., and Joshua, I. G. (1973). *Endocrinology* **92**, 853–856.

Nicolaysen, R., Eeg-Larsen, N., and Malm, O. J. (1953). *Physiol. Rev.* **33**, 424–444.

Nieto, A., Moya, F., and R-Candela, J. L. (1973). *Biochim. Biophys. Acta* **322**, 383–391.

Nieto, A., L-Fando, J. J., and R-Candela, J. L. (1975). *Gen. Comp. Endocrinol.* **25**, 259–263.

Nordin, B. E. C., Marshall, D. H., Peacock, M., and Robertson, W. G. (1975). *In* "Calcium Regulating Hormones" (R. V. Talmage, M. Owen, and J. A. Parsons, eds.), pp. 239–253. Excerpta Med. Found., Amsterdam.

Norman, A. W., Myrtle, J. F., Midgett, R. J., Nowicki, H. G., Williams, V., and Popjak, G. (1971). *Science* **173**, 51–54.

Omdahl, J. L., and DeLuca, H. F. (1973). *Physiol. Rev.* **53**, 327–372.

Pearse, A. G. E., and Carvarheira, A. F. (1967). *Nature (London)* **214**, 929–930.

Peterson, P. A. (1971). *J. Biol. Chem.* **246**, 7748–7754.

Proscal, D. A., Okamura, W. H., and Norman, A. W. (1975). *J. Biol. Chem.* **250**, 8382–8388.

Rasmussen, H. (1959). *Endocrinology* **65**, 517–519.

Rasmussen, H., Jensen, P., Lake, W., Friedman, N., and Goodman, D. B. P. (1975). *J. Cyclic Nucleotide Res.* **5**, 375–394.

Riddle, O., Rauche, V. M., and Smith, G. C. (1944). *Anat. Rec.* **90**, 295–305.

Rosenberg, T. (1954). *Symp. Soc. Exp. Biol.* **8**, 27–41.

Sammon, P. J., Stacey, R. E., and Bronner, F. (1969). *Biochem. Med.* **3**, 252–270.

Schachter, D., Dowdle, E. B., and Schenker, H. (1960). *Am. J. Physiol.* **198**, 263–268.

Schraer, H., Tannenbaum, P. J., and Posner, A. S. (1967). *J. Dent. Res.* **46**, 1072–1074.

Shane, S. M., Young, R. J., and Krook, L. (1969). *Avian Dis.* **13**, 558–567.

Shanfeld, J., Shapiro, I., and Davidovitch, Z. (1975). *Anal. Biochem.* **66**, 450–459.

Sherwood, L. M., Potts, J. T., Jr., Mayer, G. P., and Aurbach, G. D. (1966). *Nature (London)* **209**, 52–55.

Simkiss, K. (1961). *Biol. Rev. Cambridge Philos. Soc.* **36**, 321–367.

Simkiss, K., and Taylor, T. G. (1971). *In* "Physiology and Biochemistry of the Domestic Fowl" (D. J. Bell and B. M. Freeman, eds.), Vol. 2, pp. 621–640. Academic Press, New York.

Stoeckel, M. E., and Porte, A. (1973). *Arch. Anat. Microsc. Morphol. Exp.* **62**, 55–58.

Stringer, D. A., and Taylor, T. G. (1961). *Biochem. J.* **78**, 119P.

Tannenbaum, P. J., Schraer, H., and Posner, A. S. (1974). *Calcif. Tissue Res.* **14**, 83–86.

Tauber, S. D. (1967). *Proc. Natl. Acad. Sci. U.S.A.* **58**, 1684–1687.

Taylor, A. N. (1974). *Arch. Biochem. Biophys.* **161**, 100–108.

Taylor, A. N., and Wasserman, R. H. (1969). *Fed. Proc., Fed. Am. Soc. Exp. Biol.* **28**, 1834–1838.

Taylor, A. N., and Wasserman, R. H. (1970). *J. Histochem. Cytochem.* **18**, 107–115.

Taylor, A. N., and Wasserman, R. H. (1972). *Am. J. Physiol.* **223**, 110–114.

Taylor, T. G. (1965). *Proc. Nutr. Soc.* **24**, 49–51.

Taylor, T. G. (1971). *In* "Physiology and Biochemistry of the Domestic Fowl" (D. J. Bell and B. M. Freeman, eds.), Vol. 1, pp. 473–480. Academic Press, New York.

Taylor, T. G., and Bélanger, L. F. (1969). *Calcif. Tissue Res.* **4**, 162–173.

Taylor, T. G., and Moore, J. H. (1954). *Br. J. Nutr.* **8**, 112–124.

Taylor, T. G., Simkiss, K., and Stringer, D. A. (1971). *In* "Physiology and Biochemistry of the Domestic Fowl" (D. J. Bell and B. M. Freeman, eds.), Vol. 2, pp. 621–640. Academic Press, New York.

Terepka, A. R. (1963). *Exp. Cell. Res.* **30**, 171–182.

Tyler, C., and Geak, F. H. (1958). *J. Sci. Food Agric.* **9**, 472–483.

Tyler, C., and Moore, D. (1965). *Br. Poult. Sci.* **6**, 175–182.

Urist, M. R., Deutsch, N. M., Pomerantz, G., and McLean, F. C. (1960). *Am. J. Physiol.* **199**, 851–855.

Wasserman, R. H. (1963). *In* "The Transfer of Calcium and Strontium Across Biological Membranes" (R. H. Wasserman, ed.), pp. 211–228. Academic Press, New York.

Wasserman, R. H., and Corradino, R. A. (1973). *Vitam. Horm. (N.Y.)* **31**, 43–103.

Wasserman, R. H., and Taylor, A. N. (1966). *Science* **152**, 791–793.

Wasserman, R. H., Corradino, R. A., Fullmer, C. S., and Taylor, A. N. (1974). *Vitam. Horm. (N.Y.)* **32**, 299–324.

Wells, R. G. (1967). *Br. Poult. Sci.* **8**, 193–199.

Williams, G. A., Hargis, G. K., Bowser, E. N., Henderson, W. J., and Martinez, N. J. (1973). *Endocrinology* **92**, 687–691.

Wong, G. L., and Cohn, D. V. (1975). *Proc. Natl. Acad. Sci. U.S.A.* **72**, 3167–3171.

Ziegler, R., Delling, G., and Pfeiffer, E. F. (1970). *Calcitonin, Proc. Int. Symp., 2nd,* 1969 pp. 301–310.

Zull, J. E., Czarnowska-Misztal, E., and DeLuca, H. F. (1956). *Science* **149**, 182–184.

CHAPTER 9

Energy: Expenditures and Intakes

Larry L. Wolf and F. Reed Hainsworth

I. Introduction

The uses of chemicals by organisms fall into two categories: energy value and nutrient value. These can vary in relative importance but must be provided for in consumer organisms through food selection. Additionally, the importance of either will vary during the life of an individual (e.g., changing demands associated with reproduction, Ricklefs, 1974) and with the variety of life forms of different species. To place the importance of chemicals as energy sources into perspective, we will review the determinants of energy requirements of birds. Then we will attempt to view these requirements as they influence food selection and feeding behavior. Our goal is to view energy in an ecological context.

II. Determinants of Energy Requirements

Attempts to view food selection must consider what is required of the selection. Since foods can be selected for energy value, we will summarize major determinants of energetics for birds including the effects of size, temperature, and activity (flight). A number of recent reviews dealing with the energetic requirements of birds (e.g., Farner, 1970; Dawson and Hudson, 1970; Calder and King, 1974; Paynter, 1974; Peaker, 1975) should be consulted for extensive documentation.

A. Size

Birds share with all organisms a dependence of energy requirements on body size. To assess the importance of size, temperature and activity must be standardized which usually involves measuring expenditures at minimum rates, i.e., for birds at rest in the zone of thermoneutrality during the inactive portion of the daily cycle (Aschoff and Pohl, 1970). Comparisons for a variety of organisms indicate that energy expenditures per unit time increase with size raised to a fractional power, usually approximately 0.75 (Schmidt-Nielsen, 1975a; see Fig. 1A). The similarity of the slopes for birds and other groups of

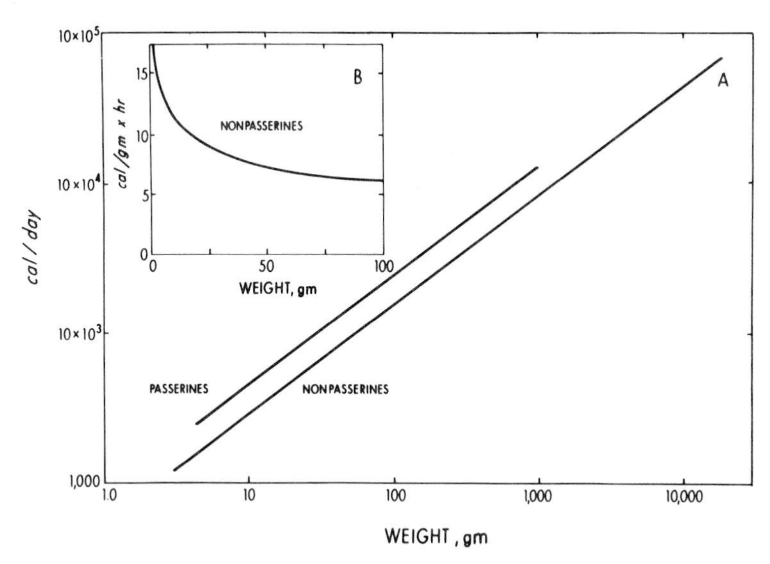

FIG. 1. (A) Standard metabolic rate as a function of body weight for passerine and nonpasserine birds (redrawn from Lasiewski and Dawson, 1967). (B) Standard metabolic rate per gram as a function of body weight for nonpasserine birds. Calculated from regression equation of Lasiewski and Dawson (1967).

organisms suggests a uniformity of constraints. This has been debated extensively (e.g., Ultsch, 1973; Calder, 1974; McMahon, 1973; Schmidt-Nielsen, 1975a).

Although regression slopes are generally the same between taxa (Schmidt-Nielsen, 1975a), the intercepts differ. Poikilotherms generally have the lowest intercepts (Schmidt-Nielsen, 1975b). Among homeotherms, the intercept is lowest for nonplacental mammals (Dawson and Hulbert, 1970), intermediate for eutherian mammals (Schmidt-Nielsen, 1975b), and highest for birds (Lasiewski and Dawson, 1967). Differences reflect energy maintenance requirements. Thus, for birds the maintenance of circulation, muscular tension, digestive functions, etc., are geared at a higher level and require more energy at a given size. In this regard, it would be interesting to examine the relative contributions of various organ systems to this maintenance level. For example, since muscle represents such a large portion of an organism, differences in its maintenance (e.g., for tonus) might contribute considerably to intercept differences.

In comparisons involving maintenance and the size of homeotherms it is common to express energy expenditures relative to weight (e.g., cal/gm × hr versus cal/hr). The slope of this relationship with respect to weight is negative (with an exponent approximating −0.25; Calder, 1974; Schmidt-Nielsen, 1975b). The differences in weight-relative energy requirements are quite dramatic when plotted on arithmetic coordinates (Fig. 1B). A great deal has been written about the increase in weight-specific metabolic rate (cal/gm × hr) with decreasing size, including: (1) considering surface area changes with size (see Calder, 1974); (2) suggesting this as a basis for a lower limit to body size in homeotherms (Pearson, 1948); and (3) demonstrating that a variety of physiological functions such as heart rates (Calder, 1968), oxygen dissociation curves (Schmidt-Nielsen and Larimer, 1958), and capillary densities (Schmidt-Nielsen and Pennycuick, 1961) parallel requirements of supplying tissues of smaller homeotherms with energy at a higher rate. For present purposes, a decrease in size for a homeotherm can be viewed as influencing temperature regulation through an increased heat loss forced by a relative increase in surface area (Calder, 1974; see Section II,B).

Some caution is required in viewing energy requirements for maintenance on a per gram basis. Both McNab (1971) and Calder (1974) stress that in an ecological context costs for the entire organism are important and not just those associated with maintaining a gram of tissue. Examining the physiology of small homeotherms can lead to insights on how energy is supplied at rates influenced by rates of energy loss, but small size ultimately translates into lower costs (cal/hr) for maintenance, which will affect food selection and processing.

B. Temperature

Ambient temperature is perhaps the most easily measured environmental variable that can influence energy expenditure, and the papers documenting its effect are legion (see Dawson and Hudson, 1970, for a recent review). To understand temperature effects it is important to consider animals at rest and to document the effects of body size.

To maintain a constant body temperature the rate of heat production must equal the rate of heat loss. Ambient temperature has its effect primarily through the dependence of heat loss through conduction, convection, and radiation on the difference between body and ambient temperature (e.g., Kleiber, 1961, 1972). As ambient temperature decreases, heat loss increases and more energy must be expended to maintain body temperature. At high ambient temperatures increased heat loss becomes a consideration since energy expenditures will increase if body temperatures increase through Q_{10} effects on chemical rate processes (Dawson and Hudson, 1970).

Data for two species of hummingbirds (Fig. 2, solid points) demonstrate increased metabolism with decreasing ambient temperature (Wolf and Hainsworth, 1972). The linear change in energy ex-

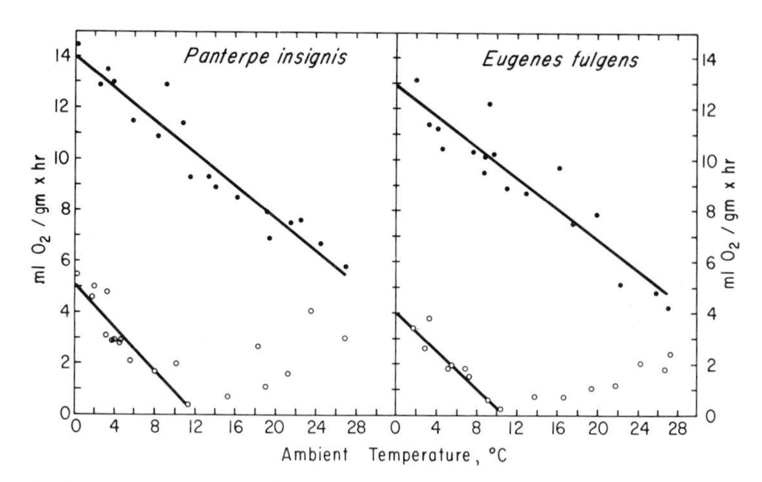

Fig. 2. Oxygen consumption (an index of metabolic rates) as a function of ambient temperature for two species of hummingbirds. Closed symbols denote birds regulating body temperatures at 41°C; open symbols denote birds in torpor (Wolf and Hainsworth, 1972).

penditure with temperature has often been interpreted relative to Newton's Law of Cooling (e.g., Yarbrough, 1971), but this has been criticized on physical grounds since homeotherms do not normally cool (Calder, 1972). Regardless, the dependence of energy expenditures on the difference between body and ambient temperature is clear and has its basis in physical factors influencing heat transfer.

The slope of the line relating energy expenditure (cal/gm × hr) to ambient temperature (slope = cal/gm × hr × °C) has been interpreted as "thermal conductance" or the reciprocal of insulation. A steeper slope indicates greater heat loss or lower insulation (e.g., Herreid and Kessel, 1967; Lasiewski *et al.*, 1967). Kleiber (1972) recently pointed out that a more physically accurate expression is in units of cal/hr × °C. An examination of this heat transfer coefficient for a variety of passerine and nonpasserine birds of different sizes indicates that it is related to approximately the square root of weight (Kleiber, 1972; Calder, 1974). When expressed per unit of surface area, the exponent on weight is -0.21 (Calder, 1974) which is a reflection of a change in insulation with size. Thus, the slope of the line for energy expenditure (cal/gm × hr) as a function of temperature will be larger for smaller species.

The dependence of heat transfer (cal/hr × °C) on size can result in energy requirements for maintenance at low temperatures that are considerably different than at thermoneutral temperatures (Calder, 1974). At a low ambient temperature a small bird must add relatively more to its basal requirements. For the small bird relative to the large bird this would raise the intercept and lower the slope of the thermoneutral relationship between energy requirements (cal/hr) and body size. Kendeigh (1969) provides evidence for this for birds from measurements of existence metabolism at low ambient temperatures.

C. Reducing Maintenance Expenditures: Torpor

One method to lower energy requirements is to lower metabolic rate through reduction of body temperature. This is most common among "poikilotherms" that regulate their temperatures only over short periods by physiological (Heinrich, 1974) or behavioral (Regal, 1967) means. Among mammals, seasonal hibernation is common in moderate sized species such as ground squirrels and is associated with seasonally low food availability and low temperatures (Fisher and Manery, 1967). Torpor on a daily basis is found in several small species of mammals and also appears related to food availability and energy expenditures (Tucker, 1966; Brown and Bartholomew, 1969; French, 1976).

Few species of birds enter torpor seasonally; rather, many migrate from areas of low food availability and low temperatures. The energetics associated with flight compared to running suggest that migration could be an efficient way to adjust expenditures on a seasonal basis (see Section II,E). However, a number of species of birds become torpid for short periods. These include Swifts (*Chaetura pelagica,* Ramsey, 1970; *Apus apus,* Koskimies, 1948; *Aeronautes saxatalis,* Bartholomew *et al.,* 1957), Poorwills (*Phalaenoptilus nuttallii,* Bartholomew *et al.*, 1957), Lesser Nighthawks (*Chordeiles acutipennis,* Marshall, 1955), and Colies (*Colius,* Huxley *et al.*, 1939), as well as all hummingbird species that have been studied (Lasiewski, 1963; Hainsworth and Wolf, 1970; Wolf and Hainsworth, 1972; Carpenter, 1974; Hainsworth *et al.*, 1977).

Energy expenditures for hummingbirds in torpor are illustrated in Fig. 2 (open circles). Oxygen consumption typically decreases with decreased temperature until a particular body temperature is reached which is then regulated at lower ambient temperatures. The temperature at which regulation in torpor occurs is not related to size or other physiological characters influencing energy expenditure, but it is related to the minimum environmental temperatures in the habitats of the species (Wolf and Hainsworth, 1972; Carpenter, 1974). The phenomenon of body temperature regulation in torpor suggests that this state is precisely controlled, and it has been observed in species from tropical (Hainsworth and Wolf, 1970; Wolf and Hainsworth, 1972; Carpenter, 1974) as well as temperate latitudes (Calder and Booser, 1973; Hainsworth and Wolf, 1978). Since regulation in torpor occurs near the minimum environmental temperatures, it should result in maximum energy savings.

Why and under what conditions does torpor occur? The function of torpor for saving energy is obvious, and its occurrence in hummingbirds suggests an importance for conserving energy in very small homeotherms. Hummingbirds must store energy during the day to meet their overnight requirements when they do not feed. Since they are small, they can store less energy relative to their requirements than larger species. Indices of energy storage capacity (such as gut and crop volumes) are related to $(\text{weight})^{1.0}$, while expenditures (cal/hr) are related to $(\text{weight})^{0.75}$ (Calder, 1974). Depending on proportionality constants, larger species could store more energy relative to their expenditures. Torpor would allow very small species to adjust expenditures to lower energy storage capabilities. Advantages of torpor for large species are limited by size constraints on rates of entry and arousal (Pearson, 1960).

Hummingbirds do not always enter torpor under laboratory conditions (Lasiewski, 1963). We have found that when hummingbirds have access to adequate food sources in the laboratory they can store sufficient energy during the day to meet their overnight requirements at moderate temperatures (24°C) and long daylengths (14L:10D) without entering torpor (Wolf and Hainsworth, 1977). When the photoperiod is shortened (9L:15D) with colder temperatures overnight (e.g., 10°C), they will not enter torpor until they have used most of their energy reserves (Hainsworth *et al.*, 1977). Also, torpor for incubating Broad-tailed Hummingbirds *(Selasphorus platycercus)* in a natural setting was related to inability to forage and to store sufficient energy (Calder and Booser, 1973). The occurrence of torpor only in "energy emergency" situations may be related to a degree of risk associated with hypothermia, such as a higher risk from predation.

D. Other Means to Conserve Energy

Organisms can employ a variety of mechanisms to conserve energy other than through physiological modifications. For example, toads *(Bufo boreas)* will select temperatures in such a way as to maximize their growth rate (Lillywhite *et al.*, 1973). Roadrunners *(Geococcyx californianus)* exhibit a degree of hypothermia and can conserve energy by basking (Ohmart and Lasiewski, 1971). Also, hummingbirds save energy by microhabitat selection (Calder, 1973a) as well as from nest insulation during incubation (Calder, 1973b; Smith *et al.*, 1974; Drent, 1975).

Detailed climatological studies of microhabitats have indicated that temperatures, although important, are not the only possible determinants of energy expenditures (Calder, 1973a). Wind velocity around an organism appears to be particularly important (Southwick and Gates, 1975). Presumably, an organism with the ability to make a choice will select those microhabitats that minimize maintenance requirements (Horvath, 1964) if no other constraints are more important.

E. Activity

1. "Resting" Metabolic Rates

A change from standard metabolic rate to an awake, resting condition produces a predictable increase in energy expenditure. This has been documented for a variety of species including: (1) mammals ranging in size from mice to dogs where resting = 1.3 to 2.1 times standard metabolic rates (Taylor *et al.*, 1970); (2) Herring Gulls *(Larus argentatus)* = 1.7 times standard metabolic rate (Baudinette and

Schmidt-Nielsen, 1974); and (3) hummingbirds where resting = 1.7 times standard metabolic rate (Wolf and Hainsworth, 1971). The increase presumably is associated with postural effects.

2. *Energetics of Flight*

The energetic expenditures for flight depend on speed, size, and morphological characteristics of the flight apparatus. The energy expenditures for flight in the Budgerigar *(Melopsittacus undulatus)*, Laughing Gull *(Larus atricilla)*, and Fish Crow *(Corvus ossifragus)* flying in wind tunnels varied with air speed (Tucker, 1968, 1972; Bernstein *et al.*, 1973) (Fig. 3). Energy expenditure was based on direct measurement of oxygen consumption. The relatively constant performance curves for the Laughing Gull and Fish Crow (Fig. 3) were the result of technical inabilities to examine expenditures at extreme air speeds for these species. Other methods for estimating flight energetics, such as weight changes (Raveling and LeFebvre, 1967) or using D_2O^{18} (Utter and LeFebvre, 1970), are restricted to longer time periods where control and measurement of the bird's activity are less precise.

It is of interest to ask why the performance curves for birds in general are U shaped. Pennycuick (1969) provided an elegantly simple treatment by considering birds as flying machines to which he applied aerodynamic theory. An important aspect of his treatment was that essentially only two variables, wing length and body weight, were

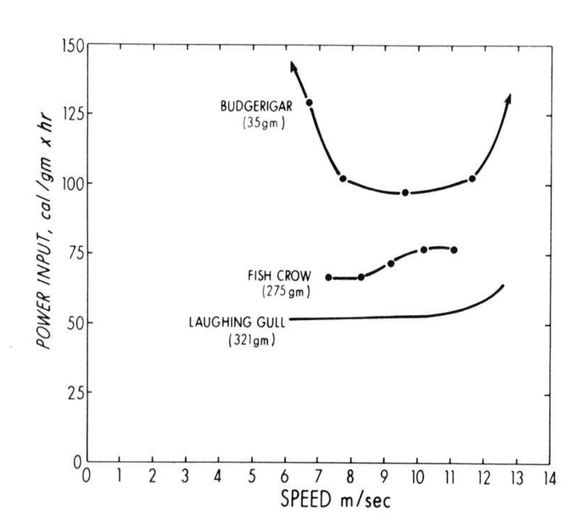

FIG. 3. Performance curves for power input for flight as a function of flight speed for the Budgerigar, Fish Crow, and Laughing Gull. Redrawn from Thomas (1975).

required to construct a performance curve. This treatment described "power output" (work done on the environment) rather than the "power input" measured from oxygen consumption which differs from power output by an efficiency term.

The U-shaped curve can be divided into three power terms (Fig. 4; Pennycuick, 1969): (1) induced power (P_i) required to produce lift; (2) parasite power (P_p) required to overcome drag on the body; and (3) profile power (P_o) required to overcome drag on the wings. At increasing flight speed the power for lift decreases while the power to overcome drag on the body increases. At zero flight speed a special situation occurs (hovering flight) which will be considered below (see Section II,G). The drag on the wings is very difficult to estimate, but Pennycuick (1969) indicates that this remains relatively constant with flight speed for the Pigeon *(Columba livia)*.

The addition of these power terms results in the U shape which has important implications for energy expenditures. As Pennycuick (1969) and Tucker (1970) point out, there is a speed which represents a minimum power for flight, (V_{mp}, where the sum of P_i and P_p is minimum) (Fig. 4). Also, there is a higher speed where power is minimum to cover a distance (V_{mr}, or speed for maximum range, Fig. 4). This latter power provides a means for comparing a variety of species since it represents a "minimum cost of transport" expressed as the minimum cal/gm × km for a particular species (Tucker, 1970). Minimum cost of transport is related negatively to body size. Running

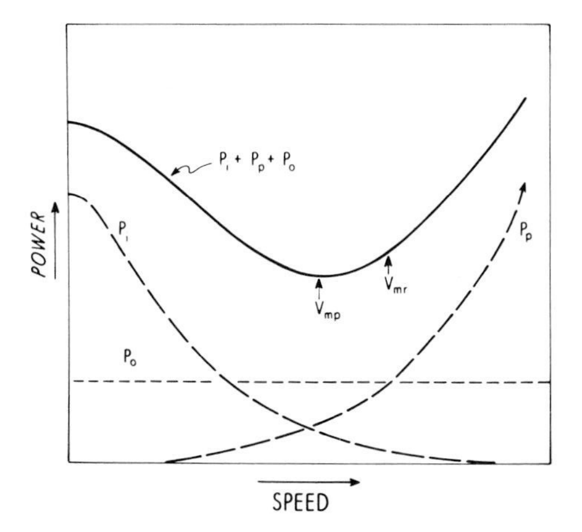

FIG. 4. Performance curve for a theoretical bird for power output as a function of flight speed. P_i = induced power; P_p = parasite power; P_o = profile power; V_{mp} = speed for minimum power; V_{mr} = speed for maximum range. Redrawn from Pennycuick (1969).

animals have the highest transport costs per gram at a given weight, flying animals intermediate, and swimming animals the lowest costs (Schmidt-Nielsen, 1972).

As was the case with comparisons for animals at standard metabolic rates, it is important to distinguish between metabolic rates and weight-relative metabolic rates for minimum transport costs. For example, Schmidt-Nielsen (1975b) stated that the significance of the negative relation between minimum costs of transport (cal/gm × km) and weight was that it cost larger animals less for transport. Also, Gaulin and Kurland (1976) improperly concluded that larger animals are "more efficient" because they have lower transport costs (cal/gm × km). However, there is a positive relationship between metabolic rate (cal/hr) and body size for birds of different weights at minimum costs of transport (Fig. 5). Also, when costs are calculated as kcal/km, there is a positive slope with respect to weight [kcal/km = $1.25(\text{kg})^{0.77}$ for the 16 species in Table 1 of Tucker, 1970]. It is these expenditures which should have the most impact on the ecology of a given species. For example, costs in cal/km together with energy storage capacity [scaled to $(\text{weight})^{1.0}$ for migrating birds?] would determine possible nonstop flight ranges for migrating birds (see Tucker, 1975, for other estimates of flight ranges).

Tucker (1973) attempted to modify Pennycuick's theoretical treatments of power output to make them correspond more closely with measured power input. Unfortunately, the most completely studied species for estimating power output is the domestic pigeon for which oxygen consumption during flight has not been measured. Regardless, aerodynamic considerations provide a rationale for differences in measured power inputs for the few species studied. For example, the differences in power input for the similar sized Laughing Gull and Fish Crow (Fig. 3) can be attributed to differences in their wing disc loading. Wing disc loading is body weight divided by the area of the disc swept by the wings ($W/\frac{1}{4}\pi d^2$; d = tip-to-tip wing length) and represents an index of the weight to be lifted relative to the morphology to lift it (Pennycuick, 1969). For the Fish Crow wing loading is 1.9 times higher than for the Laughing Gull which corresponds to the 1.5 times higher power expenditure for flight by the Fish Crow (Fig. 3).

F. Energy Savings during Flight

1. Flight at Minimum Cost of Transport Speeds

It would be interesting to determine when and if birds fly at speeds which provide minimum costs of transport. From measurements on Laughing Gulls, Tucker and Schmidt-Koenig (1971) indicate that

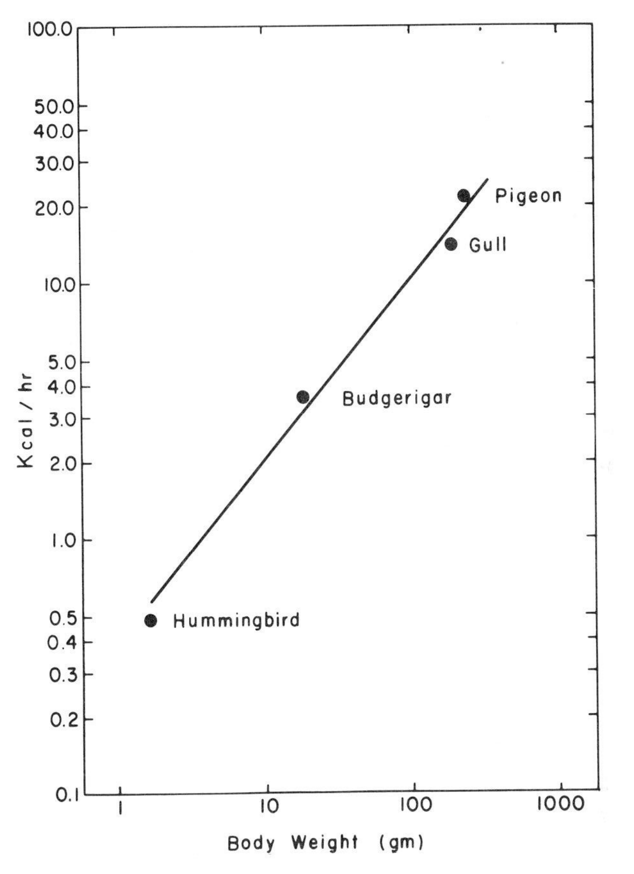

FIG. 5. Energy expenditure per unit time for flight at minimum cost of transport speeds as a function of body weight for a hummingbird *(Calypte costae)*, Budgerigar, Laughing Gull, and pigeon. Values recalculated from data presented in Table I of Tucker (1970). The equation is kcal/hr = 37.03 $(kg)^{0.71}$.

these birds moved at or near minimum transport cost speeds during migration. Also, a bat, *Phyllostomus hastatus*, may fly close to its minimum transport speed under natural circumstances (Thomas, 1975). For birds moving over short distances there may be constraints on the speeds they can achieve. For example, the Violet-ear Hummingbird *(Colibri thalassinus)* moving distances up to a few meters while foraging at flowers showed a positive, linear relationship between distance moved and flight speed (Wolf *et al.*, 1976). This suggests that it may have been constrained in speed by acceleration and deceleration such that an "optimum" speed was not achieved.

2. *Use of Winds: Soaring and Gliding*

Another method to reduce energy expenditures for flight is to use prevailing winds (e.g., Able, 1973; Tucker, 1975) which could add appreciably to flight ranges. Use of tail winds may be particularly important for migratory birds that carry increased fuel loads (fat) since an increase in weight would increase V_{mr} in proportion to $(\text{weight})^{\frac{1}{2}}$ (Pennycuick, 1969). However, Bellrose (1967) indicates that night migrants may decrease forward speed with tail winds such that they would be flying at minimum power speeds rather than at minimum cost of transport speeds. In addition to using tail winds, a number of species of relatively large birds use local wind currents (thermals, ridge and wave currents) for soaring (Pennycuick, 1971). Gliding flight involves an increase in energy expenditure for Herring Gulls of 1.9–2.4 times resting metabolic rate which represents a savings of approximately 70% relative to flapping flight (Baudinette and Schmidt-Nielsen, 1974).

3. *Formation Flight*

Finally, an intriguing possibility to reduce expenditures in flight could involve group behavior. A number of migrating birds fly in a V formation (e.g., Canada Geese, *Branta canadensis*). Lissaman and Shollenberger (1970) have treated this behavior with aerodynamic theory. The potential savings from flying in a group formation would depend on the wing tip spacing between birds since the closer individuals are, the more they can utilize the updraft produced past the tip of an adjacent wing. A V formation could distribute the savings among individuals such that the birds at the ends of the vee receive as much benefit as the bird at the apex (a consequence of Munk's stagger theorem). Depending on group size and wing tip spacing, flying in a group could result in savings of 50–60% relative to individual power requirements.

A preliminary study indicated that distances between adjacent Canada Geese in formations varied considerably and appeared too large in many cases for them to have gained lift from the formation (Gould and Heppner, 1974). However, the birds studied were not migrating, and additional studies should be interesting.

G. Hovering Flight: A Special Case

In many comparisons of flight energetics it is often necessary to consider species that may differ considerably in their mode of flight. Generalizations about minimum costs of transport can involve very different flight strategies, and considering all forms together obscures interesting differences. Hummingbirds are a good example because of their obvious specialization for hovering flight. It is also of interest to examine the energetics of flight for these birds because of the importance of this for their food selection and feeding efficiency (see Section III).

The specialization of hummingbirds for hovering is reflected in their wing lengths relative to their weights. Greenewalt (1975) pointed out that wing lengths for hummingbirds increased more with increased body weight than for other birds. This is a reflection of the very high costs associated with hovering where weight must be lifted with no lift generated by forward speed (the intercept of P_i at zero flight speed, Fig. 4). Measurements of oxygen consumption for hovering hummingbirds are the highest per gram values measured for vertebrates. Assuming a respiratory quotient of 1.0, Lasiewski (1963) indicated a cost of about 215 cal/gm × hr for a 3-gm Costa's Hummingbird *(Calypte costae)*, Wolf and Hainsworth (1971) found about 215 cal/gm × hr for a 9-gm *Eulampis jugularis*, Berger and Hart (1972) obtained a similar value for a 6-gm *Amazilia fimbriata*, and Berger (1974) obtained the same value for a 7- to 8-gm *Colibri coruscans*.

The similarity of expenditures per gram for these species suggests that metabolic rate for hovering (cal/hr) would be linearly related to size. Hainsworth and Wolf (1972) calculated power output costs for hovering for 40 species and noted that the per-gram costs were statistically independent of weight. However, much of the variation in the values calculated by Hainsworth and Wolf (1972) could be explained by variations in wing disc loading (Epting and Casey, 1973). Recently, Greenewalt (1975) independently examined power output for hovering and reached the same conclusion as Hainsworth and Wolf (1972). He used a different expression for estimating profile power and a more precise way of calculating wing spans from wing lengths.

To reexamine this problem we obtained additional data on wing lengths and body weights for hummingbirds (see legend of Fig. 6 for details). We used Greenewalt's equations to estimate wing span and profile power and calculated the relationship between power per gram for hovering and body weight (Fig. 6). A power function provides the best fit, where

$$P_{\text{hov}} \text{ (calories)/gm} = 14.6(\text{gm})^{-0.065}$$

Using Greenewalt's (1975) equations we find that the exponential relationship between power for hovering and wing disc loading accounts for much less of the variation in values of power for hovering ($r^2 = 0.19$) than was reported earlier by Epting and Casey (1973; $r^2 = 0.83$).

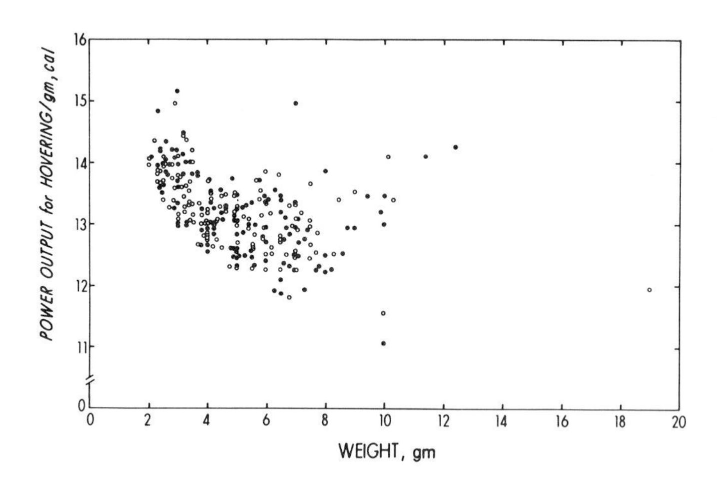

FIG. 6. Power output for hovering per gram as a function of body weight for hummingbirds. Data for males of 82 species (solid circles) and for females of 81 species (open circles) from the collection of the Louisiana State Museum with additional values for other species from Greenewalt (1975), Feinsinger and Chaplin (1975), and F. G. Stiles (personal communication) for a total of 262 values representing 129 males, 119 females, and 14 species of undetermined sex. Calculations from the equations of Greenewalt (1975).

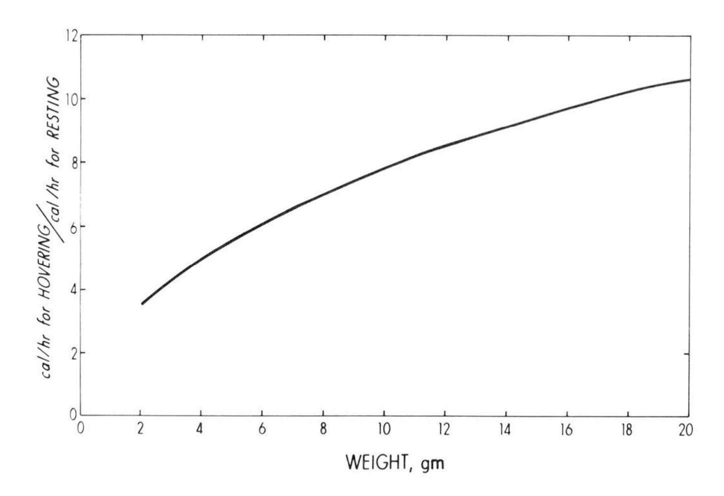

FIG. 7. Energy expenditures for hovering relative to energy expenditures for resting (standard metabolism at 20°C estimated from Herreid and Kessel, 1967 × 1.7) as a function of body size for hummingbirds. Calculations of hovering expenditures are from the regression of P_{hov}/gm versus weights (Fig. 6) assuming an output efficiency of 5.8% (Hainsworth and Wolf, 1972).

These results suggest that the per-gram energy expenditure for hovering should increase somewhat with decreasing body weight, particularly for species less than about 4 grams (Fig. 6). The variations in output per gram at any given size can be appreciable, making it reasonable to use 215 cal/gm $\times$ hr as an estimate of power input for hovering. Certainly variations in weight have more pronounced effects on expenditures for hovering (cal/hr) than wing disc loading variations or power/gm estimates of different species at a given weight (Fig. 6; Hainsworth and Wolf, 1975).

The ability to make reasonable estimates of expenditures at rest and for activity within a group of organisms allows us to examine relationships that could have importance for differences in behavior or resource exploitation. One parameter of interest in subsequent discussions is the relationship between energetic costs for hovering (i.e., for feeding) relative to costs for resting and body size in hummingbirds (Fig. 7). Despite the increase in the per-gram costs for hovering as size decreased (Fig. 6), when translated into cal/hr and related to resting metabolic rates (1.7 times standard at 20°C) the comparison indicates that smaller hummingbirds experience less increase, relatively and absolutely, in metabolic rates for hovering than larger species. Measurements of proportions of time spent flying (excluding feeding flight time) for unrestrained hummingbirds in the laboratory indicate that smaller birds spend a larger proportion of their time flying than larger birds, perhaps as a consequence of smaller energetic constraints (Fig. 8).

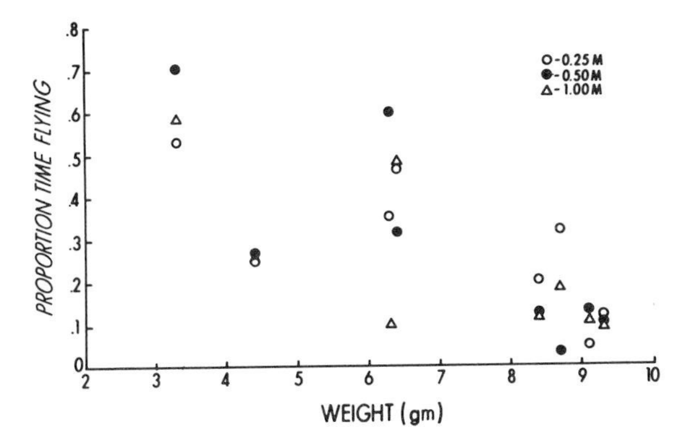

FIG. 8. Relation between proportion of nonfeeding time hummingbirds spent in flight and body weight. Data from hummingbirds in the laboratory. Different symbols represent different food concentrations.

H. Determinants of Size

In view of the importance of size for a variety of energetic relationships, biologists have been interested in the determinants of body size. The sizes of living organisms span an enormous range, 10^{21} grams, from submicroscopic organisms to the blue whale (Schmidt-Nielsen, 1975a).

One approach to this problem has been to examine the mechanical limitations that might be responsible for the scaling of metabolic rate (cal/hr) to $(\text{weight})^{0.75}$. A variety of explanations has been considered such as limits on (1) respiratory exchange area and size (Ultsch, 1973); (2) surface area for heat loss and size (Rubner, 1883); (3) scaling of rate functions relative to size (Calder, 1974); and, most recently (4) changes in bone elasticity relative to size (McMahon, 1973). Although these approaches may give some insight into why the exponent for metabolic rate relative to size is 0.75, they provide little rationale for explaining the sizes of organisms per se. The fact that organisms as large as some dinosaurs evolved indicates that mechanical constraints on size cannot explain the diversity of smaller organism sizes.

Considering the total size range of animals, the exponent of 0.75 on weight-related energy expenditure does have implications for abilities of organisms of different sizes to obtain sufficient energy to meet their requirements. Ability to store energy is generally related to $(\text{weight})^{1.0}$ rather than $(\text{weight})^{0.75}$. For example, the mass of the gut of mammals is related to $(\text{weight})^{0.94}$ (Stahl, 1962), and for birds to $(\text{weight})^{0.985}$ (Calder, 1974). Thus, larger animals could store more energy in their digestive tracts relative to their expenditures. Small organisms must feed more frequently on a given resource or resort to some mechanism, such as torpor, to reduce expenditures. In this context, obligatory reduction in expenditures of many poikilotherms at low ambient temperatures could reduce their maintenance requirements and permit a degree of size reduction.

At the other extreme, large animals could be limited by the efficiency with which they can obtain food to meet their greater energy expenditures (cal/hr). Many large animals appear to feed at relatively low trophic levels (e.g., herbivores versus carnivores). In an examination of territory size relative to body weight, Schoener (1968) noted for birds, and McNab (1963) for mammals, that feeding areas could be relatively smaller for herbivores, and they related this to the abundance and distribution of prey items. Also, Pough (1973) examined the sizes of extant lizards in four families and found that almost all large species (>300gm) were herbivores, while small species (<50–100gm) were carnivores. Although he interpreted these results such that larger lizards would be unable to meet caloric demands by eating insects, it seems more reasonable to suggest that an herbivorous diet could permit larger size within a foraging technique.

There are certainly exceptions to the rule that size is directly related to resource type (Pough, 1973). An example for birds is the large size of many raptors. An explanation may lie in a variety of factors in addition to diet that could influence the efficiency of resource exploitation. These are best illustrated by considering the range of sizes of organisms within a similar group that exploits a particular resource. Schoener (1969b) found that several major factors influenced the size pattern in West Indian *Anolis* lizards. These were (1) aggressive interactions allowing access to food; (2) daily and seasonal food production; (3) relative availability of prey of different sizes; (4) predation as influenced by size; (5) interspecific reproductive isolation; (6) thermoregulation; (7) habitat structure; and (8) stress limitations upon structure and organ systems. Similarly, the interspecific sizes of hummingbirds were suggested to influence aggressive differences (Wolf, 1970), which, in turn, should influence access to resources (see Sections III,D and III,E). Size dimorphism in a variety of birds has also been related indirectly to resource utilization efficiency (Selander, 1966; Mosher and Matray, 1974).

In conclusion, the size of an organism is likely to be the result of a variety of factors that represent compromises in terms of food utilization, competition, and predation within particular communities. These compromises can best be understood by examining physiological and ecological differences between similar groups of organisms rather than through allometric analyses of large numbers of species with very different evolutionary histories.

III. Food Intake

A. INTRODUCTION

Energy expenditures for an organism must be provided for from outside sources such that

$$\text{Energy assimilated} = \text{energy expended} + \text{energy stored}$$

This general thermodynamic balance equation provides a basic framework in which to view feeding behavior. The nutrient balance of an organism can be viewed similarly, and the two may or may not be coupled. Control of energy intake should occur relative to different time scales. The first is a short-term basis (e.g., one feeding episode to another) with an additional feedback control that influences the accumulation of energy for daily periods of nonfeeding (e.g., at night for diurnal birds). The second is a level of control that induces seasonal changes in accumulation of energy for demanding activities such as reproduction, migration, and/or molt. A number of other factors can influence short-term as well as long-term energy accumulation under natural circumstances. These include interactions among individuals, effects of resource availability, and the resource distribution. We will consider theories dealing with food selection in a natural context after we consider the control of feeding as studied in a laboratory situation where factors influencing individual feeding behaviors can be controlled. Ultimately, the two must be joined to provide a complete picture of the regulation of energy intake.

For birds, most information comes from studies of hummingbirds (Wolf and Hainsworth, 1977). In large part this is due to the ease with which expenditures can be estimated (see Section II) and the ease with which their food (primarily sugar water) can be reproduced and experimentally manipulated.

B. Feeding Onset and Termination

The short-term control of energy intake involves the two distinct activities of initiation and cessation of feeding. These may be under common control and represent separate responses to a graded signal (e.g., hepatocyte membrane potentials or other signals mediated through the brain; Russek, 1975; Friedman and Stricker, 1976). Ignoring for the moment the control of the length of a feeding episode, most laboratory studies of mammals (LeMagnen and DeVos, 1970; LeMagnen, 1975; Levitsky, 1974; Snowdon and Wampler, 1974) and birds (Duncan *et al.*, 1970; Wolf and Hainsworth, 1977) support the hypothesis that consumers use an accumulation–depletion mechanism of control of feeding onset (Fig. 9). There is a high, positive correlation for time between feeding bouts and time spent feeding on the first of two bouts, and a lack of correlation for time between bouts and time spent feeding on the second of two bouts. A more appropriate correlation is between energy expended and energy ingested, which for several hummingbird species is significantly positive for accumulation followed by depletion (Wolf and Hainsworth, 1977). Data on time comparisions from field observations of hummingbirds foraging at flowers indicate that these relationships are not laboratory artifacts (Wolf and Hainsworth, 1977).

A feeding bout probably is initiated when some internal signal is triggered (Toates and Booth, 1974). The characteristics of the signal

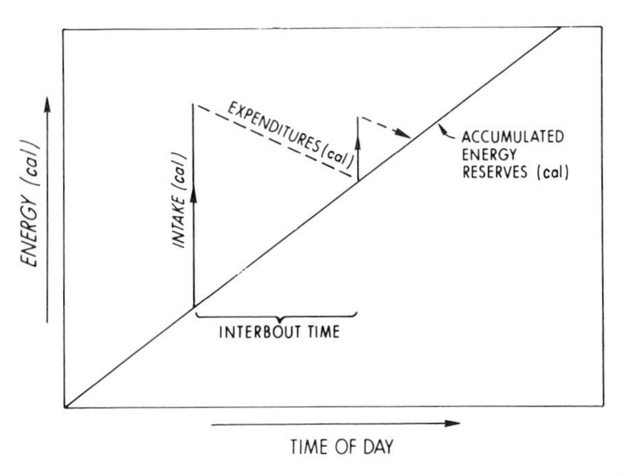

FIG. 9. General model of energy accumulation and expenditure patterns for birds during the active portion of a 24-hour cycle. Available evidence (see text) supports the hypothesis that on the long-term (= total daylight period per day) birds gradually accumulate reserves in anticipation of future negative energy budgets, while from feeding to feeding they utilize energy accumulated on the previous bout, some of which is also being shunted into the long-term stores. Figure taken from Wolf and Hainsworth (1977).

have not been explored in birds, but for mammals one or more factors related to energy availability or utilization such as blood glucose levels, ketone bodies, or fat stores have been suggested (Hervey, 1975; Epstein *et al.*, 1975; Russek, 1975; Novin *et al.*, 1976; Friedman and Stricker, 1976). Such a signal could act through several stages in a control system, such as the liver (Russek, 1975; Friedman and Stricker, 1976) and/or brain (e.g., Ritter and Epstein, 1975; Mogenson and Calaresu, 1975). In hummingbirds and chickens, there is circumstantial evidence that feeding is initiated when the volume of food in the crop reaches some lower threshold (Duncan *et al.*, 1970; Wolf and Hainsworth, 1977). The source of the actual signal is unknown as is the control system regulating crop emptying rate. For blowflies (*Phormia regina*) the signal regulating crop emptying is blood osmotic pressure which can be an index of supply of energy from the crop (Gelperin, 1966).

For hummingbirds, the rate at which the crop empties is related positively to the rate of expenditure of energy and negatively related to the molarity of the sugar solution ingested (Wolf and Hainsworth, 1977). The volume in an emptying crop is probably a negative exponential function of time, as in blowflies (Gelperin, 1966) and some other organisms (Charnov, 1976), but the exponent varies nonarithmetically with molarity for hummingbirds, e.g., a crop with 0.25 *M* sucrose empties about three times as fast as a crop with 1.0 *M* sucrose. Assimilation is essentially 100% (Hainsworth, 1974) so a hummingbird expending energy at equivalent rates can store more energy per unit time with 1.0 *M* than with 0.25 *M* nectar (Wolf and Hainsworth, 1977).

Once initiated a feeding bout continues until some control mechanism signals cessation. In the laboratory the shift in behavior from visiting the site of food to other locations and behaviors clearly indicates the end of a feeding bout. In the field it may be more difficult to define a feeding bout. However, in both situations it is possible that several feeding episodes are functionally the same bout, defined by the control of initiation and cessation of feeding. Short, incomplete feeding episodes could derive from influences other than direct feeding regulators. Slater (1974) dealt with this problem by using a survivorship curve method to arrive at a temporal definition of a feeding bout based on curve changes in the frequency of lengths of time between successive pecks.

Domestic chickens *(Gallus gallus),* domestic pigeons, Japanese Quail *(Coturnix coturnix),* and some Zebra Finches *(Poephila guttata)* in *ad libitum* feeding situations feed predominantly in a large number of very short episodes (Van Hemel and Meyer, 1969; Slater, 1974). The general shape of the frequency distributions of bout length for these species (Fig. 10A) suggests that once initiated the probability of a bout ending is constant at the end of any specific feeding period. This means that for these birds there is no characteristic meal size (Duncan *et al.,* 1970). However, as already indicated there is generally a good relation between meal size and the subsequent nonfeeding interval.

Hummingbirds and some Zebra Finches in *ad libitum* feeding situations have frequency distributions of bout lengths similar to a normal distribution with most of the values clustered around a mean (Slater, 1974; Wolf and Hainsworth, 1977; Fig. 10B). This characteristic meal size is a linear function of body weight among hummingbird species and apparently is independent of the rate of intake of either fluid or energy (Wolf and Hainsworth, 1977). For the hummingbirds a predictive model of meal size has been constructed based on the premise that the birds add to their metabolic costs as a result of the added weight of fluid stored initially in their crops (DeBenedictis *et al.,* 1978). The model reasonably predicts average meal sizes if the birds are maximizing the rate of net energy gain (or a benefit to cost function). Meal size then is thought to be a mechanism of obtaining the most excess energy per unit time within the constraints of added costs associated with feeding. It is not clear to what extent such a model of

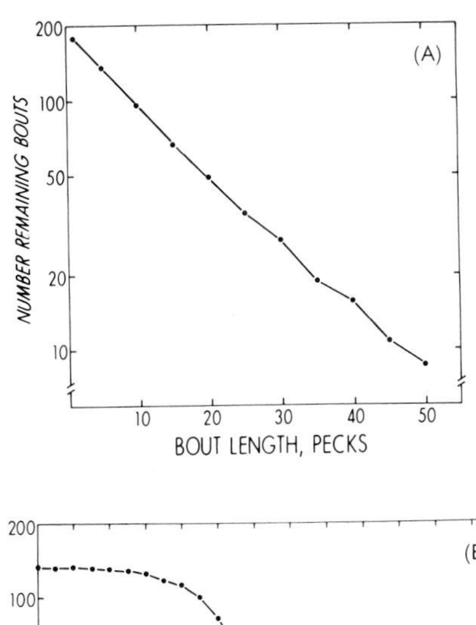

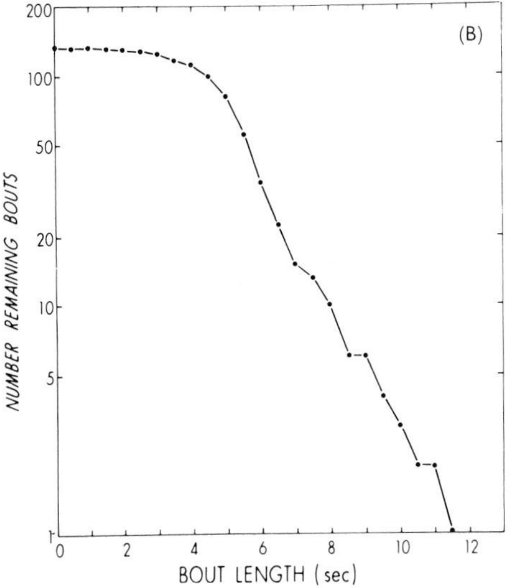

FIG. 10. Two examples of distributions of foraging bout lengths plotted as survivorship curves. The nearly straight line on semilogarithmic axes for the Zebra Finch (A) suggests a constant probability of a bout ending in the ith + 1 interval (redrawn from Slater, 1974). The curvilinear relationship (B) on semilog axes approximates a normal distribution and illustrates the characteristic feeding bout length found among hummingbirds (data from Wolf and Hainsworth, 1977).

feeding behavior would predict characteristic meal sizes for organisms that have lower energy expenditures associated with weight changes from feeding (e.g., large versus small organisms or those that must fly versus those that walk or hop).

C. Daily Energy Storage

The regulation of accumulated energy requires adjustment of short-term intake patterns to increase or decrease the storage term of the energy balance equation. Although some organisms (e.g., shorebirds, Heppleston, 1971; flamingos, Pennycuick and Bartholomew, 1973) may feed 24 hours a day under some circumstances, most have a relatively long period of enforced fast during each 24 hours; this period can be especially long for residents at high latitudes at certain seasons. This 24-hour cycle requires either replenishing nonfeeding losses or accumulating excess energy for subsequent quiescent periods. In *Zonotrichia albicollis* (White-throated Sparrow) and *Passer domesticus* (House Sparrow) energy generally is stored in anticipation of future demands as indexed by prior expenditures, although it may require several weeks to adjust appropriately energy storage rates to changing demands (Kendeigh *et al.*, 1969). Different levels of energy accumulation were reached principally by regulating the length of time early in a day that the birds were hyperphagic. Once necessary reserves were attained the birds apparently reduced their intakes to match expenditures (Kendeigh *et al.*, 1969). However, within these general patterns of energy storage there were no data on mechanisms of adjustment of short-term patterns to accommodate 24-hour cycles.

Energy accumulation can be influenced by adjustments of feeding bout lengths, frequencies, or both. Zeigler *et al.* (1971) reported for female domestic pigeons that enforced weight loss was followed by a correlated, compensatory increase in mean length of a feeding bout, but not by a correlated increase in bout frequency. In a recent revision of our model of meal size in hummingbirds (DeBenedictis *et al.*, 1978) we have incorporated variables to include both maintenance costs during the day and the rate of energy accumulation to meet the anticipated energetic costs for the next night (Hainsworth, 1978). Available laboratory data for our hummingbirds suggest that the rate of accumulation of energy during the day is dependent on the previous night's expenditures, an estimator of the next night's costs (Hainsworth, 1978). The revised model predicts that meal size should vary as overnight costs vary while meal frequency should vary with the rate of energy expenditures during the day (= maintenance costs). Preliminary laboratory data from hummingbirds kept at a constant day temperature (to hold diurnal maintenance costs approximately constant) showed a significant increase in meal size at low compared to high nocturnal temperatures. There was no significant shift in meal frequency. However, if nocturnal temperatures were held constant and the diurnal temperature was varied the birds showed a significant increase in bout frequency, but no significant shift in meal size (Hainsworth, 1978). Some minor adjustments in the rate of energy accumulation in some experimental situations suggested a competitive interaction between maintenance and storage energy pathways. In rats, Leung and Horwitz (1976) reported an adjustment to low temperature via an increased meal size at night rather than by an increased meal frequency. Rats normally store energy at night for use during the day (LeMagnen and DeVos, 1970), so although the day–night relationship is reversed the rats show a response similar to the hummingbirds.

Another possible method of shifting overnight costs is to change the photoperiod. Among sparrows Kendeigh *et al.* (1969) reported a lengthy entrainment period of up to 2 weeks to adjust accumulation rates to a changed photoperiod, but they did not report whether the adjustment was via meal size or frequency (or both). If photoperiod is decreased, hummingbirds spend more energy overnight and have less time to accumulate energy during the day. They gradually lose weight for several days during which they shift meal size. They now accumulate energy at a rate sufficient to meet the additional overnight expenditures if they were still operating under the old photoperiod regime. It may require as much as two weeks before the adjustment of meal size is sufficient to accumulate energy at a rate to meet the total new requirements in the shorter available daylength. The hummingbirds apparently respond differentially to the short term shifts in temperature likely to be encountered in nature and the normally long, gradual shifts in photoperiod. The apparent difference in the response to the two environmental variables probably is a function of the differences in predictability of the two types of environmental change in nature.

Hummingbirds fed 0.25 *M* sucrose solutions generally lose weight starting the first day (Hainsworth and Wolf unpublished). Under these conditions they normally adjust the meal size and to some extent meal frequency. The meal size shift suggests that either the control parameters associated with optimal volume intakes have been adjusted or that the factor being optimized at a specific crop volume has changed (DeBenedictis *et al.*, 1978). The compensatory increase in meal size with no shift in meal frequency suggests either that the control of bout initiation is reset at a new level, or, more probably, that the rate of reaching the old level is increased. A simple model consistent with our interpretation of meal size control in hummingbirds postulates that crop emptying is a negative exponential function of time (Fig. 11). Depending on the steepness of the curve, which is a function of the

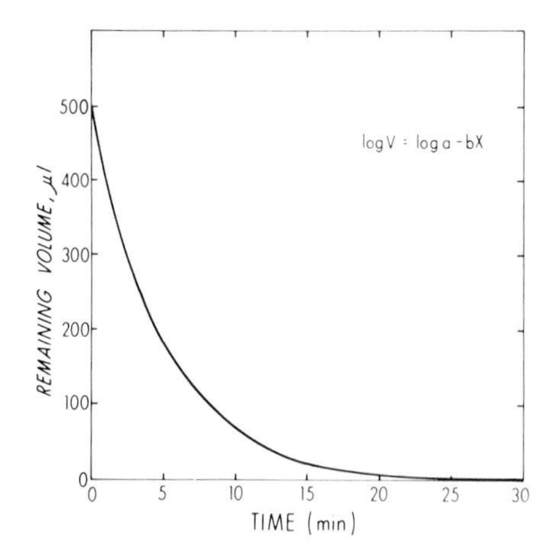

FIG. 11. A negative exponential crop emptying curve described by the equation: log $V = 2.70 - 0.09$ (minutes). This curve approximates data for a female *Lampornis clemenciae* (Blue-throated Hummingbird) feeding in the laboratory from a feeder tube containing 0.50 *M* sucrose solution.

molarity of the solution and the rate of energy expenditure, additional intake at a feeding would require very small additional increments of time between bouts. These adjustments would be sufficiently small to be obscured by other variations, such as activity levels.

In situations of reduced caloric value of food some organisms respond by increasing meal size and/or frequency essentially mimicking the deprivation situation (Levitsky and Collier, 1968; Snowdon, 1969). There is evidence that humans also compensate for caloric dilution of meals by both meal size and frequency increases, but the shift requires several days (Spiegel, 1973). If intake rate is reduced hummingbirds compensate by increasing bout length to maintain a similar volume intake per bout, unless the intake is severely reduced (to 25% or less of maximum possible in our experiments) in which case we predict that volumes taken per bout will decline (Wolf and Hainsworth, 1977; DeBenedictis *et al.*, 1978).

Some studies of daily patterns of energy accumulation by natural populations of birds have used visual estimates of energy storage (Helms, 1963) or carcass analysis of birds collected at various times (e.g., Zimmerman, 1965a; Newton and Evans, 1966; Ward, 1969a; Newton, 1969; Evans, 1969). The results generally confirm laboratory studies in demonstrating an accumulation of fat during the day with seasonal storage levels related to photoperiods and temperatures. The Bullfinch, *Pyrrhula pyrrhula* (Newton, 1969) and the Yellow Bunting,

Emberiza citrinella (Evans, 1969) deposited more fat on a daily basis in winter, while Ward (1969a) noted less seasonality in daily fat accumulated for an equatorial bird species. The fat accumulated by these birds during a day was sufficient for expenditures overnight plus part of another day but day to day survival depended on energy intake from food following nocturnal expenditures. For the Bullfinches in winter, Newton (1969) found that feeding activities could extend over the entire 8-hour day, and Evans (1969) found that Yellow Buntings roosted at night with considerable quantities of food in their gizzards. In contrast to the results of Kendeigh *et al.* (1969), Evans (1969) found no correlation between mean fat content and mean temperatures for the previous night. There was a correlation based on longer time intervals (20-year means) suggesting that long-term average temperatures rather than short-term variations may have influenced fat accumulation (Evans, 1969). However, it is not clear to what extent the considerable food consumption by these birds prior to roosting could have influenced correlations with proximate factors.

D. Seasonal Energy Storage

Seasonal patterns of energy accumulation are generally associated with seasonal changes in energy requirements for thermoregulation, molt, and migration. The seasonal periods of excess energy storage involve hyperphagia, although the short-term patterns of bout length and frequency are usually not reported. Seasonal hyperphagia can be triggered by photoperiodic shifts in some temperate zone birds (see Berthold, 1975). King (1972) reviewed seasonal fat storage in birds. However, recent work suggests that some birds do not accumulate sufficient energy in the form of fat to meet seasonally high expenditures for prolonged periods, e.g., feeding nestlings or producing eggs. At these times they may metabolize other body tissue, often pectoral muscle (Ward, 1969b; Fogden, 1972).

For a variety of species, seasonally demanding activities appear timed to availability of resources (e.g., King, 1972; Kuntz, 1974). For some species, such as mice (Millar, 1975), cave bats (Kuntz, 1974), and locusts (McCaffery, 1975) daily levels of food consumption increase during gestation. However, since most of these studies involve only measures of weight changes or daily food intake, it is not clear to what extent increased energy intake represents short-term accumulation of energy to meet demands or replace losses or to what extent longer-term energy accumulation prior to demands could be important. Several hypotheses can be proposed. If timing of demanding activities is such that resources are available in excess and time for feeding does not limit other activities (such as for mating or care of young), organisms may operate on short-term scales. However, if there were little time for feeding during demanding activities (such as during migratory flight) or food availability was reduced (e.g., hibernating in winter), long-term accumulation of energy may occur in "anticipation" of demands. These processes could be further modified by physiological constraints, such as from size differences that may influence relative abilities to store energy on a daily or seasonal basis (Calder, 1974).

1. Reproduction

Ricklefs (1974) recently viewed the area of energetics relative to reproduction in birds. His extensive review deals primarily with problems of energy allocation for growth and maintenance during development of the young. Information about effects of seasonal factors on reproduction is more extensive for lizards; the principles should be similar for birds.

Some indication of mechanisms of energy storage relative to seasonally demanding reproductive activities comes from carcass analysis of lipids for several species of lizards. For example, Hahn and Tinkle (1965) found that *Uta stansburiana* deposited sufficient quantities of fat in fat bodies to provide for the lipids deposited in the first clutch of eggs produced in the spring. From experimental manipulation of fat bodies and ovaries they found that there was an apparent feedback between ovarian hormones and fat accumulation. However, in the sagebrush lizard *(Sceloporus graciosus)* fat bodies after hibernation contain only one-third of the lipids required for the first clutch, and this species must rely on ingested energy at the time of initial reproduction (Derickson, 1974; see also Dessauer, 1955). These fat storage patterns have been related to differences in life histories and can be interpreted relative to the time and energy constraint hypotheses outlined above.

In an extensive comparative study of seasonal fat cycles in *Anolis* lizards, Licht and Gorman (1970) inferred from the inverse correlation between fat deposition and reproduction that these lizards were unable to obtain sufficient excess energy to fatten at times when reproduction was high. However, they were unable to assess the importance of changes in food availability or changes in consumption relative to expenditures. There is some indication from studies of locusts (McCaffery, 1975), based on food intake and weight changes, that egg production depends on prior ability to store energy and nutrients. Thus, locusts fed diets of variable quality show a dependence of egg production on food consumption but with a delay of several weeks in response to food quality changes.

2. *Hibernation*

Hibernation involves entry into torpor with a reduction in body temperature. The result can be energy savings during periods of low food availability. Birds that enter torpor may do so on a daily rather than a seasonal basis, but the results can be similar over different time scales. Also, hibernation for many mammals may be analogous to migration for many birds since it represents a mechanism to modify energy expenditures while migration represents a mechanism to modify energy availability on a seasonal basis. Thus, prior to considering migration it may be instructive to examine seasonal aspects of hibernation relative to energetics.

The seasonal changes in fat deposition and utilization by seasonal hibernators have been described as an example of a circannual rhythm (e.g., Pengelley and Kelley, 1966; Pengelley and Asmundson, 1970). This is based on the occurrence of an entrainable annual cycle of fat deposition (measured as weight changes) in the absence of obvious external cues. Although there may be some question whether this is a true annual cycle or a series of linked cycles (Mrosovsky, 1970), the "anticipation" of demands for energy in seasonal hibernators through fat storage is an excellent example of long-term energy utilization effects.

Because of the time intervals involved there is little information on the bout-to-bout feeding mechanism(s) underlying this process. However, data from food intake of ground squirrels during periodic arousals from hibernation suggest that some hibernators could operate under energy storage limitations either through body size effects, food availability changes, time limitations for consumption, or a combination of these. In an early discussion of periodic arousal from hibernation, Fisher and Manery (1967) suggested two hypotheses for the phenomenon: (1) arousal for replenishing energy stores, and (2) arousal to excrete metabolites accumulated during torpor. The latter hypothesis has been extensively investigated for a variety of seasonal hibernators (e.g., Pengelley *et al.*, 1971; Passmore *et al.*, 1975; Nelson *et al.*, 1973), and most experimental evidence involving manipulation of metabolites suggests they are relatively unimportant (Pengelley *et al.*, 1971).

Fisher and Manery (1967) proposed the first hypothesis from the relationship between body size and energy storage capacity and noted that those species with lower storage capacities (generally species smaller in size) would be more likely to exhibit periodic arousal. Experimental evidence for the energy supplementation hypothesis is indirect but suggestive (Mrosovsky and Barnes, 1974; Fisher and Manery, 1967; Mrosovsky and Fisher, 1970). Ground squirrels that have accumulated more fat prior to the onset of the hibernation season remain in torpor longer, arouse less frequently during the season, and, consequently, consume less food during the hibernating season than ground squirrels that have stored less energy in fat (Mrosovsky and Barnes, 1974). Thus, limitation of energy storage prior to hibernation either through availability or time constraints could influence periodic arousals. Periodic arousals in ground squirrels also occurs more fr- quently in fall and spring when food availability may be higher to permit supplementing energy reserves. This would suggest a precise monitoring of reserves relative to "expected" expenditures on a long-term basis, perhaps with a "sliding set point" as energetic demands change during a season (Mrosovsky and Fisher, 1970).

The apparent relation of torpor to energy utilization is not restricted to seasonal hibernators. Studies of the occurrence, time of entry and arousal, and duration of daily torpor strongly suggest a relationship to energy availability for the small species that show this phenomenon (Tucker, 1966; Brown and Bartholomew, 1969; French, 1976). The lower energy storage capabilities of small species that show daily torpor suggest that fat storage may be relatively less important for them, but its role has not been quantified. Most information comes from manipulation of food rations for small mice (*Perognathus californicus*, Tucker, 1966; *Microdipodops pallidus*, Brown and Bartholomew, 1969; *Perognathus longimembris*, French, 1976) that suggests adjustments in expenditures through torpor to match energy availability.

Storage of energy in species that may be restricted in fat storage ability could occur externally through deposition of food in caches, but there is also little information on the energy value of hoarded food relative to seasonal changes in expenditures. Excavations of burrow systems of kangaroo rats *(Dipodomys spectabilis)* indicate that huge quantities of seeds can be hoarded (Vorhies and Taylor, 1922). The quantity hoarded may exceed demands on a seasonal basis if relatively little cost is associated with external storage. There is little information on the total energy value of food caches relative to expenditures or on the relationships between size of food stores and losses that could occur from raids by other individuals or from long-term seasonal fluctuations in food availability. Smith (1975) suggested that the type of seeds stored or consumed by squirrels may depend on the energetic costs associated with handling. Thus, seeds of high caloric density would be consumed on a short-term basis since they would add relatively little weight and influence the cost for locomotion less than seeds of low caloric density. The latter would be more efficient to utilize on a long-term basis in a nest during periods of low activity. The evidence for this is inferential but appears to fit observations.

3. *Migration*

Fat deposition in association with migration represents another phenomenon strongly influenced by seasonal factors, particularly photoperiod (Farner, 1970). However, there is little information concerning the underlying feeding mechanisms of premigratory fattening (see review by King, 1972) or how they relate to photoperiodic, photosensitive, hormonal, and/or endogenous cycles (Farner, 1970). Fat cycles associated with migration are cited as examples of circannual rhythms (Farner, 1970), and there are other analogies with the energetics of seasonal hibernation. The initiation of migration appears related to the accumulation of sufficient fat (energy) reserves (Dolnik and Blyumental, 1967). Also, "stopover" periods for migrating birds are analogous to periodic arousals by hibernators. Stopover occurs when a bird remains in a particular location for several days prior to resuming migration. Dolnik and Blyumental (1967) noted that these periods were longer for birds that arrived lean than for those that arrived with more fat reserves. Thus, stopover periods could represent time for replenishing energy stores relative to some "expectation" of expenditures.

The technical problems associated with dealing with migratory species have limited most studies to carcass analyses, weight changes, or daily food consumption. Although this information leads to interesting relationships, such as the correlation between migrating distance and premigratory fat reserves (Dolnik and Blyumental, 1967; King, 1972) and apparent differences in rates of daily energy accumulation for premigratory and migratory birds (Dolnik and Blyumental, 1967), more detailed studies of feeding behavior would add considerably to understanding the short- and long-term mechanisms of seasonal fat reserve cycles and possible important constraints on the processes.

E. Optimal Foraging Patterns

1. *Short-Term Optimization*

In this section we will discuss generally how birds might be expected to operate in natural situations once a foraging episode is initiated. The general framework should be viewed in relation to at least three types of foraging (cf. Schoener, 1971): (1) searching followed by capture; (2) sitting and waiting for a prey item; and (3) capture that requires pursuit time, but no search time. A searching forager must incorporate some time and energy into actively moving about the habitat. Once encountered a food item will require variable energy expenditures for capture, handling, and eating. A sit-and-wait forager allows prey to move while it remains still. This requires relatively high prey abundance or relatively large prey items for organisms such as birds with high maintenance energy requirements, and sit-and-wait foragers among birds tend to be more common in tropical regions of high sustained primary productivity (Orians, 1969). The third foraging type requires that precise food locations be known in advance. Generally this third type will be found among birds utilizing food for which being eaten is a selective advantage to the "prey," e.g., nectar or fruits, which produces pollination and seed dispersal. These food items should signal clearly location and quality.

Consumers can be faced with a variety of choices once feeding is initiated. They may choose among habitats, food items, and whether to forage alone or in groups (Zahavi, 1971; Thompson *et al.*, 1974; Charnov, 1976; Davies, 1976; Gill and Wolf, 1977). Here we will explore behavioral and ecological determinants of the possible choices.

Theoretically and practically it is important to specify the basic assumptions associated with the choices. Generally it is assumed that the overt behavior has some degree of genetic control and hence is subject to natural selection acting on the individual. Selection operates to maximize fitness, measured ultimately in genetic contributions to future generations. In the case of feeding behavior, fitness normally is equated with some energy, time, or nutrient parameters that influence potential or actual reproduction. Models of foraging behavior involving choice usually are cast in benefit and cost terms. For time or energy models these terms theoretically are readily definable (Schoener, 1971); however, for nutrients, the constraints may be somewhat different. Ecologists who model food choice situations generally ignore nutrients and deal with either time or energy (Schoener, 1971; Pulliam, 1974, 1975; Real, 1975; Charnov, 1976; Estabrook and Dunham, 1976).

Models of feeding behavior generally assume that either the rate of net energy gain or the time not feeding is maximized. So far, most models have dealt specifically only with maximizing these parameters and have largely ignored possible constraints generated by parameters such as predator avoidance or competitive interactions (Keister and Slatkin, 1974; Real, 1975; Caraco and Wolf, 1975; Gill and Wolf, 1975a; Charnov, 1976), although predator avoidance is to some degree implicit in time minimizing feeding models.

a. Item Choice. A feeding organism faced with an array of potential food items that differ in caloric availability and/or cost per item should choose those that maximize its relative benefit, generally assumed to be rate of net energy gain. A number of models predict that consumers will add distinct items to the diet as total energy availability decreases and that the most energetically rewarding items per effort should be taken first and others added in the order of descending benefit minus cost (Emlen, 1966, 1968; Schoener, 1969a, 1971; Rapport, 1971; Marten, 1973; Pulliam, 1974; Werner and Hall, 1974; Charnov, 1976; Estabrook and Dunham, 1976).

The problem becomes more complex if the rate of net energy gain for a given food type is variable, unless the analysis is limited to a single food type of variable costs, e.g., a mantid eating flies of the same size that differ in distance from the predator (Charnov, 1976). Such variations could occur if each food type was not distributed uniformly so that costs to obtain one item type varied. As in the simpler models the estimates of potential benefits and costs would be a composite of previous experience with all prey types.

When a food item is encountered, a choice must be made between accepting (or attempting a capture) or continuing to forage for a different item. In this situation the principal information available is "hunger" level (Charnov, 1976) and the probability of encounters of other items based on previous experience. It is presently not clear how long previous encounter information is integrated (Schoener, 1971; Pulliam, 1974; Gill and Wolf, 1977). Relatively short-term shifts in encounter probabilities especially with clumped prey (Thompson *et al.*, 1974) can lead to the phenomenon of predator switching (Murdoch, 1969; Murdoch and Oaten, 1975), where predator preference changes with relative prey abundance. While switching may not necessarily maximize immediate rate of net energy gain, in some large temporal or spatial context it should allow estimates of encounter probabilities and the relative abilities of predators to shift prey preferences (Murdoch, 1969; Murdoch and Oaten, 1975).

Historically, the phenomenon of switching was related conceptually first to specific search images (Tinbergen, 1960) and later to profitability (Royama, 1970). However, if we assume that birds generally hunt to maximize profitability and use whatever proximate cues are available the two explanations of switching converge (Murdoch and Oaten, 1975; Hainsworth and Wolf, 1976). Birds can improve foraging profitability by using proximate cues to enhance intake rate and food quality, through either searching site specificity (Royama, 1970) or differential responsiveness to prey characteristics (Tinbergen, 1960; Croze, 1970; Hainsworth and Wolf, 1976).

Theoretically the concept of optimal choice is pleasing. However, it is difficult to obtain empirical evidence in nature, principally because of the difficulty of quantifying food availability and foraging costs. Numerous studies of foraging in birds can be interpreted relative to optimal foraging. Although most of these do not measure prey availability the birds seem to be foraging in a manner to maximize efficiency. For example, Goss-Custard (1970) found that Redshanks *(Tringa totanus)* tended to forage in areas of highest food density, while Davies (1976) found shifts in aggressive behavior and feeding sites associated with changing availability of substrate on which to forage. Murton *et al.* (1971) were able to show that peck rate was higher for birds in flocks relative to solitary individuals.

In a few cases in the field and laboratory it has been possible to measure expenditures and intakes to provide more direct tests of optimal foraging theory. Nectar-feeding birds offer the opportunity to quantify availability of diverse resource types, as well as expenditures for obtaining food (see Section II,F). Gill and Wolf (1975a) were able to show that several sunbird (Nectariniidae) species exposed to the same array of food differed in their choice as predicted on a weight-related rate of net energy gain model. Werner and Hall (1974) in more controlled conditions showed a decline in selectivity among Bluegills *(Lepomis macrochirus)* feeding on several size classes of *Daphnia* (Cladocera) as prey availability (= density) declined.

Other laboratory studies of birds and mammals have found that parameters other than rate of net energy gain appear to be maximized. For several rodents and finches handling time defined as seeds eaten per unit time or rate of intake was important in determining the seed types eaten (Rosenzweig and Sterner, 1970; Willson, 1971; Willson and Harmeson, 1973). Several species of hummingbirds feeding in the laboratory selected between two sugar water solutions on the basis of concentration, preferring the more concentrated solution rather than the solution providing more net calories per unit feeding time (Hainsworth and Wolf, 1976). In both of these examples it is possible that in nature the cues of handling time or intake rate and concentration are important in short-term assessments of the rate of net energy gain (Hainsworth and Wolf, 1976).

b. Movement Patterns. An organism beginning a foraging episode is very often faced with an unknown array of prey. Theoretically, we can ask how an organism should move through a habitat to maximize its rate of net energy gain (Cody, 1974). The theoretical maximum will depend on the relative benefits and costs associated with some large number of possible foraging movements. The question is how closely the individual in reality can approach this theoretical maximum with information available that can be used as a guide to future benefits and costs.

An organism must move following an attempted prey capture or a decision not to capture. Unless another prey item can be sighted or another location remembered, the question is where to move. The general assumption is that a reward can be used to indicate higher than average probability of another nearby reward (Cody, 1971; Smith, 1974a,b; Baker, 1974; Pyke, 1978). Organisms should remain near good reward sites and leave poor sites. The outcome from any initial, nonmobile prey distribution will be increased patchiness of prey (unless the entire foraging area is covered on each foraging episode), providing a positive feedback for foraging behavior relative to patchy distributions.

Movement patterns of many foraging organisms tend to be similar although produced by different mechanisms (Fraenkel and Gunn, 1940). Many individuals forage nearly in a straight line, generated either by small deviations from straight movements or by alternating left and right (Siniff and Jesson, 1969; Cody, 1971; Smith, 1974a; Pyke, 1978). This reduces the probability of recrossing a foraging path (Cody, 1971; Pulliam, 1974). However, many organisms, ranging from protozoa to mammals, also exhibit a characteristic change in behavior if a reward item or favorable concentration is encountered; they increase the turn angle and maintain a consistent left or right directionality (Fraenkel and Gunn, 1940; Smith, 1974a,b) resulting in circling behavior (circus movements). Theoretically, this should relate to average reward probabilities throughout the habitat and in some cases organisms might not show differential responses to rewarded and nonrewarded foraging attempts (Gill and Wolf, 1977). For organisms that are visual searchers, such as most birds, a minimum memory of movement patterns probably is required to appropriately influence subsequent directionality. At the least a foraging bird would require information on direction of arrival at the present location (Smith, 1974a,b; Pyke, 1978; Gill and Wolf, 1977). Following a nonreward the bird would move off in approximately the same direction of its arrival; if rewarded it would turn (Smith, 1974a,b).

Another possible movement pattern in response to reward is to change the distance moved, longer movements following a nonreward and shorter following a reward. This response has been found for bees (Pyke, 1978) and several birds species (Baker, 1974; Smith, 1974a,b;

Gill and Wolf, 1977). If direction and distance movement patterns are combined, the effect is for a forager to remain when the probability of a reward is high, and to leave areas of low reward probability.

c. Patch Utilization. Staying in an area of high reward would eventually deplete it. Thus, after some time the forager should shift to new areas of possibly higher quality (Tullock, 1971). The shift presumably is related to the temporal decline in the probability of reward (MacArthur and Pianka, 1966) and has been called "giving up time" (Croze, 1970; Krebs *et al.*, 1974; Krebs, 1974). Theoretically, the time an organism will stay in a foraging patch can be related to both *in situ* reward and mean and variance of rewards in previously visited patches over some time interval (Krebs *et al.*, 1974). The reward threshold for leaving a patch should increase as the probability of a good reward increases (Krebs *et al.*, 1974). In their study of Chickadees *(Parus atricapillus)* feeding on concealed meal worms, Krebs *et al.* (1974) found that giving up time tended to be negatively related to average rate of consumption of food items. In a natural situation several sunbird species feeding at flower clusters had some probability of rejecting clusters if they encountered an empty flower (increased probability of lower than average reward level per cluster), but the probability of rejecting a cluster on encountering an empty flower decreased as the average nectar per flower in the area decreased (Gill and Wolf, 1977). Similar relationships were found by Krebs (1974) for feeding Great Blue Herons *(Ardea herodias).* Gill and Wolf (1975a) presented data suggesting that giving up time, or patch transition probabilities, varied among species of sunbirds foraging at flowers of several plant species as a function of body size and hence as a function of differing benefit : cost values. Not surprisingly, the specific rules of patch transition vary among foraging techniques and patch qualities, but the general rules hold in a variety of cases and appear to depend simply on some immediately past knowledge of reward levels.

It is possible that foraging birds can incorporate past rewards into a memory of spatial locations of food sources, even up to 6 months or more after last visiting the source (Miller and Miller, 1971a). Memory would be especially practical on a short-term basis for species exploiting a renewing resource. Several authors have suggested that some hummingbirds utilize a foraging route on which they revisit specific locations at regular intervals (Janzen, 1971; Baker, 1973; Feinsinger and Chaplin, 1975; Stiles and Wolf, 1977). Similar observations have also been made for flocks of frugivores (Wolf, 1976). The length of the route would depend on energy requirements and availability. Most examples of route foraging are reported among exploitation competitors, individuals that use the resource but do not actively interact with other potential visitors (Brian, 1957; Heinrich, 1975; Stiles and Wolf, 1977). In this situation regular utilization of the resurce probably decreases the probability of regular visits by other individuals by reducing the average reward available. The revisit frequency, then, ought to be a complex interaction between maximizing the rate of net energy gain and reducing the probability of high rewards to other individuals (see also Heithaus *et al.*, 1974).

In situations of interference competition (Miller, 1967, 1969), manifested as aggressive exclusion of individuals from foraging locations, regular foraging paths could focus each feeding effort within a territory on spatial locations least recently visited. If quantity of resource available were important to rate of net energy gain (Wolf *et al.*, 1975) then this foraging would preferentially select the high end of the availability spectrum (Gill and Wolf, 1975b, 1977). Perfect territoriality makes the increased reward from this type of foraging very deterministic. The enhanced foraging efficiency should be counted as another potential benefit to be compared to costs in an economic model of territoriality (Brown, 1964; Gill and Wolf, 1975b).

d. Social Interaction. A final technique in optimal foraging involves the use of other foraging individuals to enhance the rate of net energy gain. Keister and Slatkin (1974) presented a general model of the equilibrium distribution of foraging individuals on a resource abundance gradient assuming that foraging individuals use visual cues from other feeders (called "conspecific cuing" by Keister and Slatkin, 1974) to maximize the probability of locating high quality food patches. In general, conspecific cuing predicts the highest consumer densities in areas of maximum resource abundance, but more rapid rates of arrival than for individuals whose foraging movements are determined only by patterns of food abundance. Krebs (1974) provided evidence for herons searching for foraging sites that the probability of landing increased as the number of birds already present in a feeding group increased. Thompson *et al.* (1974) developed a model of flock foraging in which the positive result, on the assumption of conspecific cuing, was not dropping below a minimum rate of intake, rather than maximizing the rate of net intake. In this model flock birds did not do better in terms of maximizing food intake rates by being in a flock but decreased the probability of not encountering prey. Various authors have suggested that groups of foraging birds enhance the probability of locating patchily distributed food and may also serve as information centers once a patch is located (Horn, 1968; Cody, 1971, 1974; Ward and Zahavi, 1973; Emlen and Demong, 1975; Davies, 1976). Birds in flocks may also use each other as "beaters," the general higher level of activity being more likely to flush prey (Rand, 1954; Meyerreicks, 1960; Heatwole, 1965). However, Goss-Custard (1970) suggested that the presence of high densities of Redshanks on a mudflat might reduce the probability of prey being at or near the surface, thus lowering food availability.

Within a feeding flock arranged in a dominance hierarchy (Sabine, 1959; Moore, 1972) it also is possible that the more dominant individual may use subordinates to locate food. Once a high quality patch of food is found by a subordinate the dominant could displace the subordinate and continually move up a quality gradient via cuing and aggressive behavior (Brown, 1975).

The outcome of interspecific dominance interactions among birds often is correlated with size with the larger species dominating the smaller (Wolf, 1970; Stiles and Wolf, 1970; Gill and Wolf, 1975a; Wolf *et al.*, 1976). When several species share the same resource larger species generally have priority. In some circumstances among both sunbirds and hummingbirds individuals of larger species regularly forage in territories being defended by smaller species (Wolf, 1970; Stiles and Wolf, 1970). The limited interaction cost of intrusion is substituted for site fixed defense thus reducing the total costs to the dominant which nonetheless suffers some loss of nectar to the subordinate resident.

2. *Long-Term Optimization*

It is possible through natural selection to evolve a system of long-term optimization of feeding in which individuals apparently "plan ahead." The degree to which these models predict behavioral phenomena depends to a large extent on environmental predictability (Slatkin and Lande, 1976). Unpredictable phenomena are much more difficult to track precisely via genetic information than highly predictable ones. Katz (1974) assumed a regularly varying environment in which the optimization "goal" was to minimize annual foraging time, even though this may at times have forced individuals to forage for longer than a predicted optimum time measured on a daily basis. Removing the daily time constraint, even to the point of allowing negative energy budgets, effectively forces an individual to accumulate energy at a maximum rate when the environment is such that energy gains exceed some threshold. Katz assumed further that low levels of potential energy gain should produce periods of nonfeeding. At a minimum weight threshold the birds would shift to a maintenance feeding level. The general model predicts fairly well the seasonal pattern of weight gains and losses in Nigerian populations of *Quelea quelea* studied by Ward (1965a,b) and could perhaps be extended to other seasonal behaviors, such as migration, molt, or hibernation that involve long-term energy storage (see Section III,D).

A potentially advantageous long-term behavior related to patch utilization is exploration of both unvisited and previously visited patches to monitor changes in resource availability (Royama, 1970; Miller and Miller, 1971b; Smith and Sweatman, 1974; Krebs *et al.*, 1974). In general, exploratory behavior might reduce rate of net energy intake below the theoretical maximum in the short term. Presumably, such behavior could be selectively advantageous in the long term if sufficient high quality patches are encountered or the depletion of know patches is rapid. Again, the foraging individual would sacrifice short term optimal foraging for long-term net benefits (see Section III,E,1).

F. Foraging: Time and Energy Budgets

1. *Hierarchical Behavior*

Foraging is only one way an organism can expend energy or time. In certain environmental situations, some birds may have to feed for 90% or more of a day simply to maintain themselves (Gibb, 1960; Newton, 1969; Goss-Custard, 1970; Heppleston, 1971; Davies, 1976). These examples generally occur in situations where low environmental temperatures force high energy expenditures and/or food availability is low. For organisms that forage on rapidly renewing resources, an alternative option at low resource levels is to stop foraging and allow resource levels to increase. For example, several territorial male Malachite Sunbirds *(Nectarinia famosa)* had negative energy budgets over short time periods as nectar accumulated in flowers, eventually leading to an increased rate of energy gain (Wolf, 1975).

While foraging time is an important class of both time and energy expenditures several other uses of time and energy may take precedence. The total time available to an individual may be divided among three behavioral categories (Schoener, 1971; Keister and Slatkin, 1974) (1) essential behaviors necessarily spent in certain circumstances, such as predator avoidance and protecting against cuckoldry (Wolf and Wolf, 1976); (2) individual interactions, including breeding season interactions among mating partners, aggressive interactions, etc.; and (3) other behaviors which include such activities as feeding, drinking, and resting.

At a specific time a choice hierarchy should prevail among these behaviors generally in the order given. An organism faced with being captured as prey should sacrifice all other activities to escape. Similarly, it is possible to conceive of long-term feeding advantages derived from sacrificing feeding for aggressive interactions (Brown, 1964; Keister and Slatkin, 1974; Gill and Wolf, 1975b; Wolf, 1975), as evidenced by the willingness of most territorial nectarivores to stop feeding and chase intruders (L. Wolf, personal observation). We have already indicated how time spent in the various categories may be reflected differentially in total energy expenditures, and several authors have suggested how a choice between two behaviors might be influenced by the relative costs and benefits (Wolf *et al.*, 1975; Gill and Wolf, 1975b).

On the other hand, the distinctness of these categories is to some extent artifical since many of the behaviors may not be mutually exclusive, especially for particular consumer types. A sit-and-wait or capture predator can announce territorial occupancy, an important component of defense, while waiting for prey or resting. Similarly, male songbirds may sing or patrol a territorial boundary while foraging (Morse, 1968; Williamson, 1973). For many male birds, rest time, especially on an exposed perch, could be important in mating probabilities (Wolf and Hainsworth, 1971). Thus, the complexity of time use precludes any simple definition of how nonexclusive categories are related to fitness. However, many of the behaviors are sufficiently narrowly defined as to be considered mutually exclusive and hence potentially hierarchical.

A formulation of the relation between foraging time and other expenditures has been attempted for nectar-feeding birds based on the assumption that an individual breaks even energetically over a 24-hour period (Wolf *et al.*, 1975). This effectively constrains an individual to meet energy demands, on average, from bout to bout of feeding while storing energy at a sufficiently high average rate to survive overnight. In general the easier food is to obtain, measured as nectar volumes per flower (Wolf *et al.*, 1972, 1976), the less time is required to feed (Fig. 12). A similar result would hold for all birds. The shape of the curve in Fig. 12 suggests that the relative gains in time from improved food quality differ depending on initial quality. At low quality a major decrease in foraging time would accompany a small increase in quality, while at initially high quality an improvement would have virtually no effect on foraging time necessary to achieve a balanced 24-hour energy budget.

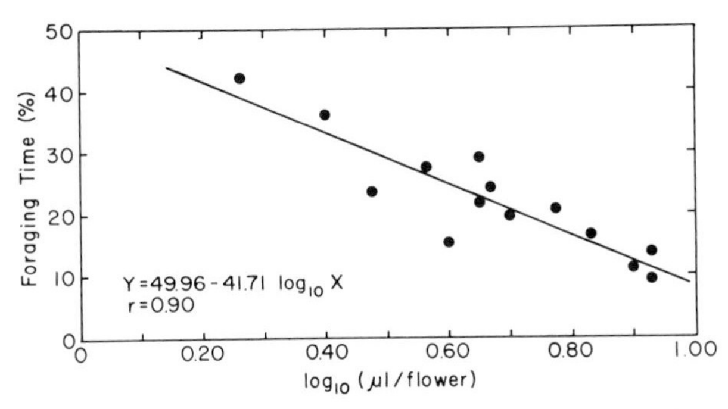

FIG. 12. The relationship between percent of time spent feeding (2–3 hour observation periods) and the volume available per average flower visited by a male Malachite Sunbird (data from Wolf, 1975).

2. *Influence of Other Behaviors on Feeding Time*

An important question in terms of time and energy budgeting is related to how activities other than feeding in the behavioral hierarchy influence feeding behavior. While antipredator or anticuckoldry behavior theoretically produce nothing more than increased total costs and a predictably raised curve in Fig. 12, a potentially critical aspect of these behaviors is their effect on realizing the theoretical maximum rate of energy intake. It is possible, for example, that antipredator or anticuckoldry behavior forces a bird to use a portion of the habitat that provides a lower average quality than other areas or that reduces average prey availability. Hyenas may force lions to hunt in larger groups than optimal (maximum energy gain per time × individual) to counter losses to hyenas once a kill is made (Schaller, 1972; Kruuk, 1972; Caraco and Wolf, 1975). Birds may be forced to hunt in suboptimal areas of the habitat to stay close to cover (T. Caraco, personal communication). Finally, to prevent cuckoldry, a male may have to leave a territory to follow his mate while she gathers nesting material, thus increasing the probability of food losses in unpredictable locations (Wolf and Wolf, 1976). The usual result of this first category of nonfeeding behavior will be to lower foraging efficiency.

An important potential outcome of aggressive interactions is to increase the rate of net energy gain above what otherwise might be available, either simply through lowered use by other birds or else through increased information on patch distribution and quality (see Section III,E,1). However, this is achieved at some cost so that the level of aggressive behavior becomes an important component of any optimality model of feeding behavior. For some birds the outcome seems to differ at high and low resource levels. At high levels of average nectar availability territoriality disappears as the potential benefit is outweighed by the added costs (Fig. 13) (Wolf *et al.*, 1975; Gill and Wolf, 1975b). At very low levels of resource availability the sunbirds also tended to reduce territoriality and reduce the area defended while resource levels increase (Wolf, 1975). Both Davies (1976) and Zahavi (1971) were able to show that the spatio-temporal display of aggressive behavior varied between territoriality and dominance hierarchies with changes in resource availability and distributions. In the case of the Pied Wagtail *(Motacilla alba)* studied by Davies this shift in aggressive behavior occurred over a time interval of less than 5 minutes.

Pulliam *et al.* (1974) have argued that the probability of aggressive

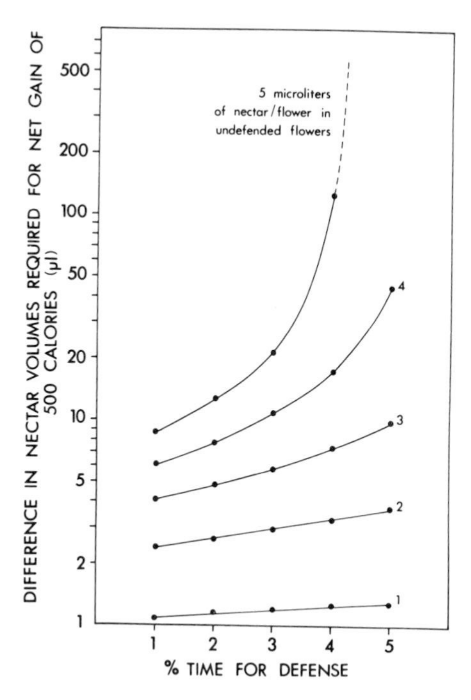

FIG. 13. The relationship between average nectar volumes per flower on a territory and the time an individual must spend on defense. Each line in the graph represents the average nectar required per flower in undefended areas (N_1 = 1, 2, . . . 5μl in undefended areas) if the bird is to realize a slightly better energy budget on the territory. If the undefended areas do not have at least the amount on the line or above relative to the amount on the territory, then the birds should be territorial (from Gill and Wolf, 1975b).

interaction among individuals in a winter junco flock is a negative linear function of ambient temperature if resources are held constant (*ad libitum* availability in the laboratory experiments). This suggests that feeding can take precedence over initiating aggressive interactions, but probably not over responding to an aggressor. However, the relative hierarchical position of aggression and feeding, in this case, is dependent on energy requirements and also food availability in nature (T. Caraco, unpublished).

3. *Food Quality*

We have already discussed mechanisms to enable a bird to select preferentially from a quality array, and we will consider here only characteristics of the food that affect quality. The principal quality characteristics are (1) item content, either energy, nutrients, or detrimental substances; (2) assimilation efficiency; (3) abundance; and (4) dispersion. Characteristics 3 and 4 principally relate to costs per item which have already been discussed in the context of food choice and energy expenditures.

a. Nutrients. Nutrients can be defined as those chemicals in a diet that are necessary for survival and reproduction but which need not be important as immediate sources of energy. Vitamins, water, and essential minerals are examples. The importance of any given nutrient will depend on its rate of utilization and its required representation in the diet.

For consumers ingesting a complex diet it is often difficult to separate selection based on energy value from selection based on specific nutrients. In a study of Icelandic Ptarmigan *(Lagopus mutus)*, Gardarsson and Moss (1970) related food selection to seasonal availability of plant species, but the preferred species were often both highest in measured nutrient levels (nitrogen and phosphorous) as well as digestibility. There were instances, however, when some plant species of relatively low digestibility but high nutrient level were present in the diet in large proportions relative to availability (Gardarsson and Moss, 1970).

Nutritional factors for food selection may be particularly important during periods of reproduction, and an extensive study related to breeding has been carried out by Moss and his co-workers (Moss, 1972; Moss *et al.*, 1975). The selection of heather by Red Grouse *(Lagopus lagopus scoticus)* in relation to nitrogen (N) and/or phosphorous (P) content indicated that the birds consumed food higher in N and P on a seasonal basis related to the growth of the heather such that N and P were at highest concentrations prior to egg laying (Moss,

1972). Also, there were differences in mean breeding success and mean density of Red Grouse on different moors related to differences in N and P content of heather. However, there was no correlation on an annual basis between breeding success and changes in N and/or P contents of heather. Variations in breeding success from year to year were associated with the length of time that heather grew prior to laying and also with heather density. Thus, the effects of maternal nutrition on breeding success appeared related to food quality, but not simply to N and/or P content (Moss *et al.*, 1975).

Calcium is an obvious chemical important to birds due to its use in eggshell. A number of intricate control mechanisms at a physiological level have been investigated for the domestic fowl (see Simkiss, 1975, and Hurwitz, Chapter 8). Included are the storage and release of calcium by medullary bone under the influence of sex hormones and parathormone (Taylor and Stringer, 1965), and an increase in efficiency of calcium absorption mediated through a calcium-binding protein stimulated by vitamin D (Wasserman and Taylor, 1966).

In the chicken, absorption of calcium from food appears necessary on a short-term basis for adequate calcium quantities for egg formation. Thus, food selection during egg formation could be based on calcium as a nutrient. Hughes and Wood-Gush (1971) noted that chickens regulate their calcium intake in relation to demands for egg formation when they are given calcium separately. Mongin and Sauveur (1974) point out that selection of calcium by chickens follows a diurnal pattern associated with egg formation, and deprivation of calcium can lead to behaviors (e.g., increased activity) to replace the nutrient (Hughes and Wood-Gush, 1973).

Hughes and Wood-Gush (1971) suggested that the selection of calcium by laying hens represents a "specific appetite." Such an appetite for a specific nutrient is not unprecedented. Mammals show a highly specific hunger for sodium that is related to salt retention mechanisms (Richter, 1956; Denton, 1965). For sodium (but not calcium) this specificity is apparently not associated with other characteristics of the food, such as novelty (Rozin, 1967), since rats will consume foods with a salt taste after an acute sodium deficiency even when those foods had been avoided previously as the result of aversive aftereffects (Stricker and Wilson, 1970).

Except for a few important chemicals (such as sodium), the selection of nutrients by many consumers does not appear to be specific to those nutrients but rather is mediated by some long-term feedback mechanism influencing food choice through a learning process and the taste novelty of the diet (see Rozin, 1967, for a review). It has long been recognized that organisms deficient in a nutrient select foods containing that nutrient, but Rozin (1967) and his colleagues (Rozin and Kalat, 1971) pointed out that deficient food is avoided and that selection occurs for food that is different from the one that produced the deficiency. These results are for laboratory rats but should be generally applicable to consumer organisms capable of modifying food choice through learning, and it provides an explanation for the selection of foods containing calcium by chickens (Hughes and Wood-Gush, 1971). In general, as long as a diet is energetically profitable and results in no adverse effects, that diet will continue to be selected. If the diet results in an impairment of function, another diet will be selected, and this would be more likely to contain the nutrient of importance (Rozin, 1967).

This phenomenon of food selection based on novelty involves learning and presents some problems for organisms that normally consume a mixed diet. Depending on the nutrient, the feedback on food selection could involve time periods that may encompass consumption of a variety of prey types (Freeland and Janzen, 1974; Westoby, 1974). To the extent that a mixed diet results in nutritious food, individual components need not be selected against. If a mixed diet was not adequate, a consumer would need to obtain information on what components were deficient, presumably through relative proportions consumed.

b. Secondary Compounds. In addition to the nutrient components of food, selection of prey can be influenced by secondary compounds evolved by prey to reduce predation (Freeland and Janzen, 1974). The general behavior of predators in response to poisons is opposite to their behavior toward novel foods when they are deficient in a nutrient since poisons can produce discrimination against novel foods (Rozin and Kalat, 1971; Kalat, 1974). The most completely studied effect of poisons on food selection by birds is the antipredator defense evolved by some species of insects in a Batesian model-mimic system (e.g., Brower *et al.*, 1970). Monarch butterflies *(Danaus plexippus)* sequester cardiac glycosides from milkweed plants (Asclepiadaceae) which induce vomiting by predators such as Blue Jays *(Cyanocitta cristata)* (Brower and Moffitt, 1974). The emetic experience can result in a learning process modifying food selection on a short-term basis. Generalization from such noxious stimuli can result in avoidance of novel foods (Coppinger, 1970; Kalat, 1974). In many natural situations where food selection may be governed both by nutrient requirements and secondary antipredator compounds, a rather delicate balance in food selection could occur based on short- and long-term learning effects.

TABLE I
ASSIMILATION EFFICIENCIES FOR AVIAN PREDATORS

Species	Food	Range of assimilation (%)	Reference
A. Herbivores			
1. *Canachites canadensis* (Spruce Grouse)	Spruce leaves	30	Pendergast and Boag, 1971
2. *Lagopus lagopus scoticus* (Red Grouse)	Heather	21–30	Moss and Parkinson, 1972
3. *Phainopepla nitens* (Phainopepla)	Mistletoe berries	49	Walsberg, 1975
B. Granivores			
4. *Richmondena cardinalis* (Cardinal)	Foxtail seed	47–80	Willson and Harmeson, 1973
	Smartweed seed	12–81	
	Hemp seed	54–84	
	Ragweed seed	68–86	
	Sunflower seed	67–85	
5. *Melospiza melodia* (Song Sparrow)	Foxtail seed	85–91	Willson and Harmeson, 1973
	Hemp seed	75–83	
	Pigweed seed	71–86	
6. *Sporophila aurita* (Variable Seedeater)	Commercial bird seed	76–83	Cox, 1961
7. *Arremonops conirostris* (Green-backed Sparrow)	Commercial bird seed	67–72	Cox, 1961
8. *Volatinia jacarina* (Blue-black Grassquit)	Commercial bird seed	76–83	Cox, 1961
9. *Sporophila nigricollis* (Yellow-bellied Seedeater)	Commercial bird seed	71–78	Cox, 1961
C. Carnivores			
10. *Asio otus* (Long-eared Owl)	Laboratory mice	87	Graber, 1962
11. *Bubo virginianus* (Great-horned Owl)	Mice	68	Duke *et al.*, 1973
	Turkey poults	71	
12. *Nyctea scandiaca* (Snowy Owl)	Laboratory rats	74–80	Gessaman, 1972
13. *Spiza americana* (Dickcissel)	Insects	68	Zimmerman, 1965b
D. Nectarivores			
14. *Archilochus alexandri* (Black-chinned Hummingbird)	Sugar water	97–99	Hainsworth, 1974
15. *Lampornis clemenciae* (Blue-throated Hummingbird)	Sugar water	97–99	Hainsworth, 1974
E. Birds Fed "Laboratory" Food			
16. *Passer domesticus* (House Sparrow)	Chicken mash	77–84	Davis, 1955
17. *Passer domesticus* (House Sparrow)	Ground chicken mash	84–94	Seibert, 1949

TABLE I (continued)

Species	Food	Range of assimilation (%)	Reference
18. *Spizella arborea* (Tree Sparrow)	Chicken mash	65–79	West, 1960
19. *Anas discors* (Blue-winged Teal)	"Duck Growena"	72–81	Owen, 1970
20. *Poephila guttata* (Zebra Finch)	Ground chicken feed	73–81	El-Wailly, 1966
21. *Acanthis hornemanni exilipes* (Hoary Redpoll)	Chicken mash	68–78	Brooks, 1968
22. *Acanthis flammea flammea* (Common Redpoll)	Chicken mash	69–74	Brooks, 1968
23. *Junco hyemalis* (Slate-colored Junco)	Ground chicken mash	77–92	Seibert, 1949
24. *Zonotrichia albicollis* (White-throated Sparrow)	Ground chicken mash	80–91	Seibert, 1949
25. *Cyanocitta cristata* (Blue Jay)	Ground chicken mash	74–76	Seibert, 1949
26. *Spizella pusilla* (Field Sparrow)	Ground chicken mash	92	Seibert, 1949

c. Assimilation of Energy. An important component of benefits and costs in foraging relates to the assimilation of energy from ingested food. Assimilation values range from essentially 100% for simple carbohydrates to less than 30% for plant material that contains large quantities of cellulose or lignin (Table I and references therein). Digestibility and assimilation values are correlated mostly with general food types although specific morphological and metabolic adaptations can produce some modifications in achieved assimilation (Walsberg, 1975).

The quality of a resource could also have a pronounced impact on the time budget of an organism (Wolf and Hainsworth, 1971). Within the limits set by abilities to extract energy from a particular food, the lower the assimilation efficiency, the more food must be processed for a given energy gain. This could place a premium on the evolution of factors influencing intake rates and bulk food processing. Variations in assimilation efficiency also could influence temporal patterns of feeding and food storage as well as body size and energy expenditures for foraging and maintenance.

IV. Synopsis

Our emphasis in this review was on ecological (i.e., natural biological) determinants of energy expenditures and intakes. We explored a variety of metabolic costs of birds and suggested possible important requirements and adaptations associated with their energy expenditures. Expenditures vary with bird size, environmental conditions, and type of activity. The partitioning of available energy or time is an important component of adaptations to ecological conditions.

The energy consumption of an organism was viewed as primarily determined by short- and long-term expenditures in natural situations where biological fitness should be dependent on ability to exploit energy or nutrient resources. Energy intake was explored from a mechanistic and evolutionary point of view. We considered energy intake as a problem in the control of initiation and termination of feeding episodes as well as the daily and seasonal control of energy intake. We also reviewed feeding as an evolutionary problem that could be modeled using theories suggesting that feeding behavior should be optimal as the result of natural selection. The optimality criteria usually are assumed to depend on time or energy constraints. Within this framework the specific foraging techniques employed by an organism, such as item choice, patch use, movement patterns, or meal size, vary with the ecological context of the feeding behavior. The specific rules within a foraging technique undoubtedly vary among most environment–organism interactions. The time period over which changes in behavior occur to maintain optimal foraging in varying ecological conditions is presently an important research area.

Finally, we viewed food intake as it is influenced by other characteristics of quality such as nutrient content, the presence of antipredator compounds, and the ranges of assimilation efficiencies of different foods consumed by birds.

ACKNOWLEDGMENTS

The authors' research has been supported by grants from the National Science Foundation. We thank Stephanie Petrarca and Georgia Ventura for typing several drafts of this manuscript. Drs. Edward Stricker and Bernd Heinrich provided helpful comments on the manuscript.

REFERENCES

Able, K. P. (1973). *Ecology* **54**, 1031–1041.

Aschoff, J., and Pohl, H. (1970). *Fed. Proc., Fed. Am. Soc. Exp. Biol.* **29**, 1541–1552.

Baker, H. G. (1973). *In* "Tropical Forest Ecosystems in Africa and South America: A

Comparative Review" (B. J. Meggers, E. S. Ayensu, and W. D. Duckworth, eds.), pp. 145–159. Random House (Smithsonian Inst. Press), New York.
Baker, M. C. (1974). *Ecology* **55,** 162–167.
Bartholomew, G. A., Howell, T. R., and Cade, T. J. (1957). *Condor* **59,** 145–155.
Baudinette, R. V., and Schmidt-Nielsen, K. (1974). *Nature (London)* **248,** 83–84.
Bellrose, F. (1967). *Proc. Int. Ornithol. Congr., 14th, 1966* pp. 281–309.
Berger, M. (1974). *J. Ornithol.* **115,** 273–288.
Berger, M., and Hart, J. S. (1972). *J. Comp. Physiol.* **81,** 363–380.
Bernstein, M. H., Thomas, S. P., and Schmidt-Nielsen, K. (1973). *J. Exp. Biol.* **58,** 401–410.
Berthold, P. (1975). *In* "Avian Biology" (D. S. Farner and J. R. King, eds.), Vol. V, pp. 77–128. Academic Press, New York.
Brian, A. D. (1957). *J. Anim. Ecol.* **26,** 71–98.
Brooks, W. S. (1968). *Wilson Bull.* **80,** 253–280.
Brower, L. P., and Moffitt, C. M. (1974). *Nature (London)* **249,** 280–283.
Brower, L. P., Pough, F. H., and Meck, H. R. (1970). *Proc. Natl. Acad. Sci. U.S.A.* **66,** 1059–1066.
Brown, J. H., and Bartholomew, G. A. (1969). *Ecology* **50,** 705–709.
Brown, J. L. (1964). *Wilson Bull.* **76,** 160–169.
Brown, J. L. (1975). "The Evolution of Behavior." Norton, New York.
Calder, W. A., III. (1968). *Condor* **70,** 358–365.
Calder, W. A., III. (1972). *Comp. Biochem. Physiol.* **43,** 13–20.
Calder, W. A., III. (1973a). *Ecology* **54,** 127–134.
Calder, W. A., III. (1973b). *Comp. Biochem. Physiol.* **46,** 291–300.
Calder, W. A., III. (1974). *In* "Avian Physiology" (R. A. Paynter, ed.), Publ. No. 15, pp. 86–144. Nuttall Ornithol. Club, Cambridge, Massachusetts.
Calder, W. A., III, and Booser, J. (1973). *Science* **180,** 751–753.
Calder, W. A., III, and King, J. R. (1974). *In* "Avian Biology" (D. S. Farner and J. R. King, eds.), Vol. IV, pp. 260–413. Academic Press, New York.
Caraco, T., and Wolf, L. L. (1975). *Am. Nat.* **109,** 343–352.
Carpenter, F. L. (1974). *Science* **183,** 545–547.
Charnov, E. L. (1976). *Am. Nat.* **110,** 141–151.
Cody, M. L. (1971). *Theor. Popul. Biol.* **2,** 142–158.
Cody, M. L. (1974). *Science* **183,** 1156–1164.
Coppinger, R. P. (1970). *Behaviour* **35,** 45–59.
Cox, G. W. (1961). *Ecology* **42,** 253–266.
Croze, H. (1970). *Z. Tierpsychol., Suppl.* **5,** 1–85.
Davies, N. B. (1976). *J. Anim. Ecol.* **45,** 235–253.
Davis, E. A., Jr. (1955). *Auk* **72,** 385–411.
Dawson, T. J., and Hulbert, A. J. (1970). *Am. J. Physiol.* **218,** 1233–1238.
Dawson, W. R., and Hudson, J. W. (1970). *In* "Comparative Physiology of Thermoregulation" (G. C. Whittow, ed.), Vol. 1, pp. 224–310. Academic Press, New York.
DeBenedictis, P. B., Gill, F. B., Hainsworth, F. R., Pyke, G., and Wolf, L. L. (1978). *Am. Nat.* **112** (in press).
Denton, D. A. (1965). *Physiol. Rev.* **45,** 245–295.
Derickson, W. K. (1974). *Comp. Biochem. Physiol.* **49,** 267–272.
Dessauer, H. C. (1955). *J. Exp. Zool.* **128,** 1–12.
Dolnik, V. R., and Blyumental, T. I. (1967). *Condor* **69,** 435–468.
Drent, R. (1975). *In* "Avian Biology" (D. S. Farner and J. R. King, eds.), Vol. 5, pp. 333–420. Academic Press, New York.

Duke, G. E., Ciganek, J. G., and Evanson, O. A. (1973). *Comp. Biochem. Physiol.* **44,** 283–292.
Duncan, I. J. H., Horne, A. R., Hughes, B. O., and Wood-Gush, D. G. M. (1970). *Anim. Behav.* **18,** 245–255.
El-Wailly, A. J. (1966). *Condor* **68,** 582–594.
Emlen, J. M. (1966). *Am. Nat.* **100,** 611–617.
Emlen, J. M. (1968). *Am. Nat.* **102,** 385–389.
Emlen, S. T., and Demong, N. (1975). *Science* **188,** 1029–1031.
Epstein, A. N., Nicolaides, S., and Miselis, R. (1975). *In* "Neural Integration of Physiological Mechanisms and Behaviour" (G. J. Mogenson and F. R. Calaresu, eds.), pp. 148–168. Univ. of Toronto Press, Toronto.
Epting, R. J., and Casey, T. M. (1973). *Am. Nat.* **107,** 761–765.
Estabrook, G., and Dunham, A. (1976). *Am. Nat.* **110,** 401–413.
Evans, P. R. (1969). *J. Anim. Ecol.* **38,** 415–423.
Farner, D. S. (1970). *Fed. Proc., Fed. Am. Soc. Exp. Biol.* **29,** 1649–1663.
Feinsinger, P., and Chaplin, S. B. (1975). *Am. Nat.* **109,** 217–224.
Fisher, K. C., and Manery, J. F. (1967). *Mamm. Hibernation 3, Proc. Int. Symp., 3rd, 1965* pp. 235–279.
Fogden, M. P. L. (1972). *Ibis* **114,** 307–343.
Fraenkel, G. S., and Gunn, D. L. (1940). "The Orientation of Animals". Dover, New York.
Freeland, W. J., and Janzen, D. H. (1974). *Am. Nat.* **108,** 269–289.
French, A. R. (1976). *Ecology* **57,** 185–191.
Friedman, M. I., and Stricker, E. M. (1976). *Psychol. Rev.* **83,** 409–431.
Gardarsson, A., and Moss, R. (1970). *In* "Animal Populations in Relation to Their Food Resources" (A. Watson, ed.), pp. 47–69. Blackwell, Oxford.
Gaulin, S. J., and Kurland, J. A. (1976). *Science* **191,** 314–315.
Gelperin, A. (1966). *J. Insect Physiol.* **12,** 331–345.
Gessaman, J. A. (1972). *Arct. Alp. Res.* **4,** 223–238.
Gibb, J. A. (1960). *Ibis* **102,** 163–208.
Gill, F. B., and Wolf, L. L. (1975a). *Am. Nat.* **109,** 491–510.
Gill, F. B., and Wolf, L. L. (1975b). *Ecology* **56,** 333–345.
Gill, F. B., and Wolf, L. L. (1977). *Ecology* **58** (in press).
Goss-Custard, J. D. (1970). *J. Anim. Ecol.* **39,** 91–113.
Gould, L. L., and Heppner, F. (1974). *Auk* **91,** 494–506.
Graber, R. R. (1962). *Condor* **64,** 473–487.
Greenewalt, C. H. (1975). *Trans. Am. Philos. Soc.* **65,** 1–67.
Hahn, W. E., and Tinkle, D. W. (1965). *J. Exp. Zool.* **158,** 79–86.
Hainsworth, F. R. (1974). *J. Comp. Physiol.* **88,** 425–431.
Hainsworth, F. R. (1978). *Amer. Zool.* (in press).
Hainsworth, F. R., and Wolf, L. L. (1970). *Science* **168,** 368–369.
Hainsworth, F. R., and Wolf, L. L. (1972). *Am. Nat.* **106,** 589–596.
Hainsworth, F. R., and Wolf, L. L. (1975). *Am. Nat.* **109,** 229–233.
Hainsworth, F. R., and Wolf, L. L. (1976). *Oecologia* **25,** 101–113.
Hainsworth, F. R., and Wolf, L. L. (1978). *Auk* (in press).
Hainsworth, F. R., Collins, B., and Wolf, L. L. (1977). *Physiol. Zool.* **50,** 215–222.
Heatwole, H. (1965). *Anim. Behav.* **13,** 79–83.
Heinrich, B. (1974). *Science* **185,** 747–756.
Heinrich, B. (1975). *Annu. Rev. Ecol. Syst.* **6,** 139–170.
Heithaus, E. R., Opler, P. A., and Baker, H. G. (1974). *Ecology* **55,** 412–419.
Helms, C. W. (1963). *Auk* **80,** 318–334.

Heppleston, P. B. (1971). *J. Anim. Ecol.* **40,** 651–672.
Herreid, C. F., II, and Kessel, B. (1967). *Comp. Biochem. Physiol.* **21,** 405–414.
Hervey, G. R. (1975). *In* "Neural Integration of Physiological Mechanisms and Behaviour" (G. J. Mogenson and F. R. Calaresu, eds.), pp. 109–127. Univ. of Toronto Press, Toronto.
Horn, H. (1968). *Ecology* **49,** 682–694.
Horvath, O. (1964). *Ecology* **45,** 235–241.
Hughes, B. O., and Wood-Gush, D. G. M. (1971). *Anim. Behav.* **19,** 490–499.
Hughes, B. O., and Wood-Gush, D. G. M. (1973). *Anim. Behav.* **21,** 10–17.
Huxley, J. S., Webb, C. S., and Best, A. T. (1939). *Nature (London)* **143,** 683–684.
Janzen, D. H. (1971). *Science* **171,** 203–205.
Kalat, J. W. (1974). *J. Comp. Physiol. Psychol.* **86,** 47–50.
Katz, P. L. (1974). *Am. Nat.* **108,** 758–782.
Keister, A. R., and Slatkin, M. (1974). *Theor. Popul. Biol.* **6,** 1–20.
Kendeigh, S. C. (1969). *Auk* **86,** 13–25.
Kendeigh, S. C., Kontogiannis, J. E., Mazac, A., and Roth, R. R. (1969). *Comp. Biochem. Physiol.* **31,** 941–957.
King, J. R. (1972). *Proc. Int. Ornithol. Congr., 15th, 1970* pp. 200–217.
Kleiber, M. (1961). "The Fire of Life." Wiley, New York.
Kleiber, M. (1972). *J. Theor. Biol.* **37,** 139–150.
Koskimies, J. (1948). *Experientia* **4,** 274–276.
Krebs, J. R. (1974). *Behaviour* **51,** 99–134.
Krebs, J. R., Ryan, J., and Charnov, E. L. (1974). *Anim. Behav.* **22,** 953–964.
Kruuk, H. (1972). "The Spotted Hyena." Univ. of Chicago Press, Chicago, Illinois.
Kuntz, T. H. (1974). *Ecology* **55,** 693–711.
Lasiewski, R. C. (1963). *Physiol. Zool.* **36,** 122–140.
Lasiewski, R. C., and Dawson, W. R. (1967). *Condor* **69,** 13–23.
Lasiewski, R. C., Weathers, W. W., and Bernstein, M. H. (1967). *Comp. Biochem. Physiol.* **23,** 797–813.
LeMagnen, J. (1975). *In* "Neural Integration of Physiological Mechanisms and Behaviour" (G. J. Mogenson and F. R. Calaresu, eds.), pp. 95–108. Univ. of Toronto Press, Toronto.
LeMagnen, J., and DeVos, M. (1970). *Physiol. Behav.* **5,** 805–814.
Leung, P. M. B., and Horwitz, B. A. (1976). *Amer. J. Physiol.* **231,** 1220–1224.
Levitsky, D. A. (1974). *Physiol. Behav.* **12,** 779–787.
Levitsky, D. A., and Collier, G. (1968). *Physiol. Behav.* **3,** 137–140.
Licht, P., and Gorman, G. C. (1970). *Univ. Calif., Berkeley, Publ. Zool.* **95,** 1–52.
Lillywhite, H. B., Licht, P., and Chelgren, P. (1973). *Ecology* **54,** 375–383.
Lissaman, P. B. S., and Shollenberger, C. A. (1970). *Science* **168,** 1003–1005.
MacArthur, R. H., and Pianka, E. R. (1966). *Am. Nat.* **100,** 603–609.
McCaffery, A. R. (1975). *J. Insect Physiol.* **21,** 1551–1558.
McMahon, T. (1973). *Science* **179,** 1201–1204.
McNab, B. K. (1963). *Am. Nat.* **97,** 133–140.
McNab, B. K. (1971). *Ecology* **52,** 845–854.
Marshall, J. T., Jr. (1955). *Condor* **57,** 129–134.
Marten, G. M. (1973). *Ecology* **54,** 92–101.
Meyerreicks, A. J. (1960). "Comparative Breeding Behavior of Four Species of North American Herons," Publ. No. 2, pp. 1–158. Nuttall Ornithol. Club, Cambridge, Massachusetts.
Millar, J. S. (1975). *Can. J. Zool.* **53,** 967–976.
Miller, R. S. (1967). *Adv. Ecol. Res.* **4,** 1–74.

Miller, R. S. (1969). *Brookhaven Symp. Biol.* **22**, 63–70.
Miller, R. S., and Miller, R. E. (1971a). *Blue Jay* **29**, 29–30.
Miller, R. S., and Miller, R. E. (1971b). *Condor* **73**, 309–313.
Mogenson, G. J., and Calaresu, F. R., eds. (1975). "Neural Integration of Physiological Mechanisms and Behaviour." Univ. of Toronto Press, Toronto.
Mongin, P., and Sauveur, B. (1974). *Br. Poult. Sci.* **15**, 349–359.
Moore, N. J. (1972). Ph. D. Dissertation, University of Arizona, Tucson (unpublished).
Morse, D. H. (1968). *Ecology* **49**, 779–784.
Mosher, J. A., and Matray, P. F. (1974). *Auk* **91**, 325–341.
Moss, R. (1972). *J. Anim. Ecol.* **41**, 411–428.
Moss, R., and Parkinson, J. A. (1972). *Br. J. Nutr.* **27**, 285–298.
Moss, R., Watson, A., and Parr, R. (1975). *J. Anim. Ecol.* **44**, 233–244.
Mrosovsky, N. (1970). *Penn. Acad. Sci.* **44**, 172–175.
Mrosovsky, N., and Barnes, D. S. (1974). *Physiol. Behav.* **12**, 265–270.
Mrosovsky, N., and Fisher, K. C. (1970). *Can. J. Zool.* **48**, 241–247.
Murdoch, W. W. (1969). *Ecol. Monogr.* **39**, 335–354.
Murdoch, W. W., and Oaten, A. (1975). *Adv. Ecol. Res.* **9**, 1–131.
Murton, R. K., Isaacson, A. J., and Westwood, N. J. (1971). *J. Zool.* **165**, 53–84.
Nelson, R. A., Wahner, H. W., Jones, J. D., Ellefson, R. D., and Zollman, P. E. (1973). *Am. J. Physiol.* **224**, 491–496.
Newton, I. (1969). *Physiol. Zool.* **42**, 96–107.
Newton, I., and Evans, P. R. (1966). *Bird Study* **13**, 96–98.
Novin, D., Wyrwicka, W., and Bray, G., eds. (1976). "Hunger: Basic Mechanisms and Clinical Implications." Raven, New York.
Ohmart, R. D., and Lasiewski, R. C. (1971). *Science* **172**, 67–69.
Orians, G. H. (1969). *Ecology* **50**, 783–801.
Owen, R. B., Jr. (1970). *Condor* **72**, 153–163.
Passmore, J. C., Pfeiffer, E. W., and Templeton, J. R. (1975). *J. Exp. Zool.* **192**, 83–86.
Paynter, R. A., Jr., ed. (1974). "Avian Energetics," Publ. No. 15. Nuttall Ornithol. Club, Cambridge, Massachusetts.
Peaker, M., ed. (1975). "Avian Physiology." Academic Press, New York.
Pearson, O. P. (1948). *Science* **108**, 44.
Pearson, O. P. (1960). *Bull. Mus. Comp. Zool.* **124**, 93–103.
Pendergast, B. A., and Boag, D. A. (1971). *Condor* **73**, 437–443.
Pengelley, E. T., and Asmundson, S. J. (1970). *Comp. Biochem. Physiol.* **32**, 155–160.
Pengelley, E. T., and Kelley, K. H. (1966). *Comp. Biochem. Physiol.* **19**, 603–617.
Pengelley, E. T., Asmundson, S. J., and Uhlman, C. (1971). *Comp. Biochem. Physiol.* **38**, 645–653.
Pennycuick, C. J. (1969). *Ibis* **111**, 525–556.
Pennycuick, C. J. (1971). *J. Exp. Biol.* **55**, 13–38.
Pennycuick, C. J., and Bartholomew, G. A. (1973). *East Afr. Wildl. J.* **11**, 199–207.
Pough, F. H. (1973). *Ecology* **54**, 837–844.
Pulliam, H. R. (1974). *Am. Nat.* **108**, 59–74.
Pulliam, H. R. (1975). *Am. Nat.* **109**, 765–768.
Pulliam, H. R., Anderson, K. A., Misztal, A., and Moore, N. (1974). *Ibis* **116**, 360–364.
Pyke, G. H. (1978). *Theor. Popul. Biol.* **12** (in press).
Ramsey, J. J. (1970). *Condor* **72**, 225–229.
Rand, A. L. (1954). *Fieldiana, Zool.* **6**, 1–71.
Rapport, D. J. (1971). *Am. Nat.* **105**, 575–588.
Raveling, D. G., and LeFebvre, E. A. (1967). *Bird-Banding* **38**, 97–113.
Real, L. A. (1975). *Theor. Popul. Biol.* **8**, 1–11.
Regal, P. J. (1967). *Science* **155**, 1551–1553.

Richter, C. P. (1956). *In* "L'instinct dans le comportement des animaux et de l'homme" (M. Autoni, ed.), pp. 577–629. Masson, Paris.

Ricklefs, R. E. (1974). *In* "Avian Energetics" (R. A. Paynter, Jr., ed.), Publ. No. 15, pp. 152–292. Nuttall Ornithol. Club, Cambridge, Massachusetts.

Ritter, R. C., and Epstein, A. N. (1975). *Proc. Natl. Acad. Sci. U.S.A.* **72,** 3740–3743.

Rosenzweig, M. L., and Sterner, P. W. (1970). *Ecology* **51,** 217–224.

Royama, T. (1970). *J. Anim. Ecol.* **39,** 619–668.

Rozin, P. (1967). *Handb. Physiol., Sect. 6: Aliment. Canal* **1,** 411–431.

Rozin, P., and Kalat, J. W. (1971). *Psychol. Rev.* **78,** 459–486.

Rubner, M. (1883). *Z. Biol.* **19,** 535–562.

Russek, M. (1975). *In* "Neural Integration of Physiological Mechanisms and Behaviour" (G. J. Mogenson and F. R. Calaresu, eds.), pp. 128–147. Univ. of Toronto Press, Toronto.

Sabine, W. S. (1959). *Condor* **61,** 110–135.

Schaller, G. B. (1972). "The Serengeti Lion." Univ. of Chicago Press, Chicago, Illinois.

Schmidt-Nielsen, K. (1972). *Science* **177,** 222–228.

Schmidt-Nielsen, K. (1975a). *J. Exp. Zool.* **194,** 287–308.

Schmidt-Nielsen, K. (1975b). "Animal Physiology." Cambridge Univ. Press, London and New York.

Schmidt-Nielsen, K., and Larimer, J. L. (1958). *Am. J. Physiol.* **195,** 424–428.

Schmidt-Nielsen, K., and Pennycuick, P. (1961). *Am. J. Physiol.* **200,** 746–750.

Schoener, T. W. (1968). *Ecology* **49,** 123–141.

Schoener, T. W. (1969a). *Symp. Biol.* **22,** 103–114.

Schoener, T. W. (1969b). *Syst. Zool.* **18,** 386–401.

Schoener, T. W. (1971). *Annu. Rev. Ecol. Syst.* **2,** 369–404.

Seibert, H. C. (1949). *Auk* **66,** 128–153.

Selander, R. K. (1966). *Condor* **68,** 113–151.

Simkiss, K. (1975). *Symp. Zool. Soc. London* **35,** 307–337.

Siniff, D. B., and Jesson, C. R. (1969). *Adv. Ecol. Res.* **6,** 185–219.

Slater, P. J. B. (1974). *Anim. Behav.* **22,** 506–515.

Slatkin, M., and Lande, R. (1976). *Am. Nat.* **110,** 31–55.

Smith, C. C. (1975). *In* "Coevolution of Animals and Plants" (L. E. Gilbert and P. H. Raven, eds.), pp. 53–77. Univ. of Texas Press, Austin.

Smith, J. N. M. (1974a). *Behaviour* **48,** 276–302.

Smith, J. N. M. (1974b). *Behaviour* **49,** 1–61.

Smith, J. N. M., and Sweatman, H. P. A. (1974). *Ecology* **55,** 1216–1232.

Smith, W. K., Roberts, S. W., and Miller, P. C. (1974). *Condor* **76,** 176–183.

Snowdon, C. T. (1969). *J. Comp. Physiol. Psychol.* **69,** 91–100.

Snowdon, C. T., and Wampler, R. S. (1974). *J. Comp. Physiol. Psychol.* **87,** 339–409.

Southwick, E. E., and Gates, D. M. (1975). *In* "Perspectives of Biophysical Ecology. Ecological Studies" (D. M. Gates and R. B. Schmerl, eds.), Vol. 12, pp. 417–430. Academic Press, New York.

Spiegel, T. A. (1973). *J. Comp. Physiol. Psychol.* **84,** 24–37.

Stahl, W. R. (1962). *Science* **137,** 205–212.

Stiles, F. G., and Wolf, L. L. (1970). *Auk* **87,** 467–491.

Stiles, F. G., and Wolf, L. L. (1977). *Am. Ornithol. Union Monogr.* (in press).

Stricker, E. M., and Wilson, N. E. (1970). *J. Comp. Physiol. Psychol.* **72,** 416–420.

Taylor, C. R., Schmidt-Nielsen, K., and Raab, J. L. (1970). *Am. J. Physiol.* **219,** 1104–1107.

Taylor, T. G., and Stringer, D. A. (1965). *In* "Avian Physiology" (P. D. Sturkie, ed.), pp. 485–501. Cornell Univ. Press, Ithaca, New York.

Thomas, S. P. (1975). *J. Exp. Biol.* **63,** 273–293.

Thompson, W. A., Vertinsky, I., and Krebs, J. R. (1974). *J. Anim. Ecol.* **43**, 785–820.
Tinbergen, L. (1960). *Arch. Neerl. Zool.* **13**, 265–343.
Toates, F. M., and Booth, D. A. (1974). *Nature (London)* **251**, 710–711.
Tucker, V. A. (1966). *Ecology* **47**, 245–252.
Tucker, V. A. (1968). *J. Exp. Biol.* **48**, 67–87.
Tucker, V. A. (1970). *Comp. Biochem. Physiol.* **34**, 841–846.
Tucker, V. A. (1972). *Am. J. Physiol.* **222**, 237–245.
Tucker, V. A. (1973). *J. Exp. Biol.* **58**, 689–709.
Tucker, V. A. (1975). *Symp. Zool. Soc. London* **35**, 49–63.
Tucker, V. A., and Schmidt-Koenig, K. (1971). *Auk* **88**, 97–107.
Tullock, G. (1971). *Am. Nat.* **105**, 77–80.
Ultsch, G. R. (1973). *Respir. Physiol.* **18**, 143–160.
Utter, J. M., and LeFebvre, E. A. (1970). *Comp. Biochem. Physiol.* **35**, 713–719.
Van Hemel, S. B., and Meyer, J. S. (1969). *Physiol. Behav.* **4**, 339–344.
Vorhies, C. T., and Taylor, W. P. (1922). *U.S., Dep. Agric., Bull.* **1091**, 1–40.
Walsberg, G. E. (1975). *Condor* **77**, 169–174.
Ward, P. (1965a). *Ibis* **107**, 173–214.
Ward, P. (1965b). *Ibis* **107**, 326–349.
Ward, P. (1969a). *Physiol. Zool.* **42**, 85–95.
Ward, P. (1969b). *J. Zool.* **157**, 25–45.
Ward, P., and Zahavi, A. (1973). *Ibis* **115**, 517–534.
Wasserman, R. H., and Taylor, R. H. (1966). *Science* **152**, 791–793.
Werner, E. E., and Hall, D. J. (1974). *Ecology* **55**, 1042–1052.
West, G. C. (1960). *Auk* **77**, 306–329.
Westoby, M. (1974). *Am. Nat.* **108**, 290–304.
Williamson, P. (1973). *Ecol. Monogr.* **41**, 129–152.
Willson, M. F. (1971). *Condor* **73**, 415–429.
Willson, M. F., and Harmeson, J. C. (1973). *Condor* **75**, 225–234.
Wolf, L. L. (1970). *Condor* **72**, 1–14.
Wolf, L. L. (1975). *Ecology* **56**, 92–104.
Wolf, L. L. (1976). *Am. Mus. Novit.* No. 2606, pp. 1–37.
Wolf, L. L., and Hainsworth, F. R. (1971). *Ecology* **52**, 980–988.
Wolf, L. L., and Hainsworth, F. R. (1972). *Comp. Biochem. Physiol.* **41**, 167–173.
Wolf, L. L., and Hainsworth, F. R. (1977). *Anim. Behav.* **25**, 976–989.
Wolf, L. L., and Wolf, J. S. (1976). *Condor* **78**, 27–39.
Wolf, L. L., Hainsworth, F. R., and Stiles, F. G. (1972). *Science* **176**, 1351–1352.
Wolf, L. L., Hainsworth, F. R., and Gill, F. B. (1975). *Ecology* **56**, 117–128.
Wolf, L. L., Stiles, F. G., and Hainsworth, F. R. (1976). *J. Anim. Ecol.* **45**, 349–379.
Yarbrough, C. G. (1971). *Comp. Biochem. Physiol.* **39**, 235–266.
Zahavi, A. (1971). *Ibis* **113**, 203–211.
Zeigler, H. P., Green, H. L., and Lehrer, R. (1971). *J. Comp. Physiol. Psychol.* **76**, 468–477.
Zimmerman, J. L. (1965a). *Physiol. Zool.* **38**, 370–389.
Zimmerman, J. L. (1965b). *Wilson Bull.* **77**, 55–70.

CHAPTER 10

Respiratory Proteins in Birds

A. G. Schnek, C. Paul, and C. Vandecasserie

I. Introduction

The primary function of the respiratory system is to facilitate gas exchange between air and blood to provide oxygen to blood and to extract carbon dioxide from blood.

The morphology of the respiratory system in birds is quite different from that in other vertebrates. The general features have been reviewed (Lasiewski, 1973; Fedde 1975). In contrast to mammals, avian lungs are not elastic. They are small rigid structures but connected to thin walled air sacs that fill large parts of the body cavity. The air sacs change volume under the action of respiratory muscles and thus help gases to move through the lungs where gas exchanges take place. The exchange of oxygen and carbon dioxide between air and blood occurs across the walls of the air and blood capillaries. The size of the exchange area of the air capillaries per unit volume of lung opposed to blood capillaries is around ten times larger in birds than in man (Duncker, 1972).

Oxygen travels through the microbronchi and reaches the blood where a slight portion remains in solution but where most is bound chemically to hemoglobin inside the red cells. As oxygen diffuses toward the blood, the carbon dioxide that is carried by blood, either in solution, bound to proteins (mostly to hemoglobin), or in the form of bicarbonate ion, is displaced to the air capillaries to be removed from the lungs by ventilation.

The blood circulates gases to and from the tissues. The circulatory system of both birds and mammals is derived from the reptilian model. Sturkie (1975) summarized its anatomic and dynamic aspects. Although the morphology of avian circulation has departed less from the reptilian pattern, its mechanism is more efficient than that of mammals. It works at a temperature which is 3°–5°C higher than the normal temperature range of mammals (Calder and King, 1974). Furthermore, the avian heart appears larger relative to body mass than that of mammals, the heart rate is higher, and blood is circulated at pressures that generally exceed those found in mammals of comparable size. Such indications of a relatively high efficiency and performance are associated with the high respiratory and energy requirements of flight (Berger and Hart, 1974).

The oxygen carrier, hemoglobin, is confined inside the erythrocytes. The hemoglobins are essential to the life of all vertebrates; they also occur in some invertebrates and in the root nodules of leguminous plants. All bear the same prosthetic heme group: iron (II) protoporphyrin IX, associated with a polypeptide chain of between 136 and 153 residues. In all the ferrous iron of the heme is linked to nitrogen epsilon of a histidine, the porphyrin is maintained in its pocket by a phenylalanine. About 35 other specific sites along the polypeptide chain are occupied by nonpolar residues. These requirements seem to be sufficient to determine the characteristic fold of the polypeptide chain which is common to the hemoglobins of all species including birds. The remainder of the amino acid sequences of the hemoglobin are variable, the number of amino acid differences between any two species rising with their distance or separation on the evolutionary tree.

From coelacanth to man, the hemoglobins of bony vertebrates consist of $\alpha_2\beta_2$ chains and exhibit cooperative ligand binding. The cooperative effects arise from an equilibrium between two structures, one characteristic for deoxyhemoglobin (T) and the other for liganded hemoglobin (R). Crystals of all liganded hemoglobins from any one species are isomorphous, regardless of the valency of the iron and the nature of the ligand, and their structures have been found to differ only in details at and around the heme.

The average life span of the nucleated avian red blood cells (RBC) is shorter than in mammals and may be related to the higher avian body temperature and metabolic rate. The avian RBC is smaller than the red cells of reptiles but larger than those of mammals. The number and volume of fowl erythrocytes are influenced by many physiological factors and can undergo considerable fluctuation. Bird erythrocytes, like those of other nonmammals, are nucleated, contain mitochondria, although fewer than other somatic cells, and the enzymes of both aerobic glycolysis and the tricarboxylic cycle (Bell, 1971). Biochemical evidence has shown that the mature avian erythrocyte slowly synthesizes hemoglobin. However, wide gaps still exist in our knowledge of the metabolism of erythrocytes and their sources of energy. There is no measurable consumption of glucose nor production of lactate.

This chapter will be mainly devoted to some molecular level problems of avian myoglobins and hemoglobins. Hemoglobin transports the bound molecular oxygen to the tissues. Myoglobin facilitates transfer from the RBC to the tissue and stores the oxygen until it is required for metabolic oxidations. It will be impossible to review the entire literature concerning these proteins. Some approaches will be detailed but others will be mentioned only briefly or even omitted. We will consider mainly erythropoiesis, chemical structures, and physiological aspects.

II. Myoglobin

Myoglobin, like hemoglobin, is a respiratory heme protein, able to undergo a reversible reaction with molecular oxygen, and is present in varying amounts in smooth and striated muscles of all birds. Myoglobin occurs as a monomer and is therefore only one quarter the size of hemoglobin and has only one heme.

Myoglobin and hemoglobin form a molecular system for oxygen transport. Hemoglobin is a carrier with high affinity for oxygen in the presence of a moderate supply and a lower affinity in an oxygen-poor environment. Myoglobin, which can be considered as a repository pigment, shows a high affinity for oxygen under conditions where the carrier gives up its supply. This relationship facilitates movement of oxygen down a diffusion gradient from blood to tissue. Properties of myoglobin are illustrated by its oxygen binding curve, which is hyperbolic and reflects a simple one-to-one association of myoglobin heme and oxygen.

If we represent myoglobin with the abbreviation Mb, one can consider $Mb + O_2 \rightleftharpoons MbO_2$ the equilibrium reaction and $K = (MbO_2)/(Mb)(O_2)$ the equilibrium saturation constant. When the fraction of saturated myoglobin molecules and the oxygen concentration is expressed in terms of the partial pressure of oxygen, then

$$K = \frac{y}{(1 - y)P_{O_2}} \qquad y = \frac{KP_{O_2}}{1 + KP_{O_2}}$$

This is the equation of the hyperbola.

The richest sources of myoglobin are the muscles of aquatic diving vertebrates such as ceatecians among the mammals and the Spheniscidae. It was from sperm whale that Kendrew *et al.* (1961) obtained myoglobin for the first protein X-ray structure analysis. One remarkable feature was that myoglobin was essentially a box for the heme group, formed from eight connected pieces of α helix. The functional site of the myoglobin molecule is the heme group, to whose iron atom the oxygen molecule binds. The purpose of the heme and of the polypeptide chain around it is to keep the ferrous ion from being oxidized giving rise to metmyoglobin, which does not bind oxygen, and giving the iron its special oxygen storage properties. Deconinck *et al.* (1972, 1975) and Peiffer *et al.* (1973) have extended the structural studies to avian myoglobins. They approached the problems with a double purpose: first, to make comparisons at the amino acid sequence level and, if possible, at the three-dimensional level with mammalian myoglobins. The second goal was to elucidate the evolution of the bird globins relative to the other vertebrate classes.

Complete amino acid sequences of chicken *(Gallus gallus)* and a partial sequence of penguin *(Aptenodytes forsteri)* (up to residue 70) have been aligned with the sperm whale myoglobin sequence (Edmundson, 1965) (Table I). On the basis of the total number of amino acid differences between chicken globins and all myoglobins of known amino acid sequence, it appears that these chicken hemoproteins are markedly different from mammalian myoglobins. However, chicken myoglobin is more closely related to myoglobins from other species than to avian hemoglobins.

An examination of the amino acid sequence of chicken myoglobins in relation to the three-dimensional structure of sperm whale shows that the replacements occur at both external and internal positions in almost all parts of the molecule. None of the substitutions observed would induce a drastic change in the spatial conformation exhibited by the sperm whale molecule. Comparison of the amino acid sequences of the chicken myoglobin with those from other species and physicochemical studies in solutions of the chicken, penguin, and sperm whale proteins confirm this conservation of structure (Deconinck, 1973; Harrison and Blout, 1965). The numerous hydrophobic interactions in sperm whale myoglobin which determine the tridimensional structure are also present in both avian proteins. Nevertheless, the greater number of differences between the two avian myoglobins in comparison to the relatively greater degree of sequence homology in all other vertebrates studied might confirm the classic taxonomic studies suggesting that the evolutionary divergence among

TABLE I

AMINO ACID SEQUENCES OF SPERM WHALE, CHICKEN, AND PENGUIN MYOGLOBINS

	10 · 20
Sperm whale[a]	Val -Leu-Ser -Glu-Gly-Glu-Trp-Gln-Leu-Val-Leu-His-Val -Trp-Ala-Lys-Val-Glu-Ala-Asp-Val -Ala
Chicken[b]	Gly-Leu-Ser -Asp-Gln-Glu-Trp-Gln-Gln-Val-Leu-Thr-Ile -Trp-Gly-Lys-Val-Glu-Ala-Asp-Ile -Ala
Penguin[c]	Gly-Leu-Asn-Asp-Gln-Glu-Trp-Gln-Gln-Val-Leu-Thr-Met-Trp-Gly-Lys-Val-Glu-Ser-Asp-Leu-Ala
	30 · 40
Sperm whale	Gly-His-Gly-Gln-Asp-Ile -Leu-Ile -Arg-Leu-Phe-Lys-Ser -His-Pro-Glu-Thr-Leu-Glu-Lys-Phe-Asp
Chicken	Gly-His-Gly-His -Glu-Val -Leu-Met-Arg-Leu-Phe-His-Asp-His-Pro-Glu-Thr-Leu-Asp-Arg-Phe-Asp
Penguin	Gly-His-Gly-His -Ala -Val -Leu-Met-Arg-Leu-Phe-Lys-Ser -His-Pro-Glu-Thr-Met-Asp-Arg-Phe-Asp
	50 · 60
Sperm whale	Arg-Phe-Lys-His-Leu-Lys-Thr-Glu-Ala -Glu-Met-Lys-Ala -Ser-Glu-Asp-Leu-Lys-Lys-His-Gly-Val
Chicken	Lys-Phe-Lys-Gly-Leu-Lys-Thr-Glu-Pro -Asp-Met-Lys-Gly-Ser-Glu-Asp-Leu-Lys-Lys-His-Gly-Gln
Penguin	Lys-Phe-Arg-Gly-Leu-Lys-Thr-Pro (Asp,Glu)Met-Arg-Gly-Ser-Glu-Asp-Met-Lys-Lys-His-Gly-Val
	70 · 80
Sperm whale	Thr-Val -Leu-Thr-Ala-Leu-Gly-Ala-Ile -Leu-Lys-Lys-Lys-Gly-His-His-Glu-Ala-Glu-Leu-Lys-Pro
Chicken	Thr-Val -Leu-Thr-Ala-Leu-Gly-Ala-Gln-Leu-Lys-Lys-Lys-Gly-His-His-Glu-Ala-Asp-Leu-Lys-Pro
Penguin	Thr-Val -Leu-Thr
	90 · 100 · 110
Sperm whale	Leu-Ala-Gln-Ser -His-Ala-Thr-Lys-His-Lys-Ile-Pro-Ile -Lys-Tyr-Leu-Glu-Phe-Ile-Ser-Glu-Ala
Chicken	Leu-Ala-Gln-Thr-His-Ala-Thr-Lys-His-Lys-Ile-Pro-Val -Lys-Tyr-Leu-Glu-Phe-Ile-Ser-Glu-Val
	120 · 130
Sperm whale	Ile-Ile-His-Val-Leu-His-Ser -Arg-His-Pro-Gly-Asn-Phe-Gly-Ala-Asp-Ala-Gln-Gly-Ala-Met-Asn
Chicken	Ile-Ile-Lys-Val-Ile -Ala -Glu-Lys-His-Ala -Ala -Asp-Phe-Gly-Ala-Asp-Ser-Gln-Ala -Ala-Met-Lys
	140 · 150
Sperm whale	Lys-Ala-Leu-Glu-Leu-Phe-Arg-Lys-Asp-Ile -Ala-Ala-Lys-Tyr-Lys-Glu-Leu-Gly-Trp -Gln-Gly
Chicken	Lys-Ala-Leu-Glu-Leu-Phe-Arg-Asp-Asp-Met-Ala-Ser-Lys-Tyr-Lys-Glu-Phe -Gly-Phe-Gln-Gly

[a] Edmundson (1965).
[b] Deconinck *et al.* (1975).
[c] Peiffer *et al.* (1973).

birds took place earlier than the divergence between mammals themselves.

III. Hemoglobin

A. Erythropoiesis

Erythropoiesis includes the biological processes of cell differentiation and proliferation that give rise to biosynthetic activities which provide appropriate hemoglobin supplies to answer the demands of the respiratory system.

Most data concerning avian erythropoiesis were obtained from studies on the chicken. As in other higher vertebrates, chicken erythropoiesis involves a series of developmental changes in the sites where the red cells grow and multiply, in the red cell populations, and in the various types of hemoglobin produced. The successive hemopoietic organs and cell types will be briefly described, as extensive reviews have been published (Lucas and Jamroz, 1961; Bruns and Ingram, 1974; Godet, 1975).

1. *Erythropoiesis during Development*

a. Successive Erythropoietic Sites. Primitive erythropoiesis starts in the area opaca, extends to the area vasculosa, and then to the entire yolk sac. Simultaneously a low erythropoietic activity takes place in some vessels and mesenchymal islands. This vitellin erythropoiesis reaches its maximum between the tenth and the fifteenth day of incubation and disappears before hatching.

A brief erythropoiesis was observed also in the embryonic liver around the eighth or ninth day and in the spleen between the eighth and the fourteenth day.

Around the twelfth day of incubation, the bone marrow displays a low erythropoietic activity which becomes significant on the sixteenth day. The marrow is the only active organ after hatching.

b. Successive Red Cell Types. Primitive erythropoiesis involves first the differentiation of megalocytes, which are large cells that develop and reach maturation progressively, going from basophilic erythroblasts on the third day of incubation to polychromatophilic erythroblasts on the fourth and the fifth days and then to mature erythrocytes from the sixth day onwards. From the ninth day on, that erythrocyte population starts to decrease.

Before the disappearance of megaloblastic erythropoiesis, normoblastic erythropoiesis begins and is characterized by smaller cells without synchronism in their maturation. It reaches its maximum between the tenth and the fifteenth day.

The first types of red cells which appear in the bone marrow between the tenth and the twelfth day are proerythroblasts followed by erythroblasts. In the next few days polychromatophilic and then mature erythrocytes appear. By hatching time, the blood contains only the latter.

During the first month after birth, erythrocytes whose nuclei are less dense than those present at hatching time are observed. After 25 days, the erythrocytes have a denser nucleus but not yet the same aspect as the adult type.

c. Successive Hemoglobin Components. The production of different hemoglobin types seems to parallel the succession of red cell forms, especially in the first stage of development. First, major and minor embryonic hemoglobins are synthesized by the immature megalocytes. When those start to decrease, the immature normocytes synthesize adult hemoglobins and embryonic hemoglobins between the sixth and seventh days. Fetal hemoglobin appears between the tenth and the twelfth days. All forms of embryonic hemoglobin disappear by day 19, followed by the decrease in fetal hemoglobin until the end of the first postnatal month.

In summary, chicken erythropoiesis follows the general scheme observed in other vertebrates. Basically, several different forms of erythropoiesis succeed one another. RBC production starts with a cell family that differs from the next population in their morphology, their synchronized maturation, and their homogeneous division and differentiation. Changes in the amount or type of hemoglobin are related to changes in erythropoietic sites. However, there is no rigorous correlation between both events. The intrinsic properties of the cells during differentiation may in some way account for the nature of the hemoglobins produced. It seems that from the data gathered to date it is not yet possible to conclude as to the monophyletic or polyphyletic origin of the blood stem cells (Godet, 1975).

2. *Erythropoiesis in the Adult Animal*

From light and electron microscopy studies, Campbell (1967) showed that the first erythroid cells, or hemocytoblasts, in bone marrow have a diameter of 10–12 mμ, large nuclei and nucleoli, and basophilic cytoplasm. They give rise to smaller cells, known as basophilic erythroblasts, with smaller nuclei and more basophilic cytoplasm. There are also smaller cells with nuclear chromatin, the polychromatophil erythroblasts, that contain hemoglobin and from which mature erythrocytes probably arise. Erythropoiesis is generally believed to be under humoral control. An erythropoietic stimulating factor has been also discovered in birds. Although very little is known about this avian erythropoietin it must be different from the mammalian type because it is ineffective in these animals (Rosse and Waldmann, 1966).

B. Heterogeneity and Structural Characterization

The hemoglobin types in circulating RBC's vary during the life cycle, and different kinds of hemoglobin may coexist at each stage of development. Actually, hemoglobin heterogeneity among vertebrates is the rule rather than the exception. Avian hemoglobin is composed of a prosthetic group, heme, and of a protein moiety, globin. The globin is formed of four polypeptide chains, two identical pairs, two α-type and two β-type chains. As the heme is believed to be identical in all hemoglobins, heterogeneity relies on chemical variations in the protein part. The heterogeneity resides in the primary amino acid sequence and is a direct result of the translation of the information encoded in the nuclear material.

1. *Embryonic Hemoglobins*

Most studies on the nature of embryonic hemoglobins have been performed on the chicken protein. Contradictory data have emerged as to the life time as well as their properties (d'Amelio and Salvo, 1961; Fraser, 1961, 1963; Wilt, 1965; Manwell *et al.*, 1963, 1966; Hashimoto and Wilt, 1966). Brown and Ingram (1974) reexamined thoroughly these hemoglobins and added more precise chemical characterizations. No information is available on the possible relation between embryonic hemoglobins and myoglobins.

After 2–5 days of embryonic life a major component (HbP) accounts for about 70% of the hemoglobin material. Two minor ones are present which represent about 20% (HbE) and 10% (HbM). Their replacement by the adult type proteins is complete by days 14–16. From polyacrylamide gel electrophoresis, immunodiffusion, amino acid compositions, and peptide maps, Brown and Ingram concluded that HbP contains different polypeptide chains from any adult hemoglobin while HbE has an α-chain identical to that of the adult major hemoglobin and HbM contains α-chains identical to those of the adult minor hemoglobin. Both minor components had common embryonic β-like chains, designated as ϵ-chains. Moreover, HbP displayed two kinds of α-like chains which differed one from the other by one or two amino acids. It is not yet known if these are the product of alleles expressed in the chicken population used or of two different gene loci. Thus, there are at least five structural genes active in the primitive erythroid cells. Three of the chains are characteristic of embryonic life and the other two are active in the adult red cell line.

Embryonic hemoglobins have also been studied in the Turkey *(Meleagris gallapavo)*, the Red-winged Blackbird *(Agelaius phoeniceus)*, (Manwell *et al.*, 1963), and the Peking Duck *(Anas domesticus)* (Borgese and Bertles, 1965). The results appear similar although differences exist as to the time of the shift to the adult hemoproteins.

2. *Fetal Hemoglobin*

A minor hemoglobin, HbH, is apparent in hemolyzates from late embryos and young chicks up to one month after hatching (Bruns and Ingram, 1974; Keane *et al.*, 1974). By electrophoretic, chromatographic, and immunological analysis, only one type of HbH could be detected during chicken development (Bruns and Ingram, 1974; Godet, 1974), synthesized either in the yolk sac or bone morrow. The only changes seem to reside in the amount of hemoglobin produced (Godet, 1974). From amino acid composition analysis of the peptide chains, N-terminal sequences, and peptide maps, it seems that HbH has an α-chain identical to that of the major hemoglobin component in the adult but a different β-chain (Moss and Hamilton, 1974).

3. *Adult Hemoglobin*

A rapid screening by paper electrophoresis of the hemoglobins of about 50 avian species from 13 orders was performed by Saha and Ghosh (1965). Others have made a more limited sampling of bird hemoglobins using several techniques such as electrophoresis on different media and under various conditions, chromatography, peptide mapping, and amino acid analysis (Manwell *et al.*, 1963; Schnek *et al.*, 1966; Brush and Power, 1970). A few species such as the Cormorant *(Phalacrocorax niger)*, Water Hen *(Amaurornis phoenicurus)*, and Barbet *(Megalaima lineata)* possess three hemoglobin components but the general pattern in most species is either a single component, for instance in *Columba livia, Ara chloroptera*, and several penguins. Most commonly two main components, a major fraction (70%) and a minor fraction (25%) accounted for the hemoglobin present. In all cases a very minor component, less than 2%, seems to be present (Fig. 1).

The general features of these investigations indicate that the greatest similarity in pattern occurred between the proteins of closely

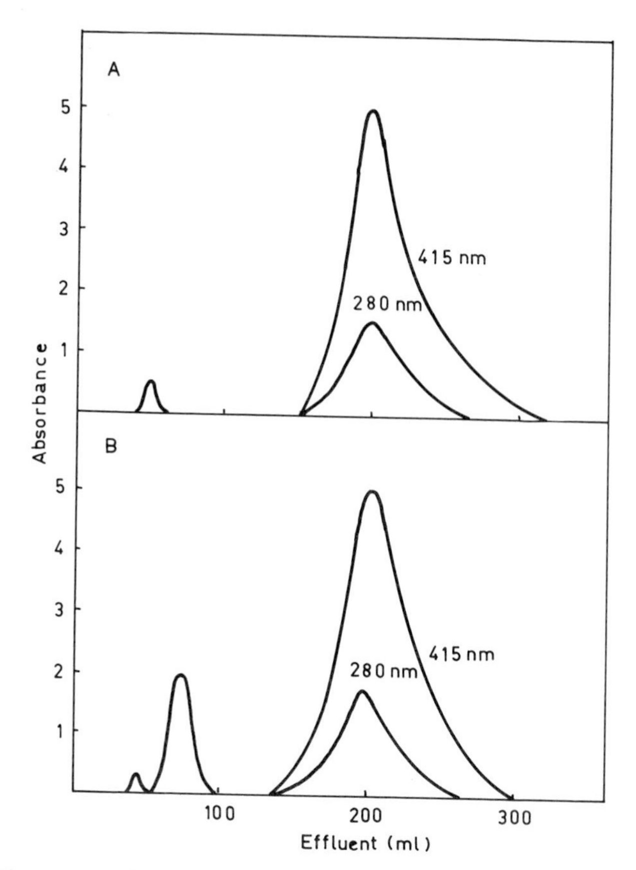

FIG. 1. Chromatographic separation of pigeon (A) and chicken (B) hemoglobin on carboxymethyl cellulose by elution with a discontinuous pH gradient. Buffer 25 m*M* NaH_2PO_4–Na_2HPO_4; pH 6.5–8. From Vandecasserie *et al.* (1971).

related species in the same order and there was more similarity between the minor components of various species or between all the major components than between the minor and the major hemoglobins within one species.

By preliminary analysis such as electrophoretic mobility, the components in species with a single band seem related to the major hemoglobin type in other species but more thorough structural studies or oxygenation curves indicate that these proteins might have an intermediate structure between the major and the minor hemoglobin types (Vandecasserie *et al.*, 1971). The most extensive studies of this sort were performed on *Gallus* hemoglobins and the various kinds of analysis mentioned above have led to confusing interpretations. Most authors (Hashimoto and Wilt, 1966; Schnek *et al.*, 1966; Moss and Thompson, 1969; Brown and Ingram, 1974; Moss and Hamilton, 1974) agree that the presence of two main components differed by the structure of their α-chains but shared very closely related, if not identical, β-chains. Other authors (Lee *et al.*, 1976) obtained a higher resolution of chicken RBC hemolyzates by isoelectrofocusing in polyacrylamide gel. Nevertheless, they were not able to reconcile the chemical nature of that heterogeneity with the current structural model.

C. Primary Structure

The complete amino acid sequences of both peptide chains belonging to the chicken major hemoglobin component were partially established by Schnek *et al.* (1970) and elucidated completely by Matsuda *et al.* (1971, 1973). The amino acid sequence of the minor component α-chain published by Takei *et al.* (1975) and an almost complete primary structure of that hemoglobin β-chain was published by Paul *et al.* (1974) and Vandecasserie *et al.* (1975a). An unexpectedly large structural divergence existed between the two α-chains (65 amino acid substitutions among the 141 residues of the chain), and the probable identity of the β-chains proved that at least three genes must operate in hemoglobin synthesis in the adult chicken. Very few other complete sequences are available for avian hemoglobins. The tryptic peptides of the α-chain derived from the goose *(Anser anser)* major hemoglobin component were aligned by homology (Debouverie, 1975). The primary structure of the penguin single component α-chain was almost completed (Monier, 1972; Monier *et al.*, 1973). Amino acid sequences of avian α-chains are indicated in Table II in comparison with the human α-chain sequence. Chicken β-chains are aligned with the human β-chain in Table. III.

The α-chains of chicken and goose major hemoglobin components are very similar and differ from the penguin hemoglobin α-chain. Some divergence could be expected from the phylogenetic relation of these birds. These three peptide chains have about the same degree of homology with the human α-chain, around 70%. The primary structure of the chicken minor component α-chain is distinct from any of the others. The two chicken β-chains also have approximately 70% of their sequence identical to the human β-chain.

Despite the differences, when examined in detail most of the amino acid substitutions are very conservative. For example, the polar or hydrophobic character of the substituted amino acid is unchanged. So the spatial conformation of all these peptide chains must be very similar. The contact between the α and β polypeptide chains which to-

TABLE II

AMINO ACID SEQUENCES OF AVIAN HEMOGLOBIN α-CHAINS COMPARED TO HUMAN α-CHAINS

	1 10 20 30
Human[a]	Val -Leu-Ser-Pro-Ala -Asp-Lys-Thr-Asn-Val-Lys -Ala -Ala-Trp-Gly-Lys-Val -Gly-Ala-His-Ala -Gly -Glu-Tyr -Gly-Ala -Glu-Ala-Leu-Glu
Chicken 1[b]	Met-Leu-Thr-Ala -Glu -Asp-Lys-Lys-Leu-Ile -Gln -Gln -Ala-Trp-Glu-Lys-Ala -Ala -Ser-His-Gln -Gln -Glu -Phe -Gly-Ala -Glu-Ala-Leu-Thr
Chicken 2[c,d]	Val-Leu-Ser-Asn-Ala-Asp-Lys-Asn-Asn -Val -Lys-Gly -Ile -Phe -Thr-Lys -Ile -Ala -Gly-His -Ala-Glu -Glu -Tyr-Gly-Ala-Glu-Thr -Leu-Glu
	(Ala)
Goose 2[c,e]	Val-Leu-Ser-Ala-Ala-Asp-Lys-Thr-Asx -Val -Lys-Gly -Val -Phe -Ser -Lys -Ile -Ala -Gly-His -Ala-Glx -Gln (Tyr,Gly,Gly,Glx,Thr) Leu-Glu
Penguin[f]	Val-Leu-Ser-Ala-Asp-Lys-Ser-Asn- Val -Lys-Ser -Ile -Phe -Ser -Lys-Leu-His-Thr-His -Ala -Cys-Gly -Tyr -Gly-Ala-Glu-Pro-Leu-Glu
	40 50 60
Human	Arg-Met-Phe-Leu-Ser -Phe -Pro -Thr -Thr-Lys-Thr-Tyr -Phe-Pro-His-Phe-Asp-Leu-Ser-His-Gly-Ser-Ala-Gln-Val-Lys-Gly-His-Gly-Lys
Chicken 1	Arg-Met-Phe-Thr -Thr-Tyr -Pro -Glu -Thr-Lys-Thr-Tyr -Phe-Pro-His-Phe-Asp-Leu-Ser-Pro-Gly-Ser-Asn-Gly-Val-Arg-Gly-His-Gly-Lys
	(Leu)
Chicken 2	Arg-Met-Phe-Ile -Gly-Phe -Pro-Thr-Thr-Lys-Thr-Tyr-Phe-Pro-His-Phe-Asp-Leu-Ser-His-Gly-Ser-Ala-Gln-Ile-Lys-Gly-His-Gly-Lys
Goose 2	Arg-Met-Phe- Glx -Ala -Tyr -Pro-Thr-Thr-Lys-Thr-Tyr-Phe-Pro-His-Phe-Asx-Leu-Ser-His(Gly,Glu,Ala)Glx-Ile- Lys-Gly-His-Gly-Lys
Penguin	Arg-Met-Phe-Glx-Thr-Tyr -Pro-Thr-Thr-Lys-Thr -Tyr-Phe-Pro-His-Phe-Asp-Leu-Ser-His-Gly-Ser-Ala-Glx-Val-Lys-Ala-His-Gly-Lys
	70 80 90
Human	Lys-Val- Ala -Asp -Ala -Leu-Thr -Asn -Ala-Val -Ala -His -Val -Asp -Asp -Met-Pro-Asn -Ala -Leu-Ser -Ala -Leu-Ser -Asp -Leu-His -Ala -His -Lys
Chicken 1	Lys-Val- Leu-Gly -Ala -Leu-Gly -Asn -Ala-Val -Lys -Asn -Val -Asp -Asn -Leu-Ser-Gln -Ala -Met-Ala -Glu -Leu-Ser -Asn -Leu-His -Ala -Tyr -Asn
Chicken 2	Lys-Val- Ala -Leu-Ala -Ile -Thr -Asn -Ala-Ile -Glu -His -Ala -Asp -Asp -Ile -Ser-Glu -Ala -Leu-Ser -Lys -Leu-Ser -Asp -Leu-His -Ala -His -Lys
Goose 2	Lys-Val- Ala -Asx -Ala -Leu-Gly -Lys -Ala-Val -Glu -His -Gly -Asp -Asp -Ile -Ser-Gly -Ala -Leu-Ser -Lys -Leu-Ser -Asp -Leu-His -Ala -His -Lys
Penguin	Lys-Val- Ala -Asx -Glx -Ile -Gly -Lys -Ala-Ile -Ala -Glx -Leu-Asx -Asx -Ile -Ala-Gly -Ala -Leu-Ser -Lys -Leu-Ser -Asx -Leu-His -Ala -Glx -Lys
	100 110 120
Human	Leu-Arg-Val-Asp-Pro-Val-Asn -Phe -Lys -Leu-Leu-Ser -His -Cys -Leu-Leu-Val -Thr -Leu-Ala-Ala -His -Leu-Pro -Ala -Glu -Phe -Thr -Pro -Ala
Chicken 1	Leu-Arg-Val-Asp-Pro-Val-Asn -Phe -Lys -Leu-Leu-Ser -Gln -Cys -Ile -Gln -Gln -Val -Leu-Ala-Val -His -Met-Gly -Lys -Asp -Tyr -Thr -Pro -Glu
Chicken 2	Leu-Arg-Val-Asp-Pro-Val-Asn -Phe -Lys -Leu-Leu-Gly -Gln -Cys -Phe -Leu-Val -Val -Leu-Val -Ala -His -Leu-Pro -Ala -Glu -Leu-Ala -Pro -Lys
Goose 2	Leu-Arg-Val-Asp-Pro-Val-Asn -Phe -Lys -Leu-Leu-Gly -Glx -Cys -Phe -Leu-Val -Thr -Leu-Val -Ala -His -Leu-Pro -Ala -Glu -Phe -Ala -Pro -Lys
Penguin	Leu-Arg(Val,Asx)Pro-Val- Asx -Phe -Lys -Leu-Leu-Ser -His -Gly -Leu-Glx -Val -Ala -Asx- Ala-Lys -Asx -Leu-Val -Arg -Glx -Phe -Thr -Pro - Gly
	130 140
Human	Val-His-Ala-Ser-Leu-Asp-Lys-Phe-Leu-Ala -Ser-Val-Ser -Thr-Val-Leu -Thr -Ser -Lys-Tyr-Arg
Chicken 1	Val-His-Ala-Ala-Phe-Asp-Lys-Phe -Leu-Ser -Ala-Val-Ser -Ala -Val-Leu -Ala -Glu -Lys-Tyr-Arg
Chicken 2	Val-His-Ala-Ser-Leu-Asp-Lys-Phe -Leu-Lys-Ala-Val-Gly -Thr-Val-Leu -Thr -Ala -Lys-Tyr-Arg
Goose 2	Val-His-Ala-Ser-Leu-Asp-Lys-Phe -Leu-Lys-Ala-Val-Gly -Thr-Val-Leu -Thr -Ala -Lys-Tyr-Arg
Penguin	Val-Thr-Ala-Ser-Leu-Asx-Lys(Ile, His, Lys,Ser,Val,Ser, Ala, Ala,His, Gln, Ala) Lys-Tyr-Arg

[a] Braunitzer *et al.* (1961).
[b] Takei *et al.* (1975).
[c] Paul *et al.* (1974).
[d] Matsuda *et al.* (1971).
[e] Debouverie (1975).
[f] Monier (1972); Monier *et al.* (1973).

TABLE III

COMPLETE AMINO ACID SEQUENCE OF THE CHICKEN MAJOR HEMOGLOBIN β-CHAIN AND PARTIAL AMINO ACID SEQUENCE OF THE CHICKEN MINOR HEMOGLOBIN β-CHAIN COMPARED TO THE HUMAN β-CHAIN SEQUENCE

	10 ... 20
Human[a]	Val -His-Leu -Thr-Pro -Glu-Glu-Lys-Ser -Ala -Val -Thr-Ala -Leu-Trp-Gly-Lys-Val-Asn-Val-Asp -Glu
Chicken 2[b]	Val -His-Trp -Thr-Ala -Glu-Glu-Lys-Gln -Leu -Ile -Thr-Gly -Leu-Trp-Gly-Lys-Val-Asn-Val-Ala -Glu
Chicken 1[c]	Cal -His-Trp -Thr-Ala -Glu-Glu-Lys-Gln -Leu -Ile -Thr-Gly -Leu-Trp-Gly-Lys-Val-Asn-Val-Ala -Glu
	30 ... 40
Human	Val -Gly-Gly-Glu-Ala-Leu -Gly-Arg-Leu-Leu -Val -Val-Tyr-Pro-Trp -Thr-Gln-Arg-Phe-Phe -Glu -Ser
Chicken 2	Cys-Gly-Ala -Glu-Ala-Leu -Ala -Arg-Leu-Leu -Ile -Val-Tyr-Pro-Trp -Thr-Gln-Arg-Phe-Phe -Ala -Ser
Chicken 1	Cys-Gly-Ala -Glu-Ala-Leu -Ala -Arg-Leu-Leu -Ile -Val-Tyr-Pro-Trp (Thr,Gln)Arg-Phe-Phe
	50 ... 60
Human	Phe-Gly -Asp -Leu-Ser-Thr-Pro -Asp -Ala -Val -Met-Gly-Asn-Pro -Lys -Val -Lys-Ala-His-Gly-Lys-Lys
Chicken 2	Phe-Gly -Asn -Leu-Ser-Ser -Pro -Thr -Ala -Ile -Leu -Gly-Asn-Pro -Met-Val -Arg -Ala-His-Gly-Lys-Lys
Chicken 1	Ala-His-Gly-Lys-Lys
	70 ... 80
Human	Val-Leu-Gly-Ala -Phe-Ser -Asp-Gly -Leu -Ala -His -Leu-Asp-Asn -Leu -Lys -Gly -Thr-Phe -Ala-Thr -Leu
Chicken 2	Val-Leu-Thr-Ser -Phe-Gly -Asp-Ala -Val -Lys-Asn-Leu-Asp-Asn -Ile -Lys-Asn -Thr-Phe-Ser-Gln -Leu
Chicken 1	Val-Leu-Thr-Ser -Phe(Gly, Asp,Ala)Val -Lys(Asn, Leu,Asp,Asn, Ile) Lys, Asn(Thr,Phe,Ser,Gln,Leu
	90 ... 100 ... 110
Human	Ser-Glu-Leu-His-Cys-Asp-Lys -Leu-His-Val-Asp -Pro-Glu-Asn-Phe-Arg -Leu-Leu-Gly -Asn -Val -Leu
Chicken 2	Ser-Glu-Leu-His-Cys-Asp-Lys -Leu-His-Val-Asp -Pro-Glu-Asn-Phe-Arg -Leu-Leu-Gly -Asp -Ile -Leu
Chicken 1	Ser, Glu, Leu, His)Cys-Asp-Lys Pro-Glu-Asn-Phe-Arg
	120 ... 130
Human	Val -Lys-Val-Leu-Ala -His -His-Phe -Gly -Lys- Glu-Phe-Thr-Pro-Pro-Cal -Gln-Ala-Ala-Tyr -Gln-Lys
Chicken 2	Ile -Ile -Val-Leu-Ala -Ala -His-Phe -Ser -Lys- Asp-Phe-Thr-Pro-Glu-Cys -Gln-Ala-Ala-Trp -Gln-Lys
Chicken 1	Asp(Phe, Thr, Pro)Gly-Cys
	140
Human	Val -Val-Ala -Gly-Val-Ala -Asn -Ala-Leu-Ala -His -Lys-Tyr-His
Chicken 2	Leu-Val-Arg -Val -Val-Ala -His -Ala-Leu-Ala -Arg -Lys-Tyr-His
Chicken 1	Leu-Val-Arg ,Cal -Val-Ala -His -Ala-Leu-Ala -Arg, Lys-Tyr-His

[a] Braunitzer *et al*. (1961).
[b] Matsuda *et al*. (1973).
[c] Paul *et al*. (1974); Vandecasserie *et al*. (1975a).

gether form either of the two dimers is called the $\alpha_1\beta_1$ contact. The residues involved in the interactions between polypeptides in the avian hemoglobin chains examined are not unlike those proposed for horse (Perutz *et al.*, 1968; Perutz, 1969). Thus, the structural correlates of molecular function are retained in avian hemoglobins.

The bonds which hold the two dimers together are not as rigid as between the dimers and allow the rearrangement of the chains in response to the presence of oxygen. This contact is known as the $\alpha_1\beta_2$ or $\alpha_2\beta_1$ contact and is vital for the heme–heme interaction. The residues involved in this association again form mainly nonpolar interactions (Perutz, 1969). As in numerous species already studied, avian hemoglobins are almost invariant in these positions. Among the amino acids in contact with the heme (Perutz, 1969), once again very few have undergone substitutions in the avian hemoglobins already sequenced. Phylogenics of both myoglobin and hemoglobins are reviewed by Holmquist *et al.* (1976).

D. Oxygen Affinity

The oxygen dissociation curve for hemoglobin has a sigmoid shape when y, the fraction saturated with oxygen, is plotted against P, the partial pressure of oxygen. Hill (1910) described this curve with the following formula: at a given pH,

$$y = \frac{KP^n}{1 + KP^n}$$

n determines the degree of heme–heme interactions that arise from the rearrangement of the molecule subunits as the two non-α-chains come near upon oxygenation. The oxygenation curves are often referred to by characteristic parameters, the n value and the oxygen half saturation pressure, P_{50}. The alkaline Bohr effect is another property of hemoglobin and occurs when a drop in blood pH due to the changed level of carbon dioxide causes a decrease in oxygen affinity. It is present down to pH 6, well below life conditions, and then gives way to the acidic Bohr effect, i.e., an increase in the oxygen affinity with a further decrease in pH.

1. Oxygen Affinity of Whole Avian Blood

The first studies performed on avian hemoglobin oxygenation were undertaken mainly in comparison to the mammalian system. Birds are able to fly at high altitudes and the formed elements of the blood differ from mammals in many respects. This raises questions on possible correlation with their hemoglobin oxygenation properties. Experiments were first carried out on erythrocytes. The data compared were the oxygen half saturation pressure (P_{50}) of whole blood to the same parameter obtained for man. Various experimental conditions were tested, most important was the effect of hypoxia by variation in altitude. The oxygen capacities of the Bolivian Goose *(Choephaga melanoptera)* and the Rhea *(Rhea americana)* were measured (Hall *et al.*, 1936). These two species live at sea level and at high altitude. The oxygen affinity of both species, measured at sea level and around 3500 m, increased with altitude. On the other hand, these birds display a lower P_{50} than related species that live at sea level. This is in contrast to the adaptive changes in humans (Lenfant *et al.*, 1968). Some adaptation may have occurred but experimental details are missing and the problem has to be reinvestigated. Later studies determined the P_{50} and Hill's coefficient for the goose blood oxygenation curve (Danzer and Cohn, 1967) and compared the same parameters obtained in man. The P_{50} measured was higher than in man, 37 and 26 Torr, respectively, but Hill's coefficient, $n = 2.7–2.8$ seems very close. The blood oxygenation properties of the Penguin *(Pygoscelis adeliae)* have also been compared to those of flying birds and of other diving animals (Lenfant *et al.*, 1969). *Pygoscelis* had a higher hemoglobin concentration and oxygen affinity, a lower Bohr effect, and a higher buffering capacity than flying birds, and resembled diving mammals in all these features.

The same oxygenation parameters were measured for many other avian species including the chicken (Christensen and Dill, 1935; Bartels *et al.*, 1966), duck (Magath and Higgins, 1934; Andersen and Lovo, 1967), pheasant *(Phaseanus colchicus)* (Christensen and Dill, 1935), pigeon (Magath and Higgins, 1934; Christensen and Dill, 1935), and sparrow *(Passer domesticus)* (Tucker, 1972). Many disagreements appear in the literature and the data on experimental conditions, pH, temperature, and measurement techniques vary and are not always detailed completely. Recent studies (Lutz *et al.*, 1973) contradict the commonly held idea that bird blood oxygen affinity was lower than in mammals. Hemoglobin deoxygenation curves at normal blood pH, pCO_2, and temperature were obtained for pigeon (Lutz *et al.*, 1973) and for other birds (Lutz *et al.*, 1974) and show half saturation pressures of bird bloods similar to those of mammals and decreasing with increasing body size. Thus it is probably related to the metabolic rate (Lasiewski and Dawson, 1967). The authors suggest that the discrepancies with earlier reports probably were due to technical differences. Oxygen pressure must be measured very rapidly, otherwise the high rate of oxygen consumption by the avian nucleated red cells leads to erroneously low P_{50} values. However Scheid and Kawashiro (1975) report that if the time of anaerobic storage of blood was short enough, no appreciable errors resulted from red cell metabolism and the source of the wide divergencies in the P_{50} published values remains to be resolved.

A high P_{50}, or a low oxygen affinity, as reported by many authors, seems incompatible with the requirements of high altitude flight. Presumably other factors of avian respiration compensate and provide the answer (Lasiewski, 1973; Berger and Hart, 1974).

Other experiments on whole blood have concerned the development of the oxygenation curve in the chicken before and after hatching (Bartels *et al.*, 1966). Whole blood showed an increase in P_{50} after hatching. However, while this change is usually continuous, a striking difference appeared in the chicken with a marked drop in affinity of 13 Torr, at pH 7.4, and 39°C, from the day before to the day after hatching. The reason for this rapid affinity change is difficult to explain. As mentioned previously, it cannot be due to a variation in hemoglobin types nor in red cells morphology; both these changes take place much earlier in avian development. A modification in the base excess or in the difference of pH between the erythrocytes and the blood may explain this difference.

2. *Oxygen Affinity of the Isolated Hemoglobins*

Chicken hemoglobins were first studied functionally in a semipure form by Huisman *et al.*, 1964; Huisman and Schillhorn van veen, 1964. Oxygenation curves of the two main fractions from the blood hemolyzate in *Gallus* were measured in phosphate buffer from pH 6.4 to 7.1. The minor component, A1, displayed a higher oxygen affinity than the major one, A2. The difference in log P_{50} was around 0.1. The Hill coefficients were 2.61 ± 0.10 for A1 and 2.35 ± 0.1 for A2. When phosphate concentration was increased or when NaCl was added to the sample, higher values for P_{50} and the n coefficient were obtained. Simultaneously, it was observed that the oxygen affinity of the isolated hemoglobins was high when compared with the blood hemolyzate that had not been either chromatographed or dialyzed. This was confirmed by the experiments performed by Chanutin and Curnish (1967) and by Benesch (1968) in mammals. When the 2,3-diphosphoglycerate (2,3-DPG) present in mammalian red cells was eliminated from the hemolyzate by gel filtration, hemoglobin oxygen affinity increased. When 2,3-DPG was reintroduced to the sample, the lower oxygen affinity was recovered. This explained discrepancies among published oxygenation curves for mammalian hemoglobins where the incomplete elimination of this cofactor by various chromatographic or dialysis procedures had occurred.

Rapoport and Guest (1941) had identified the main organic phosphate present in mammalian erythrocytes as 2,3-diphosphoglycerate (2,3-DPG). However inositol hexaphosate (IP6) predominated in birds and adenosine phosphate (ATP) in reptiles and fishes. But it was more then 20 years before the role of these molecules as allosteric regulators of oxygen binding to hemoglobins was recognized. Recently, Borgese and Lampert (1975) studied the phosphorylated compound in duck during development and reported that erythrocytes drawn from embryos contain 2,3-DPG and that IP6 becomes the major organic phosphate only in the adult. They suggested a possible relationship between the kind of cofactor and the types of hemoglobins present in the embryonic and posthatching red blood cells. Most authors accept the Rapoport and Guest (1941) identification of the avian phosphorylated cofactor as IP6. Actually, the nature of the phosphate component in pigeon and chicken erythrocytes was reinvestigated (Steward and Tate, 1969; Benzon, 1972) and was established as myoinositol pentaphosphate (IP5) rather than IP6. Nevertheless, almost all studies were pursued using IP6 because IP5 is difficult to purify. However experiments performed by Vandecasserie *et al.* (1976) demonstrate that IP5 has only a slightly lower effect than IP6 on hemoglobin oxygenation.

The influence of phosphorylated cofactors on hemoglobin oxygenation curves was demonstrated first by Benesch and Benesch (1969) who showed that the low affinity of the chicken hemolyzate when compared to the human hemolyzate was due mainly to IP6. In fact, if these two hemolyzates are stripped from their respective phosphorylated cofactor, they both respond very similarly to the addition of either phosphate. Vandecasserie *et al.* (1971) refined these studies. They showed that the unfractionated mixture of *Gallus* hemoglobins in the presence of IP6 had an oxygen affinity similar to that of the human protein. Subsequently, when the two avian hemoglobin fractions were separated, despite the increase in the P_{50} brought about by the addition of IP6, the minor component still showed a higher oxygen affinity that the major one. They concluded that it was the nature of the cofactor and the structure of the protein that together determined the value of P_{50} (Fig. 2). Further observations on pigeon *(Columba)* erythrocyte hemolyzate showed that this single hemoglobin component had oxygenation parameters intermediate between the two chicken hemoglobins, although they were more similar to the chicken major component. Vandecasserie *et al.* (1973) also measured the oxygenation

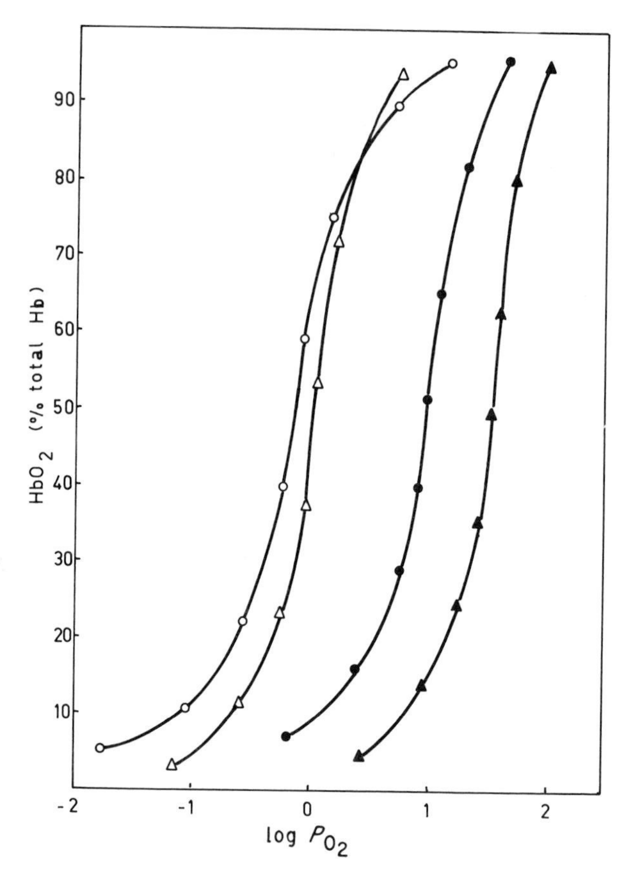

FIG. 2. Oxygen equilibria of major and minor components of chicken hemoglobin at pH 7. From Vandecasserie *et al.* (1971). Buffer was 25 m*M* Tris-HCl; temperature 20°C. Hemoglobin concentration 0.1%. ○, Minor component free of phosphate. △, Major component free of phosphate. ●, Minor component with 1 mole IP6 per mole of tetramer. ▲, Major component with 1 mole IP6 per mole of tetramer.

curves of the two hemoglobins of turkey, pheasant, and duck, and of the single component of parakeet and penguin and established the general nature of these observations. The Bohr effect for all these hemoglobins was more pronounced in the absence of IP6 and the effect of IP6 was stronger at more acidic pH's (Vandecasserie *et al.*, 1973).

The P_{50} values obtained by various methods (Vandecasserie *et al.*, 1975c; Imai *et al.*, 1970) were in agreement and, when applied to both chicken hemoglobins and to the single pigeon protein, gave equilibrium constants k_1 and k_4 corresponding to the oxygenation equilibrium of the first and the fourth subunits, respectively. The values obtained for the constant k_1 are comparable to those for human hemoglobin. The k_4 constants, on the other hand, were four times lower than the human protein, both in the absence and presence of 2,3-DPG or IP6. The global kinetic constants of deoxygenation were obtained by stopped flow for these samples (Vandecasserie *et al.*, 1975b). Preliminary results confirmed the effect due to the phosphorylated cofactors but no obvious difference between the minor and the major components explains their different oxygen affinities.

The binding of IP6 to human deoxyhemoglobin was determined directly by X-ray crystallography (Arnone and Perutz, 1974). IP6, like 2,3-DPG, lies at the entrance of the central cavity of the tetramer and is bound by ionic interactions. These interactions are more numerous between hemoglobin and IP6 than between hemoglobin and 2,3-DPG. Indeed, 2,3-DPG forms salt bridges with residues Val 1, His 2 and His 143 of both β-chains and Lys 82 of one β subunit (Arnone, 1972). Besides those four ionized groups Asn 139 may form hydrogen bonds with IP6.

The primary structure of the chicken β subunits implied that IP6 assumes the same position and hypotheses exist as to its mode of fixation in those molecules (Arnone and Perutz, 1974). The number of basic groups facing the IP6 in avian hemoglobin would be 12 instead of 8 as in human: Val 1, His 2, Lys 82, Arg 135, His 139, and Arg 143 of both β-chains, and so the energy binding would be very high. Comparative studies on a diverse set of mammals indicate compensatory structural changes to preserve functional changes in the face of changed affinity or 2,3-DPG binding ability (Scott *et al.*, 1976). No comparable body of data exists for birds.

IV. Conclusions

Hemoglobin was the first protein to be crystallized and to be associated with a specific physiological function. Reichert and Brown (1909) were convinced that the morphology of hemoglobin crystals from different animals could provide evidence for the expression of species specificity in protein structure. Hemoglobin was also the first eukaryotic protein to be synthesized in a cell-free system *in vitro*, an experiment which proved that the apparatus for protein synthesis in eukaryotes was similar to that discovered earlier in procaryotes (Schweet *et al.*, 1958). Messenger RNA for globin was the first eukaryotic messenger to be isolated (Marbaix and Burny, 1964) and to have its nucleotide base sequence determined (Proudfoot and

Brownlee, 1976). The transition of hemoglobin synthesis from the fetal to the adult form during the late prenatal and early postnatal period, is one of the simplest manifestations of cell differentiation. Ingram's discovery that sickle cell anemia was caused by the replacement of a single one of the 287 amino acid residues in the half molecule of hemoglobin showed for the first time in any organism that a point mutation in a structural gene produces a product with a single amino acid change and can have significant physiological effects. Thus studies on hemoglobins have provided insights on the molecular basis for a variety of important biological processes.

Hemoglobin is a two-way respiratory carrier, transporting oxygen from the lungs to the tissues and facilitating the return transport of CO_2. It executes this dual function by a reversible change of structure first delineated by Haurowitz (1938). The arterial blood has a high affinity for oxygen and a low one for hydrogen, CO_2, and organic phosphates. The relative affinities are reversed in venous blood. These changes are brought about by spacial relationships of the hemoglobin dimers. If heme is considered the active site of hemoglobin, oxygen its substrate, and hydrogen and phosphate ions inhibitors, then hemoglobin seems similar to biosynthetic enzymes subject to feedback inhibition and other allosteric forms of regulation. Therefore, the mechanism of hemoglobin action allows us to understand the mechanism of respiratory transport, and also to gain information into the action of other more complex systems on which metabolic regulation depends.

There are many reasons to believe that all avian hemoglobins are formed by the four iron-containing heme units characteristic of mammals, but the globin protein moieties are different and migrate electrophoretically at speeds different from the other vertebrate counterparts. With some exceptions all adult avian species investigated have at least two main types of hemoglobin, exhibiting slow (2) and fast (1) electrophoretic mobilities. The proportions of these components average about 70–80% for type 2 and 20–30% for type 1. They display amino acid differences in the α-chains but the β-chains appear identical, in contrast to the mammals where variation is observed mainly in the β-chains. IP5 replaces 2,3-DPG in the avian RBC presumably because its structure is compatible with the requirements of the oxygenation regulatory process or because some metabolic pathway specific to birds produces this organic phosphate. The possible physiological interactions between the multiple component system, IP5 and sequence changes in the globin remain unexplored. Further studies are expected which will certainly solve some problems related to a comprehensive interpretation of the respective contribution of the principal parameters involved in the avian respiratory system.

REFERENCES

Andersen, H. T., and Lovo, A. (1967). *Respir. Physiol.* **2**, 163–167.
Arnone, A. (1972). *Nature (London)* **237**, 146.
Arnone, A., and Perutz, M. F. (1974). *Nature (London)* **249**, 34–36.
Bartels, H., Hiller, G., and Reinhardt, W. (1966). *Respir. Physiol.* **1**, 345–356.
Bell, D. J. (1971). *In* "Physiology and Biochemistry of the Domestic Fowl" (D. J. Bell and B. M. Freeman, eds.), pp. 863–872. Academic Press, New York.
Benesch, R., Benesch, R. E., and Yu, C. I. (1968). *Proc. Natl. Acad. Sci. U.S.A.* **59**, 526–532.
Benesch, R., and Benesch, R. E. (1969). *Nature (London)* **221**, 618–622.
Benzon, A. (1972). *Oxygen Affinity Hemoglobin Red Cell Acid Base Status. Proc. Alfred Benzon Symp., 4th, 1971* p. 146.
Berger, M., and Hart, J. S. (1974). *In* "Avian Biology" (D. S. Farner and J. R. King, eds.), Vol. IV, pp. 416–477. Academic Press, New York.
Borgese, T. A., and Bertles, J. F. (1965). *Science* **148**, 509–511.
Borgese, T. A., and Lampert, L. M. (1975). *Biochem. Biophys. Res. Commun.* **65**, 822–827.
Braunitzer, G., Gehring-Muller R., Hilschmann, N., Hilse, K., Hobom, G., Rudolff, V., and Wittman-Liebold, B. (1961). *Hoppe-Seyler's Z. Physiol. Chem.* **235**, 283–286.
Brown, J. L., and Ingram, V. M. (1974). *J. Biol. Chem.* **249**, 3960–3972.
Bruns, G. A. P., and Ingram, V. M. (1974). *Philos. Trans. R. Soc. London, Ser. B* **265**, 225–305.
Brush, A. H., and Power, D. M. (1970). *Comp. Biochem. Physiol.* **33**, 587–599.
Calder, W. A. III, and King, J. R. (1974). *In* "Avian Biology" (D. S. Farner and J. R. King, eds.), Vol. IV, pp. 260–415. Academic Press, New York.
Campbell, F. (1967). *J. Morphol.* **103**, 405–414.
Chanutin, A., and Curnish, R. R. (1967). *Arch. Biochem. Biophys.* **121**, 96–102.
Christensen, H., and Dill, D. B. (1935). *J. Biol. Chem.* **109**, 443–448.
D'Amelio, V., and Salvo, A. M. (1961). *Acta Embryol. Morphol. Exp.* **4**, 250–258.
Danzer, L. A., and Cohn, J. E. (1967). *Respir. Physiol.* **3**, 302–306.
Debouverie, D. (1975). *Biochimie* **57**, 569–578.
Deconinck, M. (1973). Ph.D. Thesis, Université Libre de Bruxelles.
Deconinck, M., Depreter, J., Paul, C., Peiffer, S., Schnek, A. G., Putnam, F. W., and Leonis, J. (1972). *FEBS Lett.* **23**, 279–281.
Deconinck, M., Peiffer, S., Depreter, J., Paul, C., Schnek, A. G., and Leonis, J. (1975). *Biochim. Biophys. Acta* **386**, 567–575.
Duncker, H. R. (1972). *Respir. Physiol.* **14**, 44–52.
Edmundson, A. B. (1965). *Nature (London)* **205**, 883–887.
Fedde, M. R. (1975). *In* "Avian Physiology" (P. D. Sturkie, ed.), 3rd ed., pp. 122–145. Springer-Verlag, Berlin and New York.
Fraser, R. C. (1961). *Exp. Cell Res.* **25**, 418–427.
Fraser, R. C. (1963). *J. Exp. Zool.* **156**, 185.
Godet, J. (1974). *Dev. Biol.* **40**, 199–207.
Godet, J. (1975). *Annee Biol.* **14**, 129–165.
Hall, F. G., Dill, D. B., and Barrón, E. S. G. (1936). *J. Cell. Comp. Physiol.* **8**, 301–313.
Harrison, S. C., and Blout, L. R. (1965). *J. Biol. Chem.* **240**, 299–305.

Hashimoto, K., and Wilt, F. H. (1966). *Proc. Natl. Acad. Sci. U.S.A.* **56**, 1477–1483.
Haurowitz, F. (1938). *Z. Physiol. Chem.* **254**, 266–274.
Hill, A. V. (1910). *J. Physiol. (London)* **40**, IV–VII.
Holmquist, R., Jukes, T. H., Moise, H., Goodman, M., and Moore, G. W. (1976). *J. Mol. Biol.* **105**, 39–74.
Huisman, T. H. J., and Schillhorn van veen, J. M. (1964). *Biochim. Biophys. Acta* **88**, 367–374.
Huisman, T. H. J., Schillhorn van veen, J. M., Dozy, A. M., and Nechtman, C. M., (1964). *Biochim. Biophys. Acta* **88**, 352–366.
Imai, K., Morimoto, H., Kotani, M., Watari, K., Hirata, W., and Kuroda, M. (1970). *Biochim. Biophys. Acta* **200**, 189–196.
Keane, R. W., Abbott, U. K., Brown, J. L., and Ingram, V. M. (1974). *Dev. Biol.* **38**, 229–236.
Kendrew, J. C., Watson, H. C., Strandberg, B. E., Dickerson, R. E., Philips, D. C., and Shore, U. C. (1961). *Nature (London)* **190**, 666–670.
Lasiewski, R. C. (1973). *In* "Avian Biology" (D. S. Farner and J. R. King, eds.), Vol. II, pp. 288–343. Academic Press, New York.
Lasiewski, R. C., and Dawson, W. R. (1967). *Condor* **69**, 13–23.
Lee, K. S., Huang, P. C., and Cohen, B. H. (1976). *Biochim. Biophys. Acta* **427**, 178–196.
Lenfant, C., Torrance, J., Enolish E., Finch, C. A., Reynafarje, C., Ramos, C., and Faura, J. (1968). *J. Clin. Invest.* **47**, 2652–55.
Lenfant, C., Kooyman, G. L., Elsner, R., and Drabek, C. M. (1969). *Am. J. Physiol.* **216**, 1598–1600.
Lucas, A. B., and Jamroz, C. (1961). *U.S. Dept. Agric., Monogr.* **25.**
Lutz, P. L., Longmuir, I. S., Tuttle, J. V., and Schmidt-Nielsen, K. (1973). *Respir. Physiol.* **17**, 269–275.
Lutz, P. L., Longmuir, I. S., and Schmidt-Nielsen, K. (1974). *Respir. Physiol.* **20**, 325–330.
Magath, T. B., and Higgins, G. M. (1934). *Folia Haematol.* **51**, 230–241.
Manwell, C., Baker, C. M. A., Rolansky, J. D., and Foght, M. (1963). *Proc. Natl. Acad. Sci. U.S.A.* **53**, 1147–1154.
Manwell, C., Baker, C. M. A., and Betz, T. W. (1966). *J. Embryol. Exp. Morphol.* **16**, 65–81.
Marbaix, G., and Burny, A. (1964). *Biochem. Biophys. Res. Commun.* **16**, 522–527.
Matsuda, G., Takei, H., Wu, K. C., and Shiozana, T. (1971). *Int. J. Proteins Res.* **3**, 173–174.
Matsuda, G., Maita, T., Mizuno, K., and Ota, M. (1973). *Nature (London), New Biol.* **244**, 244.
Monier, C. (1972). Ph.D. Thesis, Université Libre de Bruxelles.
Monier, C., Schnek, A. G., Dirkx, J., and Leonis, J. (1973). *FEBS Lett.* **36**, 93–95.
Moss, B. A., and Hamilton, E. A. (1974). *Biochim. Biophys. Acta* **371**, 379–391.
Moss, B. A., and Thompson, E. O. P. (1969). *Aust. J. Biol. Sci.* **22**, 1455–1471.
Paul, C., Vandecasserie, C., Schnek, A. G., and Leonis, J. (1974). *Biochim. Biophys. Acta* **371**, 155–158.
Peiffer, S., Deconinck, M., Paul, C., Depreter, J., Schnek, A. G., and Leonis, J. (1973). *FEBS Lett.* **37**, 295–297.
Perutz, M. F. (1969). *Proc. R. Soc. London, Ser. B* **173**, 113–140.
Perutz, M. F., Muirhead, H., Cox, J. M., and Goaman, L. C. G. (1968). *Nature (London)* **219**, 131–139.
Proudfoot, N. J., and Brownlee, G. G. (1976). *Br. Med. Bull.* **32**, 251–256.

Rapaport, S., and Guest, G. M. (1941). *J. Biol. Chem.* **138**, 269–282.
Reichert, E. T., and Brown, A. P. (1909). *In* "The Crystallography of the Hemoglobins." Carnegie Inst. Washington, Washington, D.C.
Rosse, W. F., and Waldmann, T. A. (1966). *Blood* **27**, 654–660.
Saha, A., and Ghosh, J. (1965). *Comp. Biochem. Physiol.* **15**, 217–235.
Scheid, P., and Kawashiro, T. (1975). *Respir. Physiol.* **23**, 291–300.
Schnek, A. G., Paul, C., and Leonis, J. (1966). *Proc. Int. Symp. Comp. Hemoglobin Struct., 1966,* pp. 103–109.
Schnek, A. G., Paul, C., Monier, C., and Leonis, J. (1970). *Proc. Int. Symp. Strukt. Funkt. Erythrocyten, 6th, 1970* pp. 209–215.
Schweet, R., Lamfrom, H., and Allen, E. (1958). *Proc. Natl. Acad. Sci. U.S.A.* **44**, 1029–1035.
Scott, A. F., Bunn, H. F., and Brush, A. H. (1976). *J. Mol. Evol.* **8**, 311–316.
Steward, J. M., and Tate, M. G. (1969). *J. Chromatogr.* **15**, 400–406.
Sturkie, P. D. (1975). *In* "Avian Physiology" (P. D. Sturkie, ed.), 3rd ed., pp. 76–101. Springer-Verlag, Berlin and New York.
Takei, H., Ota, Y., Wu, K. C., Jiyoharat, T., and Matsuda, G. (1975). *J. Biochem. (Tokyo)* **77**, 1345–1347.
Tucker, V. A. (1972). *Respir. Physiol.* **14**, 75–82.
Vandecasserie, C., Schnek, A. G., and Leonis, J. (1971). *Eur. J. Biochem.* **24**, 284–287.
Vandecasserie, C., Paul, C., Schnek, A. G., and Leonis, J. (1973). *Comp. Biochem. Physiol.* **44A**, 711–718.
Vandecasserie, C., Paul, C., Schnek, A. G., and Leonis, J. (1975a). *Biochimie* **57**, 843–844.
Vandecasserie, C., Fraboni, A., Schnek, A. G., and Leonis, J. (1975b). *Arch. Int. Physiol. Biochim.* **83**, 407–409.
Vandecasserie, C., Lorkin, P. A., Schnek, A. G., and Leonis, J. (1975c). *Arch. Int. Physiol. Biochim.* **83**, 409–410.
Vandecasserie, C., Schnek, A. G., and Leonis, J. (1976). *Cong. Hematol. Fr., Club Hemoglobine,* 2nd, *1976* p. 34.
Wilt, F. H. (1965). *J. Mol. Biol.* **12**, 331–341.

Author Index

Numbers in italics refer to the pages on which the complete references are listed.

B

D

E

H

I

M

N

S

T

U

V

Y

Z

Subject Index

F

H

M

N

Q

R

U

V

A B C 8 D 9 E 0 F 1 G 2 H 3 I 4 J 5